The Tangled Origins *of* The Leibnizian Calculus

A Case Study of a Mathematical Revolution

Gottfried-Wilhelm Leibniz
1646–1716

The Tangled Origins *of* The Leibnizian Calculus

A Case Study of a Mathematical Revolution

Richard C Brown

University of Alabama, USA

World Scientific

NEW JERSEY • LONDON • SINGAPORE • BEIJING • SHANGHAI • HONG KONG • TAIPEI • CHENNAI

Published by

World Scientific Publishing Co. Pte. Ltd.
5 Toh Tuck Link, Singapore 596224
USA office: 27 Warren Street, Suite 401-402, Hackensack, NJ 07601
UK office: 57 Shelton Street, Covent Garden, London WC2H 9HE

British Library Cataloguing-in-Publication Data
A catalogue record for this book is available from the British Library.

The portrait on page ii and cover is by an unknown artist and engraved by P. Savart in 1768. The reproduction is courtesy of History of Science collections of the University of Oklahoma Libraries.

THE TANGLED ORIGINS OF THE LEIBNIZIAN CALCULUS
A Case Study of a Mathematical Revolution

ISBN-13 978-981-4390-79-8
ISBN-10 981-4390-79-8

Printed in Singapore.

As always for Phyllis and in memory of my teacher Thomas S. Kuhn (1924–1996).

Contents

Preface

The remote origins of this book may be found in a paper dealing with Leibniz's invention of calculus, no longer extant, which I wrote in a Berkeley seminar conducted in the fall of 1960 by Thomas S. Kuhn. The theme of the seminar was "Science: Evolution or Revolution?". Being then a rather immature and disorganized graduate student, who seldom paid much attention in class (a habit which I confess has carried over to most plenary talks at mathematical conferences I have attended), I have hardly any clear memories of the seminar except for some lively discussion concerning the Nixon-Kennedy debates (neither the class or Kuhn cared much for Nixon), the beginning of a political ritual which in 1960 was for the first time shown on television. But it must have been the case that Kuhn presented preliminary versions of some of the ideas to appear in his famous book *The Structure of Scientific Revolutions* published two years later. In perfect conformity with my academic habits at the time, the paper was never finished, but Kuhn was kind enough to give me a decent grade anyway. A few years later my interests changed from history of science to mathematics and after obtaining a Ph.D. in Analysis and Differential Equations in 1972, I had a conventional modestly successful career in those subjects until my retirement from the University of Alabama in 2002. During these years I continued to have a residual interest in Leibniz as well as mild but enduring feelings of guilt (doubtlessly due to a mixed Norwegian Lutheran and Massachusetts Puritan heritage). It would be nice I felt if the original paper someday could be completed or even better expanded into a book; but the demands of teaching, writing 65^+ research papers, departmental micro-politics, and all the other duties and distractions (some a waste of time) associated with academic employment at an average public university kept this project well on the back burner. I had no time to pay much attention either to the history of science or to the humanities generally. I was locked in a narrow, intense world of splines, differential operators, various classical inequalities, and weighted Sobolev-type embeddings. It was only around 2008, well after retirement, that I had the leisure to look again at interests abandoned decades earlier. The resulting book, based on the memories of my early paper, surpasses one of Leibniz's efforts in two respects: for the last thirty years of his life he was engaged in writing a history of the family (the Guelfs) of his employers, the Dukes of Hanover which at his death and to their intense irritation[1] was still unfinished. The present book in contrast has been gestating for twenty years longer than the Guelf project, but at least I can say that it has been completed.

Since the book is fairly complicated, a brief description of its goals and organization may be convenient for the reader. We pursue several distinct but interrelated themes: At the most basic level, we attempt to give a reasonably accurate account of Leibniz's creation and development of calculus both in the Paris period (1672–1676) and afterwards. This will be based on a study of his early mathematical manuscripts and some of the articles he published in the 1680s and 90s. We focus on this task mainly in Chapters 5 and 6, which are rather technical. Right at this point, one might question the need for another book on this topic given the classic work of J.E. Hofmann *Leibniz in Paris 1672–1676: His Growth to Mathematical Maturity*, published initially in German in 1949 and then in English in 1973. Hofmann's book is certainly an invaluable source for any investigation of Leibniz's mathematics in this period, including our own. But it does not avoid a difficulty which is common among historians of mathematics. The only way, it seems, that we can understand the results of an archaic mathematician like Leibniz is to "translate" them using modern notation. This may appear to be a harmless convention, especially to a historian who may know some mathematics but who does not think mathematically or has attempted to do mathematical research himself. After all, Leibniz did invent much of the calculus notation we use today, and his results are (in the main) correct and formally equivalent to our own. A mathematical result is just that: it may have been discovered in the distant past and presented in an unfamiliar way. But it remains true, and its "nature" does not change. Consequently for the sake of clarity, modern techniques can be employed to explain it provided logical equivalence is preserved. However, in this writer's opinion such an attitude does violence to history. It treats

[1] Especially to Georg Ludwig (1660–1727) who became George I of Great Britain and Ireland in 1714.

mathematics as only an ever increasing array of "results"; it is a symptom of a kind of "Whig interpretation" applied to the history of mathematics that has long since been discredited in the history of science. To give an example of the difficulty we are talking about, consider a note of Leibniz written October 25, 1675. Here, according to Hofmann, Leibniz comes up with the result[2]

$$\int_0^b yx\,dx = \frac{1}{2}b^2c - \int_0^c \frac{1}{2}x^2\,dy.$$

Although his notation is quite different, it looks like Leibniz has discovered an example of 'integration by parts" (a major result in calculus) as early as 1675. But this is a historical distortion; for as we shall see in Chapter 5, in this particular case Leibniz's reasoning has nothing to do with integration in the modern sense. What we see and write as "integrals" were for him aggregates of lines in the sense of Cavalieri. Also even in later work, he never had a concept of the "definite integral" (and therefore certainly did not use the modern notation for one!). Even though his formula is "correct" and can be derived from the modern concept of integration by parts, such a procedure would have been meaningless to him in 1675. For Leibniz this result involves mechanics and is a statement about moments.[3] Only much later, probably in the period 1680–1690 did he think of this formula as the inverse of the differential formula

$$d\left(\frac{x^2y}{2}\right) = yxdx + \frac{x^2dy}{2},$$

and since Leibniz's calculus made no use of "derivatives," he would not have grasped its connection with the differentiation formula

$$\frac{d}{dx}\left(\frac{x^2y}{2}\right) = xy + x^2\frac{dy}{dx}.^4 \tag{0.1}$$

To give another example, in a manuscript of October 29, 1675 Leibniz mentions one of his "transmutation" techniques to find certain areas. In Figure 0.1 let TO be the y intercept of the tangent TP to the curve C at an arbitrary point $P = (x, y)$. Then $T = (0, y - x\,dy/dx)$. Let $Q = (x, (1/2)(y - x\,dy/dx))$. Then according to Leibniz the area of the variable

[2]Hofmann (1974a), p. 188.

[3]To be fair Hofmann does make reference to Leibniz's use of moments, pointing $\int_0^a x(y\,dx)$ is the moment of the area $\int_0^a y\,dx$ about the y axis. But even this statement is a "modernization" of Leibniz.

[4]It is true that the notation "$\frac{dy}{dx}$" can be found on occasion in manuscripts of 1675–1676. But as Henk Bos (1973) has observed, this is not the derivative *function*, but instead a ratio of line segments allowing us to determine the "subtangent."

region bounded by the arcs $\widehat{OP}$, $\widehat{OQ}$ and line segment PQ is always the area of the triangle OXP, i.e., $xy/2$.

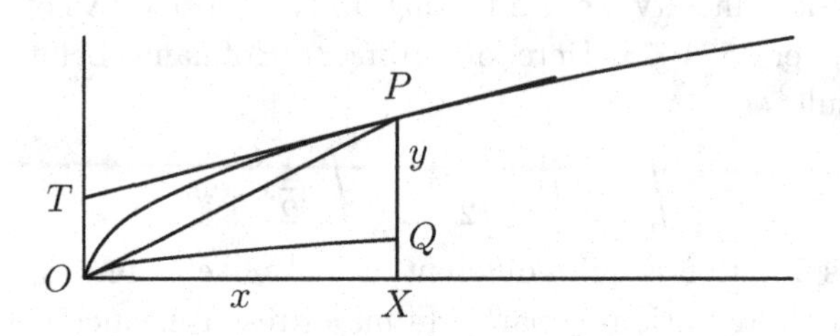

Figure 0.1

To derive this, Hofmann expresses the area by the integral

$$\int_0^x \left(y - \frac{y\,dx - x\,dy}{2\,dx} \right) dx,$$

which is readily seen to be equal to

$$\int_0^x \frac{x\,dy + y\,dx}{2\,dx}\,dx = \frac{1}{2}xy.$$ [5]

This is correct, but it cannot possibly be the same as Leibniz's argument which in fact is unknown. Aside from the illegitimate use of definite integrals in these formulas, in 1675 he probably did not know that

$$d(xy) = xdy + ydx,$$

and for the same reason as in (0.1) never would exploit the identity

$$\frac{d(xy)}{dx} = x\frac{dy}{dx} + y.$$

Instead, as we will see in Appendix A, Leibniz could have used a simple similar triangle argument combined, as in the previous "integration by parts" formula, with the ideas of Cavalieri. Unfortunately Hofmann, despite his unparalleled research, continually follows (along with most historians of mathematics) this procedure of writing Leibniz's results in modern form and treating them more or less at face value as equivalent to their modern versions. But even in its perfected form we shall find that Leibniz's calculus, even though he created our notation, is rather different from its modern descendants. And furthermore, these differences cannot be explained solely

[5] *Ibid.* p. 192.

by the fact that his calculus is less developed than ours. That is certainly true, but a more fundamental difference is that he thought about problems *differently* than we do. As we have already conceded, Hofmann's use of modern notation to describe archaic mathematics is standard, and to an extent almost unavoidable. But unless great care is taken "modern ideas" can be erroneously attributed to early modern mathematicians.

This brings us to a second theme: we will make every effort to understand the results of Leibniz and his contemporaries from their perspective. This, however, is a very difficult—perhaps an almost impossible task—even for the mathematically trained. True, we can follow (often with difficulty) the arguments, but if we think of mathematics as a creative *process* instead of a collection of "theorems" we can no longer *think* creatively in the same manner as a seventeenth century mathematician. The methods he employed and our own are too different, and we are unable to use them. There are also significant differences in mathematical "taste," values, and views of what constitutes worthwhile problems and correct solutions to them between his time and ours. As we will try to make evident in Chapter 3, not even a Field's medalist could come up with one of Isaac Barrow's (1630–1677), more complicated theorems in the form he did. We may be able to understand his proofs, but the motivation behind them is no longer accessible, even though when Barrow's results are expressed using modern notation they look very much like standard results in calculus.[6] In this sense our mathematics and that of the seventeenth century can almost said to be "incommensurable," having differences that can be psychologically analogous to those of rival scientific paradigms. Whether or not Kuhn's thesis that scientists conditioned by competing paradigms live in "different worlds" is an exaggeration, seventeenth century mathematicians definitely *do* occupy a mathematical world which is very different from our own.

Given these observations, a third theme of the book owes much both to the work of Thomas Kuhn and to that of his lesser known precursor Ludwik Fleck (1896–1961). We will attempt to fit some (but by no means all) of their theses into the history of mathematics. We have hinted elsewhere that Kuhn's book, *Structure of Scientific Revolutions* in particular, because of its relativist anti-realism ultimately had, contrary to its author's intentions and unforseen by him, a damaging effect on both the history of science and

[6]This phenomenon led J. M. Child, the editor of Leibniz (1920), to the thesis that Barrow was possibly the true inventor of calculus whose ideas heavily influenced Leibniz. We will discuss this issue in Chapter 3.

the wider perceptions of science by the humanities.[7] Fortunately, since in the writer's opinion there are no convincing arguments for the relativity of truth claims in mathematics,[8] no similar harm can be done in the history of mathematics. Yet, as we attempt to demonstrate in Chapters 2, 3, 5, 6, and 11, Leibniz's approach (together with that of René Descartes (1596–1650)) mark the beginning of a paradigm or gestalt shift or change in "thought style" (Fleck) in mathematics, but one that was not completed until at least the mid- eighteenth century. For Leibniz and Descartes began to create a new kind of mathematical analysis based respectively on a nearly self-working symbolism and algebra, both of which ultimately involve the concept of function. But although their work marked a real break with the contemporary mathematical paradigm, we shall see that they like most who begin an intellectual revolution were still heavily influenced by the past. The paradigm they began to supplant we call "geometric." Being the mathematical analogue of humanism, it was molded by the admiration and respect for the great geometers of antiquity especially Apollonius, Euclid, Archimedes, and Pappus. To help understand the novelty and limitation of Leibniz's achievement, we describe this form of early modern mathematics in Chapter 2. In Chapter 3 we show how it was totally incorporated into the work in infinitesimal analysis of just one mid-seventeenth century mathematician, Isaac Barrow. We could have selected several other candidates for this role (in particular Christiaan Huygens (1629–1695)), but Barrow is especially interesting because of the assertions Child and others have made concerning his influence on Leibniz's development of calculus.

A fourth theme in an attempt to show how a particular aspect of the seventeenth century nonmathematical cultural environment influenced the *technical* characteristics of Leibniz's mathematics. This is an unusual and rather strong claim. One one level it is clear that external social or cultural factors can affect mathematics. For instance, the need for mathematical solutions to practical problems can determine the direction of mathematical research and even generate either new subdisciplines or change existing ones. To give just two examples, the military problems of World War II created a vast new area of operations research, and the demands of signal processing led to Claude E. Shannon's (1916–2001) information theory. The nature of the problems encountered may also determine the kind of

[7]Brown (2009), especially Chapter 7. Indeed, Kuhn's book may be viewed as the opening shot in the so-called "Science Wars" of the 1990s.

[8]Attempts to the contrary, however, have been made by proponents of the "Strong Programme." See e.g., Bloor (1976).

mathematics employed and the nature of the solutions found. One can think of statistical analysis in the quality control of manufacturing or the solutions of ballistics problems which necessarily involve certain differential equations dictated by the physics. On the other hand, those parts of the prevailing Zeitgeist which are remote from scientific needs or practical problems seem to have no similar effect. A philosophical system such as Platonism or Formalism, for example, may conceivably play a role in the investigation of the foundations of mathematics but seems not to influence its technical aspects, at least in the modern period. Nor does it play a role relative to mathematics similar to Pythagorean sun worship relative to Copernicus or (very possibly) the ideology of nineteenth century capitalism relative to Darwin. The "nuts and bolts" parts of mathematics appear to be conditioned by the problems encountered or the methods previously in use, not by some nonmathematical ideology. However, a fascinating aspect of Leibniz's calculus is that it appears to be a nearly unique exception to this rule.[9] Leibniz was, of course, a very talented mathematician. But his strictly mathematical talents *per se* were no greater than those of the upper ranks of his contemporaries. The "depth" exhibited in his mathematical manuscripts cannot be compared with that of Newton, the supreme mathematician of his age. Leibniz came late to mathematics and never matched the dazzling technical virtuosity of an Isaac Barrow or Christiann Huygens.[10] And as we shall see his *specific* mathematical results were either no more demanding than what had been proven by others or (such as his "arithmetical quadrature" of the circle) original, but not new. Leibniz's calculus is not distinguished by mathematical ingenuity or technical virtuosity. Its novelty lies in its ability to generalize and simplify previous work in infinitesimal analysis. But Leibniz accomplishes this by adding a new "nonmathematical" ingredient: we shall see that he imports into the substance of his mathematics a feature of the wider cultural milieu. This was the search for a *Lingua Philosophical*, or "Universal Characteristic," a universal language or logical scheme which would allow all disputes whether religious, scientific, philosophic, or political to be "mechanically" settled in

[9]Another exception may be the Intuitionism developed by L. E. J. Brouwer (1881–1966). This both relates to the foundations of mathematics, but also has profound technical implications for the nature of proof and the allowable subject matter and technique in mathematics. On the other hand, Intuitionism at its core is a *mathematical* philosophy. It is not a product of some philosophical quest having little or nothing to do with mathematics.

[10]For the most part Leibniz's geometrical arguments are routine exercises involving similar triangles.

a way which would be satisfactory to all parties involved. This project, which obsessed Leibniz throughout his life, was very fashionable in the seventeenth century, and its origins go back to the Renaissance, and in part even further to the work of the Catalan philosopher and mystic Raymond Llull (1232–1315); and its importance for many aspects of Leibniz's thought (including mathematics) was noted many years ago by Dame Francis Yates and Paolo Rossi, and more recently by Umberto Eco[11] We will describe this Baroque intellectual environment in Chapter 8 and sketch its history in Chapter 9. Chapter 8 also argues that the grand project of creating the Characteristic guided Leibniz's every step in mathematics. From the very beginning of his exposure to infinitesimal analysis in 1672, he was trying to create a "calculus" rather than finding particular solutions to a collection of isolated problems. It is this goal rather than the quality or depth of his mathematical arguments that constitutes Leibniz true originality. The same motive is present in his work in logic, which in certain respects anticipates the development of symbolic logic in the nineteenth century, and in an entirely novel but unsuccessful symbolic geometry which he called the *Analysis situs.* Chapter 7 is devoted to Leibniz's logic, and Chapter 8 contains a sketch of some of his other mathematical efforts including the *Analysis Situs.*

Our fifth and final theme concerns Leibniz's infamous priority dispute with Newton and certain aspects of his character. When I began the book, I resolved that I would stay away this issue. After all it is an unedifying and distasteful subject which over the centuries has been written into the ground. What could be possibly new to say about it? The received opinion of modern historians of science is that Newton and Leibniz developed calculus independently, Newton in 1665/6 and Leibniz in October/November 1675, when in unpublished manuscripts he introduced his "$\int$" and "d" notation. This was well before Leibniz's second trip to London in October 1676 and his receipt of two mathematically important letters from Newton. Therefore Leibniz could not have known about or been influenced by Newtons's work. It follows that the subsequent accusations of plagiarism against him by the Royal Society, as exemplified by the *Commercium Epistolicum* of 1712 and Newton's anonymously written *Account* written in 1713, could not be true and reflected either Newton's pathological paranoia, British nationalism, or Hanoverian politics connected to the accession of George I (previously Leibniz's employer as Elector of Hanover), to the British throne,

[11]See e.g., Yates (1954), (1966), (1982); Rossi (2000), and Eco (1995).

or some unsavory mixture of these ingredients. As I continued my research, however, I gradually was forced to the conclusion that although Leibniz had been thinking of a "calculus" ever since 1672/3, the parts of it for which he is remembered today were only formulated *after* the London trip, and some respects were not perfected until the 1680s. Newton's suspicions, therefore, cannot be dismissed as "paranoia"; he had some rational grounds for them and his suspicions were fueled in part by aspects of Leibniz's character as well as by the Iago-like insinuations of Newton's colleagues. As we shall see, Leibniz had a habit of making inflated claims, promising more than he actually could deliver, and sometimes put himself in a position where he could be perceived (wrongly or not) as "improving" the results of others (including Newton and Barrow) without proper acknowledgement.[12] Yet the evidence of any direct plagiarism from Newton concerning calculus is ambiguous. What really happened (if anything) between the two will probably never be known. The priority controversy is not a major part of the book, but a brief discussion of it will be given in Chapters 6 and 10. The latter chapter also sketches Leibniz's life after Paris.

We hope that two global conclusions will emerge from our treatment of Leibniz. The first is that although the invention of calculus was one of the greatest achievements of the human mind and a necessary condition for the tremendous scientific progress made in the hard sciences between the mid-seventeenth century and the present, it has a tangled history: it was not just an easily predicted consequence of previous infinitesimal analysis, but in the form Leibniz gave it benefited from a contemporary nonmathematical intellectual climate in a way seldom seen before or since. More fundamentally, his invention shows that mathematics is a contingent, evolving, perishable historical product like any other part of human culture. It is a much more complicated enterprize than an array of theorems majestically unrolling in time according to logical necessity. Large portions are constantly being born and forgotten, and past and present mathematics can be "incommensurable" in a quasi-Kuhnian sense.

A few remarks of a pedagogical nature may be helpful: Chapter 2, 3, 5, and 6 contain a lot of mathematics and require a good knowledge of geometry or calculus to be understood. Also, in order to drive home our point that the "geometric paradigm" influencing seventeenth century mathematics—especially as relating to infinitesimal analysis—is *different*

[12]This may especially be the case in regard to some of the central ideas of the *Principia* and to Leibniz's various methods of "transmutation" which may have been influenced by Barrow's work. We will discuss this issue in Chapters 3, 5, 6, and 10.

from (not necessarily less "advanced" than) modern approaches, we have omitted the proofs of some of the archaic theorems we state (particularly those due to Isaac Barrow in Chapter 3). Instead the reader is encouraged to prove them by showing their logical equivalence to certain results in calculus. In some cases this will be a demanding task (at least it was for this writer) and should reenforce a conclusion that the motivation and thought patterns behind these results were quite remote from our own, and consequently *how* and *why* early modern mathematicians thought as they did is mysterious. Such a view may still not be quite evident to those whose mathematical experience is confined to classroom work, but it may be more so for those who have carried out real mathematical research: they will realize how dependent our reasoning and modes of discovery are on contemporary practice and how difficult it is to productively exploit a truly different mind-set. The reader of Chapter 7 should also have some background in elementary logic.[13] Fortunately, the remaining chapters deal primarily with biographical or historical issues, are independent of the mathematically demanding parts of the book, and summarize our main conclusions.

In closing, we mention that standard easily obtained information about Leibniz's life or other matters will not usually be footnoted. For biographical details we recommend the books of Antognazza (2009), Aiton (1985), Mackie (1845), or Guhrauer (1846—reprinted 1966). We will, however, give references to information that we judge more recondite, to interpretation or judgements, and to all quotations (translated versions if possible). Since this policy is a matter of judgement, we apologize if it seems arbitrary to some. The book is also by no means a study of Leibniz's metaphysics. We deal with it very briefly and only when it helps us understand his general intellectual state (Chapter 4), or impacts some feature of his science or mathematics.[14] There is, of course, a vast literature on Leibniz's metaphysics containing quite different interpretations. We recommend the classic (if now controversial) study of Russell (1900) as well as more recent works by Mates (1986), Mercer (2001), and Garber (2009). Also we should point out some possible confusion concerning dates, especially in relation to Leibniz's correspondence. The Gregorian calendar was used on the continent since Pope Gregory XIII introduced it in 1582. However, Britain did not adopt it until 1752. The two calendars in the seventeenth century and

---

[13]We give, however, a simplified sketch of traditional syllogistic logic in Appendix C.

[14]Such as logic (Chapter 7), or dynamics–the *vis viva* controversy in particular (Chapter 6 and Appendix D.

until March 10, 1700 are related by the formula

$$\text{Julian date} = \text{Gregorian date} - 10 \text{ days.}$$

After this date and until March 11, 1800 the difference increases to 11 days. We will follow the policy of using the date written on the letter. Unless otherwise indicated, letters written in Europe will use the Gregorian date, and those written in England the Julian one. For clarity we will use the abbreviations "O.S." (old style) for the Julian date and "N.S." (new style) for the Gregorian date. Finally, we have included many geometric diagrams. These for the most part follow the originals, but occasionally I have altered the labeling of points and lines. For example, Leibniz often uses the notation C, (C), $((C))\ldots$ or ${}_1C$, ${}_2C$, ${}_3C\ldots$ to refer to sequences of points or to the same points which changes its location. We will usually replace this by writing C_1, C_2, C_3, etc. This, of course, is a minor contradiction of our declared policy of historical authenticity, but we feel that the improvements in appearance and clarity excuse it.

Acknowledgments

The author is grateful to several people for their support and encouragement in writing this book. Thanks are especially due to C.C. Gillispie, Max Hocutt, James Cook, David Edmunds, Frank Roehl, and Don Hinton. He is also grateful to his beloved wife Phyllis for her careful proofreading of large sections of the manuscript, for her toleration of a messy house, littered with books and reprints, and of my prolonged absences upstairs at the computer. Appreciation is also due to the staff of Gorgas Library of the University of Alabama, especially to Pat Causey who helped via interlibrary loan to procure material for my research and who forgave many late returns. Finally, special thanks to his editors Lai Fun Kwong, He Yue, and their technical advisor Rajesh Babu of World Scientific Publishing Co. for their assistance in preparing the manuscript for publication.

Chapter 1

Evolution or Revolution in Mathematics: The Case of Leibniz

There is a conventional interpretation of Leibniz's mathematical achievements, specifically his discovery of the calculus, found in histories of mathematics like Boyer (1959), Burton (1985), or the more technical monographs by Baron (1969), Edwards (1979), or Hofmann (1974a). It goes roughly as follows: Leibniz's contribution was a predictable synthesis–mainly through improvements in notation–of earlier results of many seventeenth-century mathematicians such as Fermat (1601–1665), Cavalieri, Descartes, Barrow, Sluse (1625–1685), and others. Calculus in the form of contemporary investigation of tangent or quadrature problems was "in the air":

> Few new branches of mathematics are the work of single individuals... Far less is the development of calculus to be ascribed to one or two men... The time was indeed ripe, in the second half of the seventeenth century, for someone to organize the views, methods, and discoveries involved in infinitesimal analysis into a new subject characterized by a distinctive method of procedure.[1]

The inevitability of its creation is shown by Newton's independent and prior discovery of equivalent concepts in the 1660s. In fact, had these two giants not existed, someone else would have taken the same unifying steps and subsequent mathematical history would have been much the same.

Nevertheless over the centuries there have been many major and minor disputes over the exact nature of the Leibnizian synthesis and just what features in it were "original". All commentators agree on the importance and novelty of the Leibniz notation which is still preferred today (except in particle dynamics) over the less advanced symbolism of Newton's flux-

[1] Boyer, *op. cit.*, p. 186.

ional calculus. There is, first of all, the wonderfully compact and suggestive symbol for the derivative dy/dx, denoting the slope function of the tangent line to the curve defined by $y = f(x)$, which puts the adept of calculus in mind—by its analogy to a fraction—to the ratio of the infinitely small vertical and horizontal sides of what Leibniz called the "characteristic triangle." Likewise the definite integral symbol $\int_a^b f(x)\,dx$ which when f is nonnegative represents the area under the curve $y = f(x)$ from $x = a$ to $x = b$ cannot fail to remind the student that area is really the sum of infinitely many infinitely thin rectangles of "width" dx and height $f(x)$. (Figure 1.1)

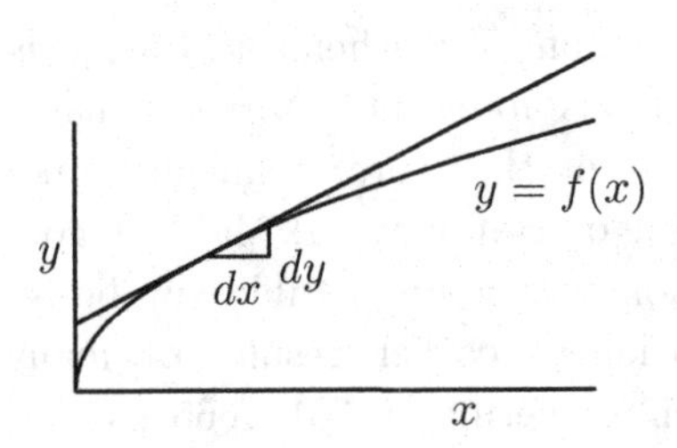

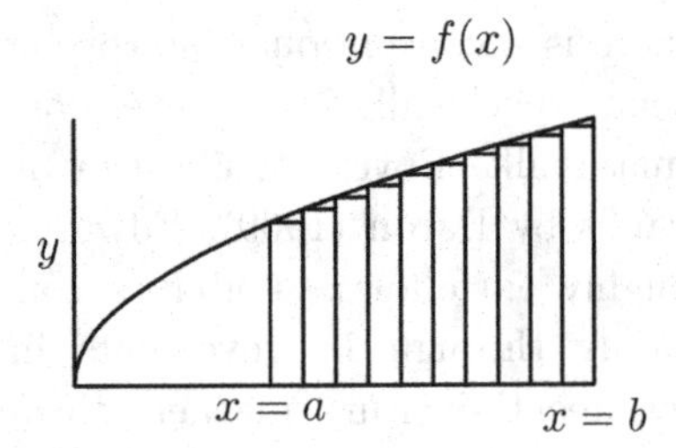

Figure 1.1

The scope of this notation is illustrated by several common formulas which we list below.

$$\frac{dy}{dx} = \frac{dy}{du}\frac{du}{dx} \tag{1.1}$$

$$\frac{dx}{dy} = \frac{1}{\frac{dy}{dx}} \tag{1.2}$$

$$y(x) = \frac{d}{dx}\int_a^x y(t)\,dt \tag{1.3}$$

$$y(b) - y(a) = \int_a^b \frac{dy}{dx}\,dx \tag{1.4}$$

$$\frac{d}{dx}f(x)g(x) = f(x)\frac{dg}{dx} + g(x)\frac{df}{dx} \tag{1.5}$$

$$\frac{d}{dx}\frac{f(x)}{g(x)} = \frac{g(x)\frac{df}{dx} - f(x)\frac{dg}{dx}}{g(x)^2} \tag{1.6}$$

Equation (1.1) is called the "chain rule"; it enables us to compute the derivative of the composition $y = g(u(x))$ by expressing it in terms of the derivatives of the functions g and u, while (1.2) allows us to take the derivative of an inverse function and express it in terms of the derivative of the original function. Formulas (1.5) and (1.6) are respectively called the "product" and "quotient" rules for differentiation. But the most important formulas in this list which lie at the heart of the achievement of both Newton and Leibniz are (1.3) and (1.4). Together they comprise what modern texts call the "Fundamental Theorem of Calculus" (henceforth abbreviated as "FTC"). Essentially they show the reciprocal relationship between integration and differentiation or, to express this differently, that finding areas under curves—in seventeenth-century language the problem of "quadrature"—and the problem of finding a tangent are equivalent problems. For instance, if we think of graphing the "area function" $A(x) = \int_a^x y(t)\,dt$, then the slope of the tangent line to this curve at a given x is $y(x)$. Conversely, if A is any function such that $dA/dx = y$, then by (1.4) the area $\int_a^b y(t)\,dt = A(b) - A(a)$. This last formula makes it possible to determine quadratures without the difficult method of viewing the area as some kind of infinite sum of "indivisibles" or of infinitely "thin" elements whose area can sometimes be determined, which was a standard procedure in the mid-seventeenth century. Formulas (1.1)–(1.6) and their correct (or incorrect) manipulation summarize the essentials of at least a semester of college calculus. It does not matter that the interpretations indicated in the above diagram of dy/dx or $\int_a^b f(x)\,dx$ are "wrong" or that it took until the time of Cauchy or Weierstrass to get rid of "infinitesimals" and to create precise concepts of limit processes.[2] The formulas work as if by magic! Permissible manipulations with them are suggested by their very structure. For example,

$$\frac{dx}{dy} = \frac{1}{\frac{dy}{dx}}$$

[2] In the case of the derivative this involves defining dy/dx as the limiting value of the difference quotient $h^{-1}(f(x+h) - f(x))$ as h remains finite and approaches but never equals 0 (for then the difference quotient is undefined). In this interpretation dy is a *finite* quantity called a "differential" defined by $dy = (dy/dx)dx$ where dx is also finite. It is interesting, however, that infinitesimals have again become legitimate, reappearing in the framework of non-standard analysis created by Abraham Robinson (1918–1974) in Robinson (1996). This involves the use of model theory (a branch of logic) to create a new number system $\mathbb{R}^*$ containing the ordinary real numbers $\mathbb{R}$ such that $\mathbb{R}^*$ has nonzero elements either "smaller" or "greater" than anything in $\mathbb{R}$. The small elements can be regarded as infinitesimals.

just as in elementary arithmetic. Engineers and physicists, in particular, find the intuitive thought processes behind them familiar and comforting. (A more rigorous attitude is viewed as "theory", and is to be avoided at all costs!) It is hard for non-mathematicians who have no direct experience with calculus to appreciate the mindless power that the system brings. Even a beginning student can show that

$$\int \frac{\sqrt{x+4}}{x}\,dx = 2\sqrt{x+4} + 2\log\left|\frac{\sqrt{x+4}-2}{\sqrt{x+4}+2}\right| + C; \qquad (1.7)$$

even better, he need not understand *what* he is doing or exactly what an integral is. Calculus as conventionally taught becomes a "turn the crank" process of mechanical calculation. Aside from minimal geometric intuition and some facility for high school algebra, no mathematical talent whatever is required; in fact such talent may be a disadvantage as it can be repelled by the boredom of calculus manipulations.

Leibniz himself was quite aware of these features of his invention and its consequent novelty. In the *Historia et origo calculi differentialis* (c. 1713), his (unpublished) answer to the *Commercium Epistolicum* published by a committee of the Royal Society in 1712 defending Newton's claim of priority, he writes:

> Now it certainly never entered the mind of any one else before Leibniz to institute the notation peculiar to the new calculus by which the imagination is freed from a perpetual reference to diagrams as was made by Vieta and Descartes in their ordinary or Apollonian geometry ...[3]

This is not a late judgment. He says in a letter written to his skeptical friend Tschirnhaus (1651–1708) in May 1678 "Such in fact are the signs applied by me to the calculation of tetragonistic equations, by which I am able to solve problems with little difficulty."[4] But it is also just these characteristics of the calculus which have been distrusted by mathematicians through the centuries beginning with Tschirnhaus and Huygens down to contemporary university research mathematicians who as a necessary condition of their employment have to teach it as a "service course" to an often captive audience without, however, regarding it as "real" mathematics.

As to other aspects of Leibniz's originality beyond the advantages of his notation opinion is divided. Quite aside from the Newton-Leibniz priority controversy whose passions are now mostly spent, much ink has been

[3]Leibniz (1920), p. 25.

[4]"Talia vero sunt signa a me in calculo aequationum tetragonisticarum adhibita, quibus problemata saepe difficillima paucis solvo." [Couturat (1961), p. 86].

spilt trying to explore the intellectual debts he owed to one or another of his mathematical precursors. This is an easy exercise since the results of most seventeenth-century mathematicians can be translated into modern symbolism with startling results. For example, J. M. Child who edited and published some of Leibniz's early mathematical manuscripts believed that Leibniz was decisively influenced by (or may have even consciously borrowed a version of the characteristic triangle as well as several other results, especially concerning "transmutation") found in the *Lectiones Geometricae* of Newton's mentor Isaac Barrow.[5] On the other hand, the characteristic triangle was fairly familiar, and there was no necessity for Leibniz to borrow it from Barrow. As J. E. Hofmann has noted:

> The characteristic triangle—to take up a particular point—was known already to Fermat, Toricelli, Hugyens, Hudde, Heuraet, Wren, Neil, Wallis and Gregory long before it was made public by Barrow.[6]

Leibniz himself acknowledges influences concerning the triangle from Huygens, Wallis (1616–1703), and especially Pascal (1623–1662). In tangent theory generally, Hofmann claims that Leibniz "received the decisive impulse not from Barrow and Newton but from Huygens, Pascal, Grégorie de Sainte-Vincent, Mercator, Gregory, and Sluse."[7] There are similar issues of priority or influence for most of the other features of the Leibnizian calculus. If we ask who really discovered the equivalent of the FTC, the chain rule, or differentiation rules for one or another class of functions, we will run into an exceedingly tangled parentage. Efforts to answer such questions have fueled a large monograph literature.

The view of Leibniz which grounds him exclusively in his seventeenth-century mathematical environment and finds disagreement only in his sources is linked to an equally common interpretation of the historical development of mathematics whose ramifications and alternatives we discuss in the remainder of this chapter: While political or historical, cultural circumstances, or the micro-sociology of mathematicians may determine whether or what mathematics is done or stimulate some branches at the expense of others,[8] the *content* of a given field seems independent of its social or

[5] Leibniz (1920), pp. 15f, 136, 173, 175.
[6] 1974a, p 74.
[7] *Ibid.*
[8] For example, as we have mentioned in the Preface, the needs of the military in relation to the development of operations research. Also NSF by its funding policies, which reflect the consensus of elite or at least politically influential mathematicians, encourages some fields, but not others.

intellectual context in a way is not true of other areas of human intellectual endeavor. In this way mathematics apparently differs from philosophy, art, political ideology, or even (possibly) other branches of science. One cannot think of any extraneous intellectual currents, for example, that would inspire theorems of algebraic topology similar to the way Malthus or the ruthless British nineteenth century climate of nascent capitalism provided metaphors that may have stimulated Darwin's concepts of natural selection and the struggle for existence. Leibniz was a Baroque mathematician, and we may call his work "Baroque mathematics" if we like; but the quotient or chain rule and, say, Baroque architecture, painting, or philosophy seem to have nothing in common aside from being contemporaries of each other. Mathematics also seems to be essentially cumulative and progressive; it develops logically out of its own prior content. The discoveries of one period lead in a natural way to those of the next period. If we disregard false starts, accidents of history, or social catastrophe, the progress of mathematics is as predictable as the progression of theorems in the Euclid's *Elements* or in Bourbaki. How can it be otherwise? By its very nature—at least to working mathematicians, if not always to intuitionist philosophers—mathematics is "true", the same in Bucharest, Cambridge, and Pyongyang. This judgment is accepted by the representatives of all epistemological schools and political persuasions.

To show this universality of mathematics, we take a brief look at some representatives of three somewhat extreme positions: Platonism, Social Constructionism, and the subordination of mathematics to political (Nazi) ideology. Roger Penrose has been perhaps the most articulate defender of an uncompromising Platonism.[9] He believes that the "Mandelbrot set" really exists "out there" as an "object," as real as any ordinary physical object but existing in a different realm. Mathematics, therefore, is independent of the human mind. We gradually discover its fantastically complicated and beautiful structure and certainly do not invent it. On the other side, Lancelot Hogben (1895–1975) wrote *Mathematics for the Million* (1940) which was a popular introduction to mathematics in its day and still worthwhile reading for students. Its point of view is that of pre-war British Marxism and exhibits an intense hatred of Platonism. Hogben's criticism, however, is more social than epistemological. He has many diatribes against Platonic "priesthoods" whom Hogben blames for making mathematics inaccessible to the masses. Hogben anticipated the more sophisticated discussions of

[9]Penrose (1989), especially Chapter 3.

contemporary "social constructivists" such as Reuben Hersh (1997) or Paul Ernest (1998). Like Hogben they primarily attack Platonism for its bad effects on education. They accuse it of authoritarianism, fostering a nasty elitism, and discouraging the average student. For them mathematics is entirely invented like any other human cultural product.[10] Turning to politics, even mathematics became partially Nazified in the Third Reich. In 1936 the eminent (and very able) mathematician and SA member Ludwig Bieberbach (1886–1982)[11] founded the journal *Deutsche Mathematik* dedicated to "Aryan mathematics". An early contributor Edward Tornier claimed that Jewish mathematics was infected by excessive abstraction or "objectless definitions" as contrasted to "German" mathematics which "answered concrete questions which concern real objects like whole numbers or geometric objects"[12] In general they simply didn't like modern "formalistic" mathematics which they identified primarily as "non-Aryan" as contrasted for example with the mathematics of Gauss or Klein. (There were probably also some real issues of mathematical substance involved here intertwined with Nazi politics.) At a subconscious level they probably also feared the competition of Jewish mathematicians for jobs and recognition; "in the future we will have German mathematics"[13]

The remarkable feature, however, which separates such quarrels from those in other realms of human culture and politics, is that none of these representatives of different mathematical philosophies would claim that the mathematics of their opponents is *false*. Bieberbach or Tornier, for example, may have found Jewish mathematics for the most part "ugly," but they never claimed that it was false or logically mistaken; indeed, sometimes they even praised it. Likewise, Hogben, Hersh, or Ernest would accept the truth of the theorems discovered by their most arrogant Platonist opponents. There are of course furious debates over just *why* this universality (which is quite unlike politics, religion, or philosophy) should be so; but these debates are mainly the business of academic philosophers concerned with "foundations", whose ruminations are far from the activities or interests of working

[10]Hersh goes so far as to deny that two dinosaurs meeting two other dinosaurs seventy million years ago, yielding (perhaps for a short time) four living dinosaurs, reflects the timeless Platonic truth that $2 + 2 = 4$. See Hersh (1997), p. 15.

[11]Bieberbach's famous function theoretic conjecture of 1916 was solved only in 1985 by de Branges.

[12]Maclane (1995).

[13]Tornier, as quoted in *Ibid.*. For an account of the elaborate psychological theory of mathematical "types" which Bieberbach in particular constructed to justify his prejudices, see the penetrating studies of Mehrtens (1987) and also Segal (1986).

mathematicians. A corollary of the apparently objective nature of mathematics is the continuing production of ever more mathematics—"200,000 mathematical theorems of the traditional handcrafted variety produced annually,"[14] contained in tens of thousands of papers bound in thousands of journal volumes in hundreds of university libraries (whose budgets and perhaps even structural integrity in an era of deferred maintenance are menaced by the accumulating bulk).

All this indicates that the development of mathematics differs not only from nonscientific areas of culture, but even from other sciences. In particular, we can ask if any of the historical ideas concerning scientific change developed by Thomas Kuhn in his famous book *The Structure of Scientific Revolutions* are really applicable to mathematics. Let us recall some of these ideas: Kuhn saw the development of science as a kind of "punctuated equilibrium." There are (often long) periods of what he called "normal science." Here scientists share a common "paradigm," that is, a set of common standards and conventions that prescribe how a particular scientific discipline is to be done, what problems are to be solved, what fundamental theories are to be employed, and what counts as a valid solution. Paradigms are often motivated by some "exemplar" or fundamental work in the field such as Ptolemy's *Almagest* or Newton's *Principia.* Normal science, then, amounts to "puzzle solving," or trying to force nature into the often seemingly Procrustean bed of the paradigm. Anomalies or apparent mismatches between nature and the paradigm are frequently observed, and the business of normal science is to somehow account for them. Scientific revolutions or radical changes in the paradigm can only happen, according to Kuhn, when this activity becomes impossible. Anomalies multiply and become intractable. The best efforts to fit them into the paradigm fail. In consequence, a sense of "scientific crisis" eventually emerges. Only then can a new paradigm be introduced, often by younger researchers having a tenuous connection with established doctrine. Kuhn further argues that the old and new paradigms are "incommensurable," in the sense that is no "rational" means to decide between them. Each paradigm holds to different standards of evidence and interpretation. Observations are "theory laden," and their meaning is relative to the paradigm. Consequently scientists on each side tend to talk past each other. Hence the process of transformation from one paradigm to another is fundamentally "sociological," akin in fact to a political revolution. As a whole, Kuhn believes that new paradigms

[14]Davis and Hersh (1981), p. 24.

arise from factors "internal" to the disciplines, namely the need to explain previously intractable anomalies. But he is willing to admit that external intellectual currents may supply metaphors helpful in the construction of new paradigms or reinforce them. One thinks of Pythagorean ideas in the case of Copernicus, Mathematical Platonism with respect to Galileo's belief in the efficacy of geometry, or Malthus in the case of Darwinism. Kuhn's views also have epistemological consequences: While he admits paradigms grow more sophisticated and can solve more and more elaborate puzzles, none of them are "truer" in the sense of correspondence with reality than any other. In fact, Kuhn doubts that "correspondence with reality" makes any sense. His is a coherence theory of truth. What scientists call reality is the interaction of a noumenal, unknowable world and the paradigm that gives it structure and meaning. In this sense when the paradigm changes so does the world. (Modern scientists actually live in a *different* world, for example, than medieval philosophers.)

One can view Kuhn's work in part as a refinement and extension of earlier ideas due to Ludwik Fleck, a little known Polish physician, survivor of Auschwitz, and medical researcher specializing in serology and bacteriology. In his book *Entstehung and Entwicken einer wissenshaftlichen Tatsache: Einführung in die lehr Denkstil and Denkkollective* published in 1935 introduced the notions of "thought style" and "thought collective." The latter consists of the set of individuals holding a thought style which according to Fleck consists of the readiness:

> ... *for directed perception, with corresponding mental and objective assimilation of what has been so perceived.* It is characterized by common features in the problems of interest to a thought collective, by the judgement which the thought collective considers evident, and by the methods which it applies as a means of cognition. The thought style may also be accompanied by a technical and literary style characteristic of the given system of knowledge.[15]

Furthermore:

> A thought style functions by constraining, inhibiting, and determining the way of thinking. Under the influence of a thought style one cannot think in any other way. It also excludes alternative modes of perception.... A thought style functions at such a fundamental level that the individual seems generally unaware of it. It exerts a compulsive force on his thinking, so

[15]Fleck (1979), p. 99. Emphasis in original.

> that he normally remains unconscious both of the thought style as such and of its constraining character.[16]

Clearly, what Kuhn calls a "paradigm" is an example of a thought style, although Fleck's concept is the more general one.

To this writer's knowledge, no one has attempted to apply the concept of "thought style" to mathematics. Yet Kuhn's views have attracted considerable attention among historians of mathematics. Opinion, however, is divided as to their utility. What are we to make, for instance, of the distinction between "normal" and "revolutionary science" if it is applied to mathematics? Michael Crowe (1975) has asserted as the last of his "ten laws" of mathematical change that: "Revolutions never occur in mathematics." Seconding this position Mehrtens (1976) rejected the concepts of "revolution" and "crisis" in mathematical development, although he has pointed out the importance of "paradigms" which "are shared examples that structure the mathematicians' perception and guide their research"[17] and "anomalies" such as the status of Euclid's fifth postulate "which eventually led to new geometries and to the overthrow of the 'metaphysics' of geometry."[18] Although Mehrtens feels that Kuhnian concepts "centering around the sociology of groups of scholars are of high explanatory power," he thinks that Kuhn has ignored external factors influencing mathematics which can range from developments in other disciplines to "the general material conditions of society" and the social status of mathematicians.[19]

In the last few decades, however, there have been renewed efforts to fit Kuhnian analysis into the history of mathematics. Crowe (1992) has since changed his mind and now accepts the existence of mathematical crises and revolutions. Joseph Dauben (1984) has suggested that one candidate generating an anomaly leading to a mathematical crisis was the invention or discovery of transfinite numbers by Georg Cantor (1845–1918). Combined with the paradoxes of the naive set theory of Frege (1848–1925), this caused a foundational crisis that has not been resolved. Because of it, at the philosophical level at least, there is yet to be a universally accepted "paradigm" to replace nineteenth-century confidence in the foundations of mathematics. Other examples pointed out by Dauben include the Pythagorean discovery of incommensurability, and the dissatisfaction among mathematicians concerning the status of the Euclidian Parallel Pos-

[16] *Ibid.*, p. 159.
[17] p. 309.
[18] *Ibid.*, p. 304.
[19] *Ibid.*, p. 312f.

tulate. And suggested candidates for mathematical analogs of scientific revolutions have included Abraham Robinson's development of nonstandard analysis, the symbolic logic of Boole (1815–1884), Frege, and Jevons (1835–1882), Weierstrass' (1815–1897) ϵ, δ rigorization of analysis, the structural changes in nineteenth century algebra culminating in van de Waerden's (1903–1996) *Moderne Algebra*, or the transition from empirically derived mathematical principles to rigorous deductive proof in the sixth century B.C. But possibly also, revolutions are limited to the "images of mathematics" and do not involve mathematical knowledge itself.[20] However, just what phenomena constitutes a mathematical revolution or "paradigm change" is very much in the eye of the beholder and depends on the semantics of the word "revolution."[21]

It still seems, however, that the consensus of most investigators is that at least the "body of mathematical knowledge"[22] grows in a cumulative non-revolutionary way. Even if there are mathematical correlates of normal and revolutionary science, Kuhn's epistemological pessimism seems ill-placed. If we are formalists the question of whether or not mathematics reflects some underlying reality does not even arise. On the other hand, if we are Platonists mathematics by its very nature mirrors this reality. One cannot argue whether or not later mathematics is "truer" than earlier mathematics. It is all "true," and the only difference is that more of this reality has been discovered. A reasonable conclusion, therefore, is whatever the debatable applicability of Thomas Kuhn's ideas, their later radicalization by Paul Feyerabend, Richard Rorty, and others simply have no place in the history of mathematics. The same seems true of the "sociology of knowledge" or cultural anthropology approaches recently advocated by David Bloor, Steven Shapin or Bruno Latour.[23] It is certainly true (although not always obvious to students) that mathematicians are human. Although as a group perhaps more badly dressed and less colorful than Art or English Professors, they are members of the subspecies *homo academicus*. Hence they may and often do hate each other, turn down each other's grant applications, form cliques and factions, conspire or rant in the hallways over Departmental micropolitics, write unflattering tenure evaluations, and so forth.[24] But even so, the triumph of a mathematical theory seems not in any way to resemble a

[20]See Correy (1993).

[21]There are several interesting articles on these issues in Gillies (1992).

[22]Correy (1993).

[23]See e.g., Bloor (1976), Shapin and Schaffer (1985), or Latour (1985).

[24]In fiction at least tensions among mathematicians have led to murder. We recommend the remarkable novel *The Calculus of Murder* [Rosenthal (1986)].

Bolshevik revolution—or even a change in taste from Augustan to Romantic poetry in eighteenth century England, analogies which Richard Rorty suggests apply to the development of science.[25] Similarly, there can be no convincing claim that Newton's *Principia* was somehow a product—like has been claimed for the controversy between Robert Boyle (1627–1691) and Thomas Hobbes (1588-1679) over the air pump— of political ideologies of Restoration England.[26] Nor does it explain much to derive Newtonian mechanics from the economic or navigational needs of early capitalism and/or imperialism. In summary, mathematical ideas appear not to be "historical" or "socially constructed" at least in the same way as those of poetry, philosophy, or perhaps Darwinian evolutionary theory, and except in the vaguest sense they are resistant to "external" political or cultural influence. This implies that Kuhnian revolutions (assuming they exist) in mathematics arise within the field itself according to technical needs and motivations by the community of mathematical practitioners. A further distinction between the development of mathematics and that of other cultural fields including science (assuming a Kuhnian rather than Whig interpretation) is that old mathematics may be forgotten or reduced in status, but it is not false.[27] It is still is part of the mathematical corpus although it may be replaced by new theories of greater power and abstraction. Indeed, its valuable parts may be incorporated into a new synthesis. This characteristic differs radically from the fate of, say, Ptolemaic astronomy, other defunct scientific theories, or ideas such as the Divine Right of Kings or Transubstantiation. It is evident even to non-Platonists that mathematical ideas are the enduring possessions of the human mind. Being immaterial they

[25]For comparisons between scientific and political revolutions see Kuhn (1962), Chapters XII–XIII or Rorty (1979). According to Rorty (p. 331):

> But what could show that the Bellarmine-Galileo controversy 'differs in kind' from the issues between, say, Kerensky and Lenin, or that between the Royal Academy *circa* 1910 and Bloomsbury?

For the suggestion that the process of scientific change may not be far different than "the shift from the *ancien régime* to bourgeois democracy, or from the Augustans to the Romantics" see *Ibid.* p. 327.

[26]But for an unconvincing claim see Grinell (1973). Steven Shapin and Simon Schaffer in their well known book *Leviathan and the Air Pump* have argued that a famous quarrel between Thomas Hobbes and members of the Royal Society over exactly what Robert Boyle's air pump demonstrated reflected and was decided against Hobbes in great part because of the congruence of the Royal Society's scientific value system and the ideology of the Restoration political elite. But see the devastating refutation of this thesis in Gross and Levitt (1994), pp. 63–68. Hobbes, to put it bluntly, was a mathematical crank.

[27]Dauben (1975), p. 84.

cannot be destroyed like Art or Literature. An extreme opinion of this kind is due to G. H. Hardy (1877–1947), perhaps the most eminent British mathematician of the twentieth century:

> Archimedes will be remembered when Aeschylus is forgotten because languages die and mathematical ideas do not. 'Immortality' may be a silly word, but probably a mathematician has the best chance of whatever it may mean.[28]

It follows, therefore, that long after the Shakespearean canon has perished mankind will contemplate the Hardy maximal function or Hardy's joint work with Srinivasa Ramanujan (1887–1920) on the theory of partitions. The historiographical consequences of this view—which would have been comforting to William Whewell (1794–1866) or George Sarton (1884–1956), if not to Thomas Kuhn—are obvious and still influential. In many textbooks and in the popular literature we get hagiographic (often inaccurate) chronicles[29] of glorious discoveries; on the more serious monograph level there is a careful tracing of real or imaginary mathematical influences of one mathematician upon another or the analysis of anticipations, antecedents, and missed opportunities relating to the "discoveries" we judge significant today. In any case, there have been until recently few attempts to connect the mathematics of a period to the surrounding culture (this again may be an impossible task given the timelessness and abstraction of the discipline) or to consider it in all its peculiarities on its own terms. As a result, the history of mathematics can become a species of antiquarianism, neither of great interest to professional historians (who lack the necessary technical background) nor to mathematicians who are too busy proving new theorems and whose interest in history (never strong in school—there were just too many "facts" to memorize) stops with last year's preprint circuit.

---

[28] Hardy (1993), p. 81.

[29] A notorious example is Bell (1965).

Chapter 2

Some Issues Raised by Seventeenth Century Mathematics

The previous description of mathematical change may be a caricature, but much of it is undoubtedly true. Whether we are Platonists, believing that mathematics discovers a timeless realm of super-sensible abstract truth, or think of mathematics as some sort of "social construction" conforming to the standards of a community of mathematicians, who are themselves affected by external cultural factors, the actual theorems of mathematics seem more usefully characterized as "discoveries" than "inventions" or "constructions," since they are initially unknown and follow (often only after a great deal of hard work and ingenuity) from a more general conceptual structure. We may or may not "invent" this structure, but much mathematical activity consists of the discovery of its consequences. For example, if we cast aside the problem of the ontological status of the system of positive integers, it is a fact that both the Lagrange Four Square Theorem stating that every positive integer can be expressed as the sum of at most four (not necessarily different) square integers[1] and the theorem that every even number is the sum of at most 300,000 primes[2] are true consequences of that system. Before their discovery we were just as unaware of them as medieval Europeans were of the New World. On a more elementary level once we have discovered (or invented) the concept of the derivative y' of a function y, then the quotient rule

$$\left(\frac{f}{g}\right)' = \frac{gf' - fg'}{g^2}$$

follows *necessarily.* In many cases we may also conjecture that certain mathematical statements are true without having discovered whether or

[1] For example, $459 = 15^2 + 9^2 + 12^2 + 3^2$. A proof of this theorem is given in any elementary text in number theory. See, for instance Burton (1998), Chapter 12.

[2] According to Dunham (1990), p. 83 this fact was proven by the Soviet mathematician Schnirelman in 1931.

not they actually are. In these circumstances our position is analogous to that of Columbus who believed, but did not know, that he might be close to China or that of a scientist conjecturing yet unverified properties of an unfamiliar virus. For instance, we believe that there is an infinity of twin primes (like, (3,5), (5,7), (11, 13), (17, 19), etc.) or that the Goldbach conjecture that every even number greater than or equal to 4 is the sum of only *two* primes (much better than 300,000!) is true,[3] but we could be wrong and in the same position as Columbus concerning the location of the New World. In all these cases and in countless others human wishes, ideology, or social circumstances seem irrelevant. In this sense the actual content of mathematics is certainly "internal," developing out of its past substance, and representing cumulative knowledge. We know far more mathematics today than, say, a century ago, and the same was true in 1911 as compared to 1811.

Furthermore, as we will repeatedly emphasize, past mathematics remains *true*. That there is an infinite number of primes has been known since the proof of this fact was recorded by Euclid about 330 B.C. No one, if they are acquainted with primes, believes otherwise. While the arithmetic relation $7+5 = 12$ had a different epistemological significance for Plato and Kant, neither they or anyone else have ever claimed that $7 + 5 = 13$. The identity $2 + 2 = 4$ is true for everyone (except for O'Brien in his final interview with Winston). While the mathematics of different cultures may be "different" in respect to subject matter, values, or procedures, it is (barring a few mistakes) still correct. The same constancy of truth through time is not found in the other intellectual productions of the Museum of Alexandria (where Euclid worked) or for that matter in *any* of the theses of Plato, Kant, or any other philosopher. Because of these characteristics mathematics seems quite different from areas of intellectual human activity such as art, philosophy, political ideology, or even other sciences; and consequently the claims of the Strong Programme and other forms of relativism are simply false.

Yet as has been conceded, a variety of non-mathematical political or cultural factors can indirectly determine mathematical content by simply

[3]As of August 2009 the largest known pair of twin primes was $5516468355 \times 2^{333333} \pm 1$. The Goldbach conjecture was stated by the mathematician Christian Goldbach (1690–1764) in a 1742 letter to Leonhard Euler (1707–1783). We do know many approximations to the conjecture, including the one stated above, as well as the fact that it true up to $2n = 12 \times 10^{17}$. But the conjecture itself remains unproven. Short but informative discussions of the twin prime and Goldbach conjectures can be found in Wikipedia or Burton (1998).

encouraging or discouraging certain areas, the investigation of certain problems, and so forth. But at the risk of contradicting some of the judgements of the previous chapter, we can probe more deeply by asking if these kind of causes can also *directly* help to shape the style, goals, standards, aesthetic values, or form and content (we are a bit vague here) of the overarching mathematical structure from which the true consequences are derived. To give some cultural analogies: everyone realizes that the art of Meissonier differs from that of Damien Hirst, that the novels of Bulwer-Lytton differ from those of Kurt Vonnegut, that the operas of Mozart differ from those of Philip Glass, and that these differences perhaps have more to do with profound cultural shifts than the personal talents and characteristics of the individual artist or author. Can there be similar culturally determined differences between mathematicians, especially mathematicians of different cultures or different periods? In other words, is mathematics, after all, in a sense similar to other products of human culture such as art, literature and music? Can it change fundamentally with respect to the aspects mentioned above in ways that are not reducible to mere cumulative "progress" (while remaining "correct"), so that the mathematics of one period is not merely the sum of previous mathematics plus undiscovered theorems, but is also essentially *different* and while not falsifying past mathematical results (as later science often does previous science) is psychologically incompatible with it?

To express these questions in a stronger form, can some modification of the ideas of Ludwig Fleck or Thomas Kuhn—despite the reservations we have raised—be applicable to the development of mathematics? If we agree that scientists are sometimes locked into "incommensurable paradigms" or "thought styles" which can only be changed in a revolutionary way, can there be analogies (at least on occasion) in the development of mathematics? And if this is so what are the causes? Are they "internal" (proceeding from felt inadequacies or anomalies in the structure of previous mathematics) or "external" (proceeding from cultural or nonmathematical intellectual changes)? Also does this mean that in some sense aspects of mathematical change may be "irrational" as Kuhn argued is often the case in the change of scientific paradigms? To be specific, was Leibniz's invention of the calculus analogous to a paradigm change or revolution in science? And if so, are there cultural factors in the seventeenth century (or before) which, interacting with the fertile mind of Leibniz, *in addition to* previous work on tangents and quadratures helped to cause this revolutionary shift? In the present essay we are going to suggest a qualified affirmative answer to both questions by sketching an interpretation of Leibniz's achievement which differs from the conventional one summarized in the previous chapter and which separates him from any simple cumulative, non-revolutionary historical model. Perhaps a reexamination of Leibniz and his period will suggest a less straightforward and more contingent picture, at least in one instance, of the process of mathematical change and will show how peculiar and long forgotten cultural phenomena can have mathematical consequences. Leibniz will emerge in our view as a problematic, complicated figure whose relation to previous mathematical trends is not at all clear. Perhaps it is even the case that in some important aspects, his version of the infinitesimal calculus is not derived from previous mathematics at all.

But before arguing this thesis by looking at Leibniz's own work, we need to get a feeling for the late sixteenth to mid-seventeenth century mathematical "paradigm" or "thought style" which Leibniz helped to destroy. We cannot present here a thorough history of early modern mathematics or of infinitesimal methods before Leibniz; that job has been done by the authors mentioned at the beginning of the previous chapter. We want only to highlight the salient features of the period in order to see in what ways Leibniz's work was revolutionary.

We begin by admitting that the standard view of the discovery of calculus given in Chapter 1 is true—up to a point. There is no doubt of a tremendous acceleration of mathematical development from the mid-sixteenth century onward. A mathematical version of Renaissance humanism paralleling the rediscovery and veneration of Cicero, Tacitus, and Virgil revived interest in the mathematics of antiquity especially the geometry of Euclid (c. 300 B.C.), Archimedes (d. 212 B.C.), Apollonius (c. 262–c. 190 B.C.), Pappus and the number theory of Diophantus (f. 250 A.D.). The major architect of this process "was Federigo Commandino (1509–1575), who singlehandedly prepared Latin translations of Euclid, Apollonius, Archimedes, Aristarchus, Autolycus, Hero, Pappus, Ptolemy, and Serenos."[4] There were certainly many types of mathematical practitioners before Commandino such as Italian or German "cossist" algebraists, "applied mathematicians, mystics, artists and artisans, and the analysts."[5] But his work had the same kind of impact on mathematics as the rediscovery of Aristotle had on the medieval intellectual world in the twelfth and early thirteenth centuries. It offered exposure to some extremely impressive mathematics (which in rigor and mathematical beauty is still unsurpassed). It guaranteed that the field

[4] Mahoney (1994), p. 3f.
[5] *Ibid.*

of highest prestige would be geometry as practiced by the ancients. For almost all mathematicians well into the seventeenth century the works of Euclid, Apollonius, Archimedes, and Pappus functioned as a kind of Kuhnian exemplar. They indicated the way mathematics should be done, what standards it should meet, the general type of problems that were of interest, and what should constitute a precise solution to them. Until—and even well beyond—the introduction of algebra by François Viète (1540–1603) towards the end of the sixteenth century the work of the most elite mathematical practitioners consisted of developing and refining it. To their minds real mathematics *was* geometry. Great emphasis was placed on solutions to problems by geometric constructions, especially problems related to the three classical problems of antiquity: the squaring of the circle, the trisection of angles, and the duplication of the cube (equivalently the finding of two mean proportionals between two line segments a and $2a$, or more generally between given line segments a and b).[6] There was a consuming interest in how such problems should be solved based on a classification made by Pappus (*ca.* 320 A.D.) in Book III of his *Mathematical Collection.* He had divided construction problems into three types: "plane," "solid," and "line-like." The first consisted of problems whose solution required only lines and circles, the second conic sections,[7] and the third–or "line-like problems" more complicated "mechanical" curves introduced in antiquity, such as the cissoid of Diocles, spiral of Archimedes, conchoid of Nicomedes, or quadratrix.[8] The desired goal was to solve a particular problem using the simplest means possible. One should not, for example, use a conic section on a plane problem such as the Apollonian three circle tangent problem[9] or some esoteric mechanical curve on what was really a solid problem. Most mathematicians (with some exceptions) agreed with the ancients that the three classical problems were not plane and required either conics or mechanical curves for their solution; indeed Pappus and other Greek geometers had already proposed solutions of this type for these and related problems. This basic division of the types of problems one might encounter and its

[6]That is, finding line segments x and y such that $a : x = x : y = y : 2a$. It follows that $x^3 = 2a^3$.

[7]These problems were called "solid" since a conic section is a section of a solid, i.e., a cone cut by a plane.

[8]See Appendix E for the definitions and properties of some of these curves.

[9]That is, given three circles fixed in position, it is required to draw all circles that are tangent to all three. In 1596 Adriaan van Roomen (1561–1615) published a "solid" solution to this problem using hyperbolas, while Viète was able to show that this could be done using ruler and compass only, and therefore was a "plane" problem.

consequences was probably the most fundamental theoretical construct of late sixteenth and early seventeenth century. It stimulated an enormous volume of commentary and many original solutions which often demonstrated that what some ancients had considered "line-like" problems like angle trisection or doubling a cube were really "solid"[10] or that an apparently "solid" like the previously mentioned Apollonian circle tangent problem was in fact plane. Most were reworkings of facts that had already been discovered in antiquity, but they sometimes contained valuable generalizations and improvements in technique.

A related area of research well into the seventeenth century consisted of attempts to restore works, especially by Pappus, Euclid and Apollonius, that had either been lost or existed only in fragmentary form, perhaps as summaries by Pappus or other ancient authors.[11] Also, even when they had survived, the results were presented in a polished and opaque form. The result and its proof were often quite elegant, but the motivation was concealed. How were the incredible theorems of, say, Archimedes and Apollonius actually discovered? It was felt that the ancients had some method of discovery which they had deliberately kept from vulgar eyes, and that the task of mathematicians was to rediscover it. Here is a quotation from Descartes which expresses a typical attitude. He is referring to a now invisible "true mathematics" which lay behind their published work. He says that the ancient geometers have:

> ... with a kind of pernicious cunning subsequently suppressed this mathematics, as we know many inventors to have done in the case of their discoveries. They may have feared that their method, just because it was so easy and simple, would be depreciated if they had divulged it; so in order to gain our admiration, what they presented us with as the fruits of their method were some sterile truths demonstrated by clever arguments, rather than giving us the method itself, which would have dispelled our admiration.[12]

[10]Pappus had used the quadratrix possibly following Hippias of Ellis (mid to late 5th-century B.C.), a Greek sophist and geometer who invented it to trisect or more generally to divide an angle in a given ratio (see Heath, I, (1981), p. 226–229). Diocles (2nd century B.C.) had used the cissoid to find two mean proportionals, although by the time of Pappus it was realized that this problem could be solved using conic sections. See Bos (2001) for an excellent discussion of these issues.

[11]This was the case, for instance, of large portions of the *Collections*, the *Conics*, *Porisms*, *Surface Loci*, or a treatise on mechanics by Euclid, and several works of Apollonius dealing with various construction problems including the three circle tangent problem. Francois Viète published his restoration of the latter work in 1600. See note 9.

[12]Quoted in Gaukroger (1998), p. 99f.

In an obscure passage at the beginning of Book VII of his *Collection* Pappus had written of two geometric methods "synthesis" and "analysis." The former amounted to reasoning from the premises to the conclusion or giving a construction using the known facts to obtain the desired result. Analysis was more interesting (and mysterious). It was of two kinds "theorematic" and "problematic," the former relating to theorems and the later to constructions. In either case we assume the theorem to be true or the construction done and explore the consequences. If one of them is false, then the theorem is false or the construction impossible. If we arrive at a known theorem or property, then by reversing the reasoning (which Pappus seems to assume is always possible) we can prove the theorem or carry out the construction. Early modern mathematicians were fascinated by this definition and assumed that the ancient geometers had some hidden method of realizing it.[13]

From hindsight one might attribute the beginning of a change (or perhaps breakdown) in the geometric paradigm cultivated by sixteenth and early seventeenth century mathematicians to the introduction of algebraic methods, previously mostly limited to the work of Italians such as Tartaglia 1499/1500–1557), Cardan (1501–1576), Bombelli (1526–1572) and others and confined to arithmetic or "word" problems. One of the earliest applications of algebra to geometry is found in the work of François Viète (1540–1603), principally in his *In artem analyticem* (1591) and *Supplementum geometriae* (1593). Perhaps the leading European mathematician of his time, Viète was a Catholic lawyer who worked first for a female Huguenot aristocrat Antoinette d'Aubeterre (1532–1580), the wife of an important Huguenot military leader, by supervising the education of her daughter, and later became Privy Councillor and code breaker to Kings Henry III and IV. Somehow he survived the vicious civil strife between Catholics and Calvinists that tore France apart in this period, although he was accused by Catholics of showing Protestant sympathies. Viète had ambitious mathematical aims. His slogan expressed at the end of his *In artem analyticem* was "*Nullum non problema solvere*," i.e., "To leave no problem unsolved."[14] It is related that in a few minutes he solved an equation of degree 45 due to the Dutch mathematician Adriaan van Roomen which was presented to

[13]A translation of the passage is given in Bos (2001), p. 95f. See also Pappus (1986), I, pp. 82–85. For an analysis of various issues and difficulties in the passage, especially relating to Pappus' apparent assumption of logical reversibility in analysis, see Mahoney (1968).

[14]Klein (1972), p. 154, 353. Klein in the Appendix to his book includes a translation of *In artem analyticem*.

Henry IV by the Dutch ambassador as a problem which no French mathematician could possibly solve.[15]

The originality in Viète's use of algebra in geometry lay in its application to construction problems. Given such a problem, his method was to find an algebraic equation in one unknown relating the known and unknown line segments, solve for the unknown, and translate the solution back into a formal construction. In this way he could solve equations up to the fourth degree. If the equation was quadratic, following the terminology of Pappus, the problem was said to be "plane," if cubic "solid." For example, in the case of the angle trisection problem if X is the radius of the circle, E the chord subtending the third of a given angle with chord B, Viète finds that "*X quadratum in E ter minus E cubo, aequetur X quadrato in B*" or in modern notation:

$$3X^2E - E^3 = X^2B.^{16}$$

Viète identified this process with the lost method behind the ancient Greek procedure of analysis, but divided it into three instead of two types as Pappus had done. What Pappus had called "problematic" analysis to Viète amounted to finding an "equation or proportion between the magnitude that is being sought and those that are given," and the "theorematic" kind is an "art by which from the equation or proportion the truth of the theorem ... is investigated." Finally, the third kind which Viète calls "rhetic or exgetic" consists of producing the magnitude sought from the equation or proportion.[17]

At this point we need to mention some complications of early modern mathematics that separates it from modern approaches which are present in most of the mathematicians of the period, including Leibniz. In the first place, the ancients had no concept of a real number. The fact that the hypotenuse of an isosceles right triangle is incommensurable with its side, indicated to them that line segments rather than numbers should be a fundamental measure of quantity in geometry or what Euclid called a "magnitude." They also allowed areas and angles to be magnitudes. This view had a number of consequences: In the first place, a ratio was, in general, not a quotient of two numbers as it is today. A ratio of two line

[15]The equation expressed the relation between $\sin(x/45)$ and $\sin(x)$.

[16]Bos (2001), p. 151.

[17]Klein (1972), p. 320f. Viète also called problematic and theorematic analysis "zetetic" and "poristic." Further information concerning Viète's applications of algebra to geometry can be found in Alvarez (2008), Freguglia (2008), or Mahoney (1994), Chapter II.

segments, for example, could not be represented as a common fraction unless the segments were commensurable (i.e., integer multiples of a common segment). Secondly, ratios could only hold between geometric entities of the same kind. Two line segments had a ratio, but not a line segment and a rectangle, or a line segment and an angle. It followed that because real numbers were not available to represent magnitudes of disparate types of objects they could not be compared with each other; such comparison was only possible between magnitudes of the same kind. However, the Greeks also recognized that two ratios could be compared; it would make sense to compare the ratio of two segments to that of two areas, for instance. What then were ratios and how can they be compared? In Book V of the *Elements*[18] Euclid gives the following definitions which were probably due to Eudoxus of Cnidus (c. 410 – c. 347 B.C.): First, a ratio is a "sort of relation with respect to magnitudes of the same kind." Magnitudes A and B then are said to have a ratio if multiplied by some integers k, l, $kA > B$ and $lB > A$. Secondly, magnitudes A and B are said to have the same ratio as C to D if for any two positive integers k and l one of the following possibilities hold: $kA > lB$ and $kC > lD$, $kA = lB$ and $kC = lD$, or $kA < lB$ and $kC < lD$. In a similar way, one can define what it means for the ratio A and B to be greater or less than the ratio of C to D. From these definitions all the standard facts about ratios that we now write in fractional form and take for granted can be proven, although the proofs in some cases are laborious. For instance, in Proposition 16 Euclid shows (using ":" to denote ratio) that if $a : b = c : d$, then $a : c = b : d$ (provided the four magnitudes are of the same kind) and in Proposition 22 that if $a : b = d : e$ and $b : c = e : f$, then $a : c = d : f$. This doctrine was held by the overwhelming majority of early modern mathematicians although there were some exceptions.[19] Even a physicist such as Galileo (1564–1642), who was also a well trained mathematician accepted the Euclidean doctrine of ratios. It seriously complicated propositions concerning motion in a dialogue he began to compose late in his life.[20]

The Euclidian concept of magnitude also implied both that algebraic operations could only be applied to magnitudes of the same type and that

[18] Euclid, II, (1956).

[19] John Wallis (1616–1703) was one. He considered arithmetic was the foundation of algebra not geometry and he gives an arithmetic interpretation of Euclid's ratio theory arguing that ratios are essentially quotients. Representing geometric entities by numbers, moreover, makes it possible to give numerical meaning to geometric operations. Dimensional homogeneity (see below) is no longer required.

[20] Heilbron (2010), p. 352.

there was a problem of interpretation concerning the operations of "multiplication" and "division" applied to magnitudes. While geometric objects of the same dimension such as line segments, rectangles, or cubes could be "added" to or "subtracted" from each other, in Greek geometry the addition or subtraction of objects of different dimension, e.g., a line segment and a rectangle, had no meaning. Also while the "product" of two or three segments could be viewed as a plane rectangle or rectangular box, there was no geometric object corresponding to the product of four segments or two rectangles.

Viète and others who followed him in applying algebra to geometry avoided some of these problems by being willing to accept higher powers than the third as abstract entities or "magnitudes" denoted by capital letters with no geometric interpretation. These magnitudes they viewed as occupying a dimensional ladder. The first rung of which Viète called "side," the second "square," and the third "cube." Higher rungs were indicated by writing "A square cube side" and so forth. But, following Greek precedent, every term of an equation was required to have the same dimension. Furthermore, only magnitudes on the same rung could be added and subtracted, while multiplication of, for example, "A-square" by B-cube which Viète wrote as "A-square in B-cube gave a result on the appropriate higher rung.[21] This demand for homogeneity sometimes introduced extreme awkwardness into the algebraic analysis of geometric problems.[22] We will see these conventions being rigidly applied by most mathematicians down to and including Leibniz.

Despite his use of algebra in geometry without which calculus would have been impossible, in many respects Viète was even more conservative than his contemporaries. For him algebraic operations in themselves as well as those designating geometric operations had to preserve dimensional homogeneity since algebra is prior to and applies to either geometric magnitudes or numbers[23] He also entirely rejected solutions of non-plane problems by the intersection of curves whether they were conics or mechanical curves. He felt that these had no place in geometry. Instead he wished to add an additional postulate to Euclid. This was that it was possible to draw a straight line L_1 from a point P to two given lines L_2, L_3, so that the intercept of L_1 AB between L_2 and L_3 is some given line segment CD. (Figure 2.1)[24]

---

[21] Mahoney (1994), Chapter II.
[22] Bos (2000), p. 151 and Chapter 8.
[23] Klein (1972), p. 138.
[24] Bos (2000), p. 168 and Klein (1972), p. 352 (translation of *In artem analyticem* in Appendix).

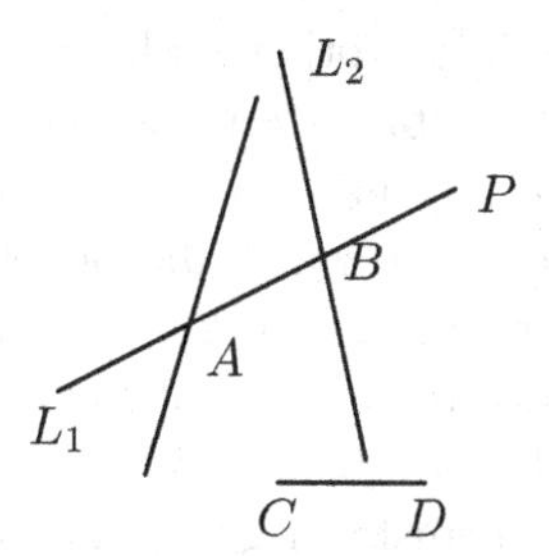

Figure-2.1

This idea was not original with Viète; versions which Pappus called "neusis" constructions are found in the *Collection.*[25] Here the given lines might be two straight lines, or a circle and a given line. They can also be found in Archimedes, especially in his *On Spirals*, and in Apollonius.[26] By using this postulate, Viète was able to give original solutions to two of the classic problems of antiquity: duplication of the cube and angle trisection[27] in addition to the construction of the roots of various cubic and quartic equations which he reduced to either angle trisection or to the construction of two mean proportionals.[28]

Viète and those who pursued related approaches such as Christopher Clavius (1538–1612) and Marino Ghetaldi (1568–1626) had enormous mathematical talent.[29] Considered qualitatively, their achievements certainly rival those of later mathematicians who are celebrated today. However, their work which exemplified the classical geometric paradigm or thought style just as it began to be penetrated by algebra has been essentially forgotten. But although their constructions were original, many of their techniques were not. Neusis constructions or the intersection of conics had, as we have indicated, already been used frequently in antiquity for cube duplica-

---

[25] "Neusis" means "verging" in Greek. The name comes from the fact that the neusis idea was often expressed as the possibility of "producing" or extending a line from the given point so that the intercept criterion was satisfied.

[26] Heath II (1981), pp. 65–68, 182–192, 401, and 412f.

[27] Earlier neusis constructions to solve these problems had also been given by the Greeks. See Appendix E.

[28] *Ibid.*, pp. 172–175. Viète's angle trisection appears to be a rediscovery of a construction attributed to Archimedes (*Ibid,* p. 172). See Appendix E.

[29] Clavius was a German Jesuit mathematician and astronomer who played a central role in the creation of the modern Gregorian calendar. Ghetaldi who was born of a noble family in what is now Croatia was a respected mathematician who was a friend of Viète and who corresponded with Galileo. He constructed a large parabolic mirror which is now kept in the National Maritime Museum in London.

tion, angle trisection, and various other problems requiring the geometrical equivalent of the solution of cubic or quartic equations. Algebraic analysis of geometric problems actually had not been anticipated by the Greeks, but early modern geometers as a whole did not realize this. They tended to identify their new algebraic techniques as a tool of analysis described by Pappus or, more generally, a restoration of the method of discovery which had been so carefully hidden by the ancients.

Algebra only began to transform geometry, rather than merely being an aid to traditional constructions, in the work of René Descartes. His *Géométrie* published in 1637 made two central advances. First, by introducing a unit line segment, Descartes was able to generalize the notion of algebraic multiplication and division to line segments in a new way which avoided the complexities of dimension. Instead of producing a rectangle the "multiplication" of two line segments gave just another line segment. This was accomplished by a simple similar triangle argument. In Figure 2.2 let AB be of unit length and BD, BC given line segments, then $BC : BE = AB : BD$ by similarity. Descartes regarded the product of BD and BC as BE. Similarly, BC can be regarded as the quotient of BE divided by BD. If in the second semicircular figure, having center K and radius KF, FG is taken as unity, then the square root of GH should be GI since this segment is the mean proportional between FG and GH.

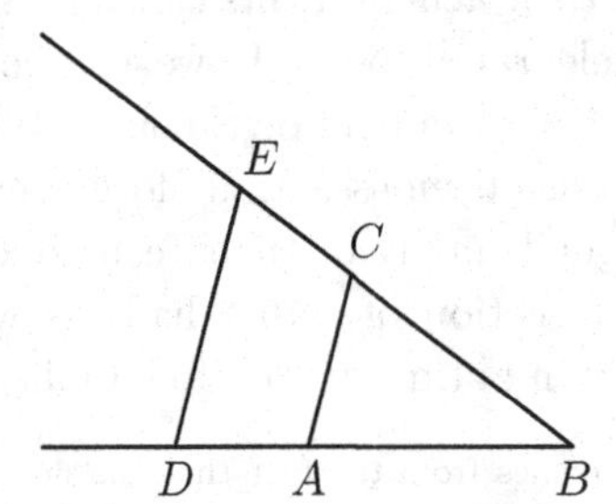

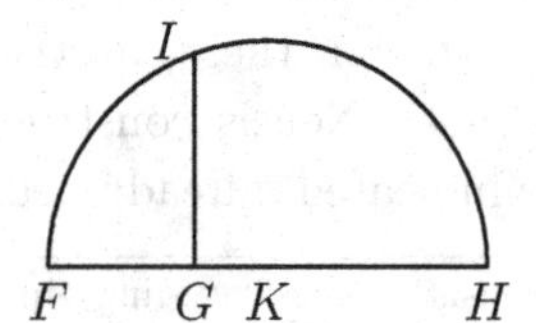

Figure 2.2

In this way equations involving geometric quantities no longer had to be homogenous, for if a, b, and c represented line segments $ab+c$ had meaning where this was not true if ab was a rectangle. However, Descartes' definition introduced a certain relativity in line segment multiplication. While the rectangle with sides a and b was uniquely determined, the line segment representing ab depended on the choice of the unit segment. This was felt to introduce an arbitrary numerical element into what should be a purely geometric argument. Consequently Descartes hedges a bit. He requires homogeneity when "unity is not determined by the conditions of the problem," but remarks that given a unit line segment one can always think of the c in $ab + c$, for example, as multiplied by as many unities as needed to preserve homogeneity.[30] Secondly, while earlier mathematicians had found roots of equations via the intersection of curves, Descartes showed how to associate the curves themselves with equations in two variables. In particular, he was able to demonstrate that equations of the form

$$ax^2 + by^2 + cxy + dx + ey + f = 0$$

corresponded to one of the conic sections or a line.[31] Descartes algebraic method reached a wide European audience and was quite influential in great part due to a commentary on and a Latin translation of his *Géométrie* by Franscisus van Schooten (1615–1660), a professor of mathematics at Leiden. It was through this book that the young Newton, for example, received his primary introduction to modern mathematics.

The long term impact of Descartes approach was certainly revolutionary. Together with Leibniz's calculus it was an essential factor in the decline of the geometric paradigm. Yet it must be emphasized that in many ways Descartes was still a transitional figure. His role in mathematics can be compared to that of Copernicus in astronomy. Both men instituted profound conceptual changes in their disciplines, but in important respects they were still locked into a fundamentally classical paradigm. Copernicus' planetary orbits were constructed using all the technical tools such as epicycles in the Ptolemaic tradition, and Descartes shared many traditional mathematical values. To be sure, he went further than anyone else in uniting algebra and geometry by relating curves to equations, but his curves are not described via a "cartesian coordinate system" which associates points in the plane with ordered pairs of numbers and vice-versa. From our vantage point this seems implicit in his work, but it almost certainly did not occur to him; instead he is still thinking of line segments instead of real numbers. For Descartes (see Figure 2.3) a given curve $\mathcal{C}$ is a locus of points defined by drawing a line PQ parallel to a fixed given line RS (our y axis) through

[30] *Ibid.*, p. 6.

[31] In this, however, Descartes had been anticipated by Fermat in his *Introduction to Plane and Solid Loci* of 1636.

a point P on the curve OPC and intersecting another given line ST (our x-axis), non-parallel to RS. Then the line segment PQ is described by an equation in terms of a parameter (our x) which is the segment OQ between the intersection of PQ and ST and some convenient fixed point O on ST.[32]

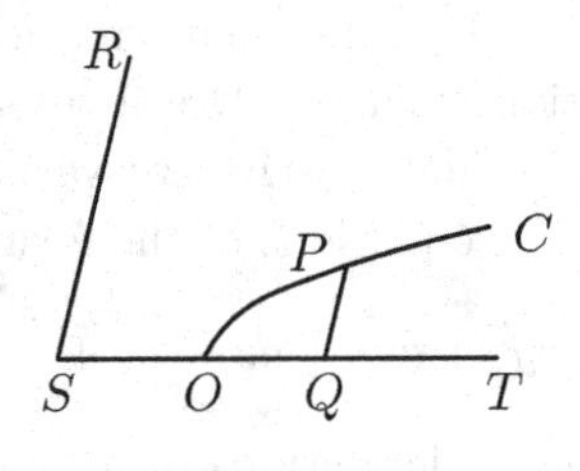

Figure-2.3

As it stands this scheme is not so different from the algebraic analysis given by Viète and others for some geometric diagram.[33] And as Emily Grosholz has observed, Descartes had no interest in using his algebra to study new curves for their own sake. His curves are all adjuncts to construction or root finding problems.[34] In Descartes' eyes a primary illustration of the power of its new technique lay in its application to a peculiarly difficult ancient locus problem recorded in Book VII of Pappus' *Collection* which had also been studied by Euclid and Apollonius. The solution of this problem occupies a major portion of the *Géometrié*. The problem is the following: Given three or four fixed lines L_1, L_2, L_3, L_4 in the plane, find the locus of points P such that in the case of three lines if line segments PP_1, PP_2, and PP_3 intersect the three lines at P_1, P_2, and P_3 with given angles θ_1, θ_2, and θ_3, then the square with side PP_1 has a given ratio to the rectangle with sides PP_2 and PP_3 (Figure 2.4).

[32]The notation here is my own.

[33]It is also nearly equivalent to a way of describing curves given in Fermat's *Introduction to Plane and Solid Loci* that was circulating in unpublished form in Paris in late 1636 or early 1637. [Mahoney (1994), p. 51.] Fermat intended his work as a restoration and extension of Apollonius' lost *Plane Loci*. But there are subtle conceptual differences between Fermat and Descartes. Fermat conceives L_1 as moving parallel to itself along L_2 so that $\mathfrak{C}$ is a *locus* described by motions. Fermat also accepted the Viétean need for homogeneity. See Mahoney (1994), pp. 76–92.

[34]Grosholz (1991).

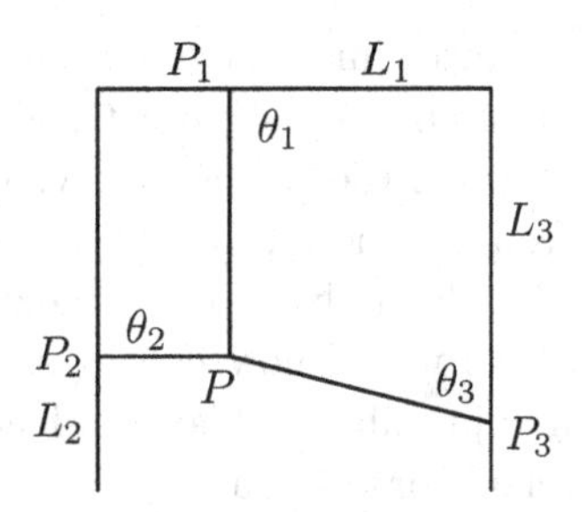

Figure-2.4

For four lines we require that the ratio be between the rectangle R_1 with sides PP_1 and PP_2 and R_2, the rectangle R_2 with sides PP_3 and PP_4 (where as before PP_4 is a line segment intersecting L_4 with fixed angle θ_4). Pappus states that in these two cases the locus is a conic (In Figure 2.4 where $\theta_1 = \theta_2 = 90°$ it is an ellipse). The problem he says may be generalized to five or six lines; but here the loci are curves "whose origins and properties are not yet known."[35] Beyond six lines the Greeks had no natural geometric interpretation of the problem, but Pappus speaks of a generalization to n lines by requiring that the line segments PP_i PP_{i+1}, $i = 1, \ldots n-1$ have given ratios to each other if n is even, or if n is odd PP_n have in addition a given ratio to a given line segment.[36] In the three or four line case Descartes uses his method to give a new proof that the locus is a conic section (or in degenerate cases a straight line or a circle). For more than four lines Descartes is able to give the natural generalization because he is able to interpret the product of more than three line segments. For example, if there are seven lines we require that the product of four line segments have a given ratio to the product of the other three; for eight that the product of four segments have a given ratio to the product of the other four, etc. In the general case he is also able to find the equation of the curve and show that it is given by a polynomial equation of degree $i + 1$, $i \geq 1$, if $2i + 1 \leq n \leq 2(i + 1)$.[37] When the locus was a conic and Descartes had derived its equation, he does not *graph* the curve. Instead, he finds certain key parameters such as the latus rectum of a parabola and then appeals to the relevant theorem in Apollonius to determine it.

Descartes is also conservative in accepting the ancient concept of the

[35] Pappus (1986), I §36, p. 120f.

[36] *Ibid.*

[37] Descartes (1954), Book I, pp. 24–37 and Book II. The notation is mine. Descartes just states (p. 25) the result up to $n = 12$ "and so on to infinity."

essential intractability of "mechanical" curves, although in an extended form. He widens the class of curves his geometrical analysis can consider. He states that all curves beyond the conics were viewed by the ancients as mechanical and barred from geometry, and that sometimes they did not even fully accept the conics themselves. In his new system, he will allow not only conics but what we would now call algebraic curves defined by polynomial equations while prohibiting transcendental curves such as the spiral of Archimedes, cycloid, or quadratrix "which truly belong only to mechanics and are not among the number that I think should be included here."[38]

Book III of the *Géométrie* is more purely algebraic. A large preliminary portion consists of what we would call the "theory of polynomial equations." Descartes answers a series of questions such as "how many roots each equation can have" including "true" (nonnegative) or "false" (negative) ones,[39] "how to make false roots true without making true ones false," etc. However, here also a root for Descartes is a line segment not a number, and the polynomial expression defining the equation as for Viété consists of operations applied to line segments (the only novelty is that homogeneity is not required due to the presence of a unit segment). The payoff of Descartes theory is its application to solid or super-solid problems that lead to cubic or quartic equations. Like earlier geometers going back to the Greeks who considered specific cases which arise from angle trisection or the finding of two mean proportionals, Descartes shows that the solutions can be found via the intersection of conics. Specifically, after noting that the general quartic equation can be reduced (by a transformation which removes the cubic term) to the form

$$z^4 = \pm pz^2 + \pm qz + \pm r = 0$$

where $p, q, r \geq 0$, he shows that the roots amount to what we would call the x-coordinates of the points of intersection of the parabola $y = -x^2$ and the circle with radius $HE = \sqrt{(\mp p - 1))^2/4 + q^2/4 + \mp r}$ and center

[38] *Ibid.*, p. 191. Descartes thinks that in the non-polynomial case the motions describing the curves cannot be exactly compared with each other; in the case of the spiral of Archimedes, for example, the spiral is described by a uniform motion from the center combined with a uniform motion along the circumference of an increasing circle. But such motions cannot be exactly compared. See also the interesting discussions of Molland (1976) and Mancosu (1992) on these issues.

[39] Part of the answer is given via his well known Rule of Signs: "An equation can have as many true roots as it contains changes of sign, from + to − or from − to +; and as many false roots as the number of two + signs or two − signs are found in succession." *Ibid.*, p. 373.

$(\pm q/2, (\mp p - 1)/2)$. Descartes treats all the different sign cases on the coefficients separately. Figure 2.5 show the construction for $-q$ and $+p$.

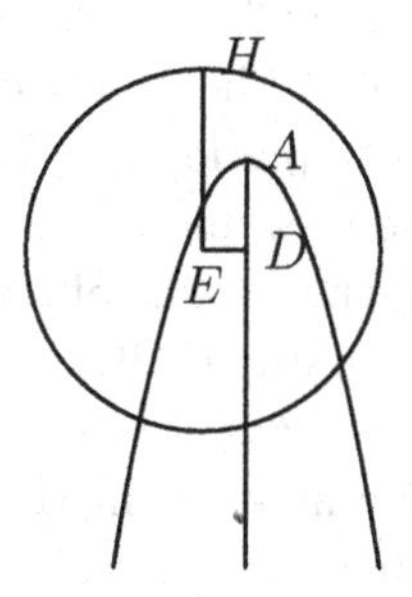

Figure-2.5

Like Apollonius or Archimedes he gives the construction and demonstrates that it works, but without any motivation or derivation. However, the analysis that led Descartes to his result was probably similar to the following. For clarity we will use the "modern" version of Descartes' mode of analysis where points are represented by coordinate pairs of real numbers. However, Descartes almost certainly would have thought solely in terms of line segments. Let a be the horizontal distance from the center of the circle E to the axis of the parabola, and let $b = AD$, and R the radius HE. Then if (x, y) signifies a point on the circle

$$(x - a)^2 + (y - b)^2 = R^2.$$

Expanding we get

$$x^2 - 2ax + y^2 - 2yb = R^2 - (a^2 + b^2).$$

If (x_r, y_r) is a point of intersection between the circle and the parabola, then

$$x_r^4 + (2b + 1)x_r^2 - 2ax_r + (a^2 + b^2) - R^2 = 0$$

which is the original equation if $b = (\mp p - 1)/2$, $a = \pm q/2$, and

$$R = \sqrt{q^2/4 + p^2/4 + \pm p/2 + 1/4 + \mp r}.$$

We turn now to problems in infinitesimal analysis whose solutions became steadily became more prominent in the seventeenth century. Like other aspects of early modern mathematics, their origin lay in the classical geometric corpus. Quadrature and tangent problems had been extensively studied by the Greek geometers, principally Archimedes and Apollonius.

As is well known, Archimedes had expressed the area of a circle in terms of its circumference and by approximating the circumference by the perimeters of regular polygons with an increasing number of sides was able to achieve any degree of accuracy.[40] He also determined the tangents to and the area cut out by part of the spiral of Archimedes (that is, the spiral we would express by the equation $r = k\theta$),[41] the area bounded by a parabolic segment, the surface area and volume of both a sphere or any of its segments, and showed that the volume of a paraboloid of revolution was half that of the cylinder that contained it.[42] Additionally Apollonius had found the tangents to all the conics.[43] Neither Archimedes or Apollonius, however, had explicitly used infinitesimal methods in their published proofs. In his so-called "method of exhaustion" Archimedes would first observe that the desired area, circumference, surface area, or volume was between the area, circumference, etc. of an "inner" and "outer" collection of n elements, both of which could be computed. For instance, in the case of the circle we can inscribe and circumscribe the circle with similar polygons having n equal sides. Call what we want to find V and the equivalent quantity of the two collections V_{in} and V_{on} so that $V_{in} < V < V_{on}$. Suppose also that the difference $V_{on} - V_{in}$ becomes as small as we please as n becomes sufficiently large. Then Archimedes proves that $V = V_{lim}$ where V_{lim} is the limiting value of either V_{on} or V_{in}. To see this, assume that $V > V_{lim}$, and let the difference between the two be δ. Then since n can be selected large enough so that, say, $V_{on} - V_{in} < \delta/2$, it follows that $V - V_{lim} < V_{on} - V_{in} < \delta/2$, which is a contradiction. A similar argument works if we suppose that $V < V_{lim}$. In the case of a circle the perimeters of the systems of circumscribed and inscribed polygons both approach the circumference of the circle, so that the area becomes half the circumference times the radius, or in geometric terms the area of a right triangle with one side the circumference and the other the radius.

Tangents were defined by generalizing Euclid's definition of a tangent to a circle to other curves by requiring a tangent line to "touch" the curve without "cutting" it. Therefore it was necessary to show (1) that the line did touch the curve at a point and (2) that every other point on it, at least in some neighborhood of the point of tangency, was on one side of

[40] In his *Measurement of a Circle* Archimedes expressed the area of a circle as half the radius times the circumference and showed that the ratio of the circumference to the diameter was between $3\frac{10}{71}$ and $3\frac{1}{7}$. [Archimedes (2010), pp. 91–93].

[41] Archimedes found the areas for $0 \leq \theta \leq 2\pi$ and $0 \leq \theta \leq 4\pi$

[42] All these results may be found in Archimedes (2010).

[43] Apollonius (1961), pp. 22–30.

the curve. This was usually done by a *reductio ad absurdum* proof. The tangent, furthermore, was viewed as an object to be *constructed.* To do this what was usually sought was the "subtangent" which in Figure 2.6 is the line segment QR between the intersection of the tangent T at the point P and some convenient line (usually what we would call the x axis or abscissa) and the intersection of another convenient line PR from the point of tangency with the line.

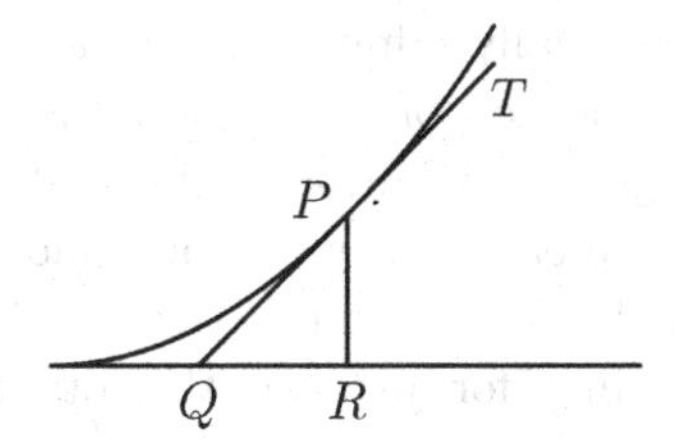

Figure 2.6

This technique mirrored Apollonius' treatment of the tangent to a parabola where he showed that (Figure 2.7) that the subtangent $TV = 2PV$.[44]

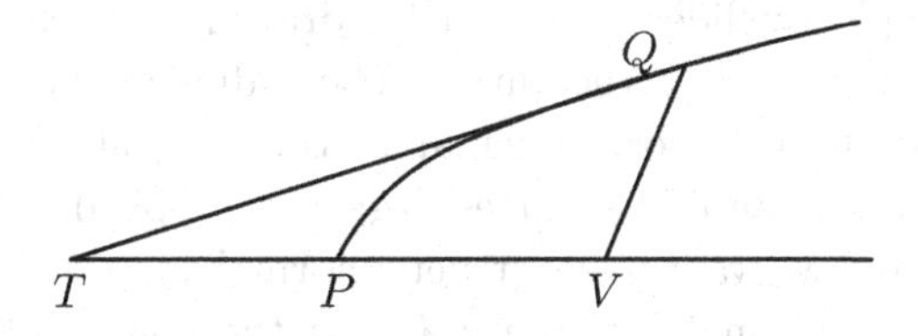

Figure 2.7

As usual Apollonius gives no indication of how this result was discovered. Instead he proves it by showing that if the candidate for the tangent has a

[44] Apollonius (1961), Proposition 12. Note that the ordinates QV, $Q'V'$ need not be perpendicular to the axis PV (which is not the case unless the cone defining the parabola is right-circular). With modern techniques, the verification of this is a nasty computation unless oblique coordinates are used. However, Apollonius' demonstration is simple and elegant.

point "inside" the parabola, there is a contradiction. We will give Apollonius's proof in some detail later.

Beginning in the late sixteenth century it was natural for mathematicians to reconsider, as they had with other material, the tangent and quadrature problems found in the Greek canon, and by the mid-seventeenth century attempt to apply the classical theory to the many new curves such as the cycloid or the "higher" parabolas given by $y = x^n$, $n > 2$, that were being discovered. Early work by Commandino, Francisco Maurolico (1494–1575), Simon Stevin (1548–1620), and others was very much in the style of Archimedes. They especially admired his treatise *On the Equilibrium of Planes* and used the method of exhaustion and *reductio ad absurdum* proofs to find centers of gravity of various solids and areas.[45] In some ways Stevin, who was a practical engineer, was more "modern" than the others. He sometimes worked with *numbers*, a practice unthinkable to a classically oriented geometer, by finding, for instance, the total water pressure on the vertical side of a tank. However, as Stevin wrote in Dutch, he had little influence until Latin and French editions of his work were prepared about fifty years after the Dutch originals.

Most early modern treatments of tangency, following the Greek model, made no mention of infinitesimal methods; in particular, there is nothing resembling a characteristic triangle or approximation of a tangent by an infinitesimal secant to the curve. If Descartes' *Géométrie* offered a unified algebraic method to analyze construction problems, this was also true of his approach to tangents. He will reduce a tangent problem for an arbitrary curve to the simple Euclidean method for drawing tangents to circles by finding a circle "tangent" to the curve. The radius of the circle from the point of tangency then is perpendicular to the tangent. The following is a very simplified version of Descartes argument applied to a parabola.[46] Suppose $y^2 = x$ and we want to construct a normal and tangent at the point $(1, 1)$. Suppose the normal cuts the x axis at the point $(p, 0)$. Consider a circle with center $(p, 0)$ and passing through $(1, 1)$. (See Figure 2.8.) The circle has equation $(x-p)^2 + y^2 = (1-p)^2 + 1$. In general the circle will cut the parabola at a second point unless the circle is "tangent" to the parabola at $(1, 1)$ in which case the two points of intersection coincide. To find the second point eliminate y. We obtain the equation

$$(x - p)^2 + x = (1 - p)^2 + 1.$$

[45] Detailed discussion of this early work may be found in Baron (1969), Chapter 3.
[46] After Burton (1985), p. 352.

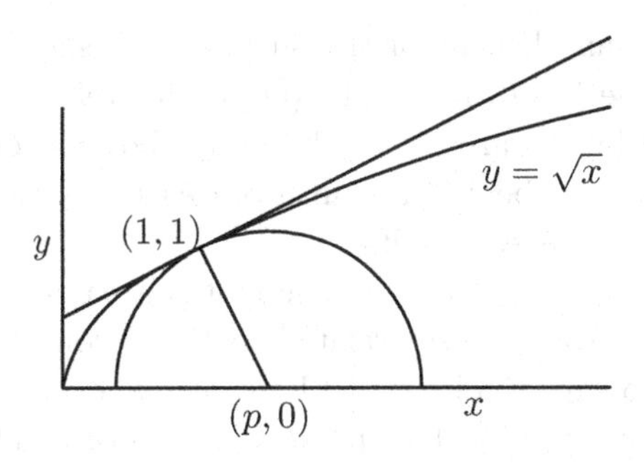

Figure 2.8

The second root of this equation (besides 1) will be the x coordinate of the second point of intersection. If therefore the equation has a double root, the circle will be tangent to the curve, the segment joining $(1, 1)$ and $(p, 0)$ will be normal to the parabola, and the perpendicular to the normal at $(1, 1)$ will be tangent to both the circle and the parabola. But it is easily shown that a double root occurs when $p = 3/2$. Descartes applies this technique to the ellipse (quoting at an appropriate point a theorem of Apollonius), the "first conchoid of the ancients," and to a variety of ovals.[47]

As we have just seen, the arguments behind the rectification, quadrature, and volume problems solved by Archimedes all followed a common pattern. The circumference, area or volume was "exhausted" or approximated as closely as we please by a collection of elements, the length, area or volume of which could be found. Then a *reductio ad absurdum* argument was applied. In an effort to streamline this technique it was but a short step to imagine (as even Archimedes and Apollonius almost certainly did to discover (but not prove) their results) the desired quantity as "made up" of infinitely many, infinitely small or thin elements and attempt to add them up. It was recognized that this a method was merely an aid to discovery which could always be modified to give an rigorous Archimedean proof if desired. One of the first examples of such an approach is found in Johann Kepler's (1571–1630) *Nova stereometria* of 1615. Using infinitesimal methods he found the volumes of a large number of solids of revolution (including wine barrels). He also derived the relation between the surface area of a sphere and its volume, by imagining the sphere to be the union of infinitely many cones with vertices at the center of the sphere and whose

[47] The technique, several examples, and applications to optics may be found in Descartes (1954), pp. 342–369.

bases approximate an element of the surface. He showed that since the volume of each cone is 1/3 its height (approximately the radius r of the sphere) multiplied by the area of the base, it follows at once that the volume of a sphere should be 1/3 its surface area times its radius, which is consistent with Archimedes' results.

An alternative technique was to imagine the area of a region or volume of a solid to be (in some sense) related to the union or "sum" of "indivisibles," that is to say, of all parallel lines or planes of a convenient type cutting it. This was the method pioneered by Bonaventura Cavalieri, a Jesuat priest and a mathematics professor at the University of Bologna from 1629 until his death in 1647.[48] In his nearly 700 page *Geometria indivisibilibus continuorum nova quadam ratione promota* (1635) and his 500 page *Exercitiationes geometricae sex* (1647) he used indivisibles to find the areas enclosed by the conic sections or Archimedean spiral and the volumes of the cone, pyramid, and several solids of revolution. These, of course, were not as a whole new discoveries, but he also made a fundamental advance by giving explicit formulas for the areas under the higher parabolas $y = x^n$ for $1 \leq n \leq 9$. Although Cavalieri's works were almost unreadable and were criticized on various philosophical and mathematical grounds, his methods proved to be very influential (even, as we shall see, on the young Leibniz). For this reason we give a very simplified description of some of them.[49] Consider a plane perpendicular to a given plane containing a figure F moving parallel to itself, say, from left to right of the figure and cutting it in a series of line segments l (Figure 2.9).

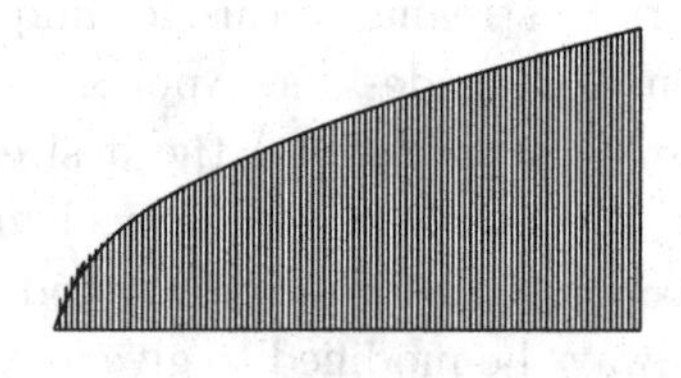

Figure 2.9

The set S of all possible line segments generated in this way Cavalieri called "all the lines" in the figure F which we will abbreviate (using Cavalieri's

[48]Cavalieri had also been the student of Benedetto Castelli (1578–1643) who had studied under Galileo.

[49]We follow the analysis of Cavalieri given in Andersen (1984), pp. 292–367.

Latin expression "*omnes*" as "$\text{omn}_F(l)$" where l is a member of S. It is important to stress that he did not identify this entity with the area of the figure (which would amount to claiming that a continuum was the union of indivisible elements). Rather omn_F was an abstract "magnitude" of its own kind. He then attempted extend Euclid's theory of ratios expressed in Book V of the *Elements* by proving that two such magnitudes could form a ratio with each other and that this ratio was the same as the ratio of the areas of the corresponding figures. The aggregate of all the cross sections made by parallel planes cutting a solid would be related to its volume in a similar manner. As a consequence of this idea he proves what is now called "Cavalieri's Principle":

> If two plane (or solid) figures have equal altitudes, and if sections made by lines (or planes) parallel to the bases and at equal distances from them are always in the same ratio, then the plane (or solid) figures are also in this ratio.[50]

In his first work, the *Geometria*, Cavalieri begins by applying his "omnes" concept to find areas of triangles and parallelograms in order to show how the method works and that it will give correct results in simple cases. Then he applies it to find the areas of conic sections. We give a simplified summary of his method (using modern notation) to find the area under the parabola $y = x^2$, $0 \leq x \leq a$. Let $P \equiv ABCD$ be a parallelogram with altitude h and $m \equiv EG$ an arbitrary line parallel to the side AB (Figure 2.10). Consider the triangle $\Delta \equiv ABC$. Let l denote the line segment FG.

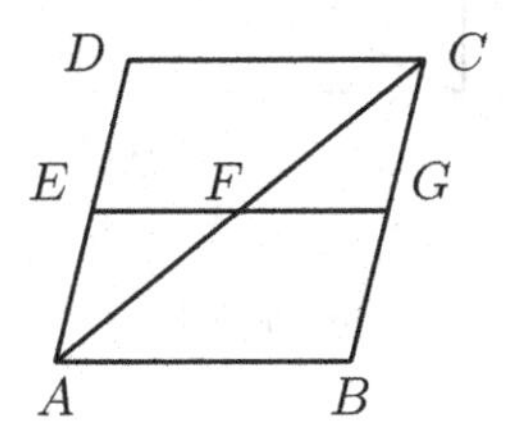

Figure 2.10

Cavalieri first proves that

$$\frac{\text{omn}_\Delta(l^2)}{\text{omn}_P(m^2)} = \frac{1}{3}.$$

[50]Quoted from the *Geometria*, p. 115, in *Ibid.*, p. 316.

One way to see why this is so is to observe that the ratio of $\text{omn}_\Delta(l^2)$ to $\text{omn}_P(m^2)$ is the same as the ratio of the volume of a pyramid of height h having a square base with side AB and square cross sections parallel to the base to the volume of a prism with the same base and height h as the pyramid and vertical cross sections congruent to $ABCD$. From classical theory or Cavalieri's previous results this ratio is $1 : 3$.

Let us see how this result can be used to find the area under a parabola. In Figure 2.11 let F denote the region under $y = x^2$, $0 \le x \le a$, and R the rectangle $ABCD$. Then

$$\frac{\text{Area}(F)}{\text{Area}(R)} = \frac{\text{omn}_F(y)}{\text{omn}_R(a^2)} = \frac{\text{omn}_{[0,a]}x^2}{\text{omn}_{[0,a]}(a^2)}$$

which is 1/3 if in the previous result we take P to be the square of side a.

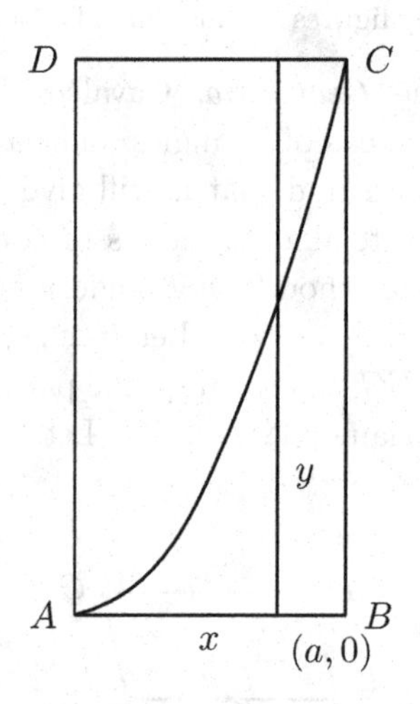

Figure 2.11

In his second work the *Exercitiationes* he extends these arguments to the higher parabolas $y = x^n$, $3 \le n \le 6$ and $n = 9$ and states but does not prove the general quadrature formula.[51] Consider the case $n = 3$. Again our account will be very condensed and simplified. As before we let P be a square with side $AB = a$. We keep the same notation as in Figure 2.11 for the previous case except that now $y = x^3$, and the upper right hand corner

[51] *Ibid.*, p. 140.

of R has coordinates (a, a^3). Then, as when $n = 2$,

$$\frac{\text{Area}(F)}{\text{Area}(R)} = \frac{\text{omn}_F(y)}{\text{omn}_R(a^3)} = \frac{\text{omn}_{[0,a]}x^3}{\text{omn}_{[0,a]}(a^3)}$$

In Figure 2.12 let Δ denote one of the congruent triangles AFE, ABF in $P = ABFE$.

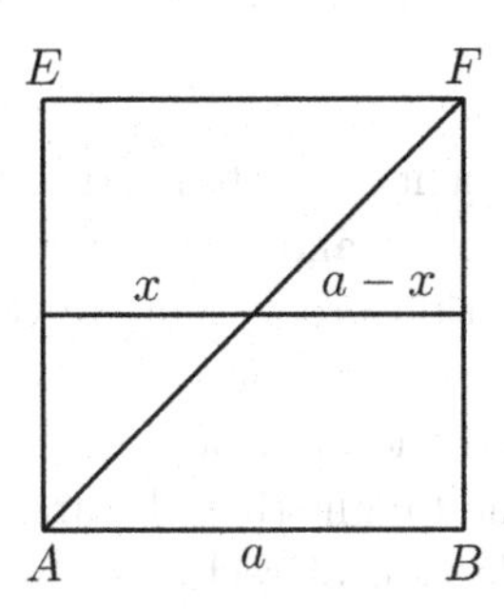

Figure 2.12

Then

$$\text{omn}_\Delta(x^3) = \text{omn}_\Delta(a - x)^3$$

which implies upon expanding $(a - x)^3$ that

$$2\text{omn}_\Delta(x^3) = \text{omn}_\Delta(a^3) - \text{omn}_\Delta(3a^2x) + \text{omn}(3ax^2).$$

Since $\text{omn}_\Delta(kl) = k\text{omn}_\Delta(l)$ the above equation becomes

$$2\text{omn}_\Delta(x^3) = \text{omn}_\Delta(a^3) - 3a^2(\ \text{omn}_\Delta(x)) + 3a(\text{omn}(x^2)).$$

Dividing both sides by one of the equivalent aggregates

$$a^2(\text{omn}_P(a)) \equiv a(\text{omn}_P(a^2)) \equiv \text{omn}_P(a^3)$$

gives that

$$\frac{\text{omn}_\Delta(x^3)}{\text{omn}_P(a^3))} = \frac{1}{2}\left(1 - 3\frac{\text{omn}_\Delta(x)}{\text{omn}_P(a)} + 3\frac{\text{omn}(x^2)}{\text{omn}_P(a^2)}\right).$$

But Cavalieri has already found in the *Geometria* that

$$\frac{\text{omn}_\Delta(x)}{\text{omn}_P(a)} = \frac{1}{2}$$

$$\frac{\text{omn}(x^2)}{\text{omn}_P(a^2)} = \frac{1}{3},$$

and hence

$$\frac{\mathrm{omn}_{\Delta}(x^3)}{\mathrm{omn}_P(a^3))} = \frac{1}{4},$$

from which the result that we would express by

$$\int_0^a x^3\,dx = \frac{a^4}{4}$$

follows immediately.

While Cavalieri's argument was often criticized, it or Kepler's basic strategy rapidly became popular and began to supplant the earlier emphasis on Archimedean exhaustion proofs or pure construction problems. In fact, the two approaches began to merge. Cavalieri's careful distinction between "all the lines" and the area of a figure was increasingly passed over, and it was felt that he taught that the area, for example, was either a "sum" of lines or a collection of rectangles with infinitely small bases to be somehow summed. Gilles de Roberval (1602–1675) in fact, inspired by Cavalieri, found the area of parabolic segments by treating Cavalieri's "*toutes les lignes*" as the sum of infinitely thin rectangles. His procedure was basically the same as that inflicted on modern calculus students to evaluate $\int_0^a x^2\,dx$ by Riemann sums before they are introduced to the FTC.[52] A similar approach was used by Fermat who generalized it to find the volumes of various solids of rotation.[53]

In addition to quadrature and volume problems early modern mathematicians attempted to extend the results of ancient geometers concerning tangents. As has been already noted, their conceptual framework concerning tangents was quite different than the modern, and we shall find that this is in part true even for Leibniz. Recall that in the Greek tradition a tangent was a line touching a curve at a unique point without "cutting" it, and so to show that a line is tangent to a given point, we need to verify both that the line passes through that point and that it does not touch the curve at any other point in some neighborhood of the given point. Here, for example, is how Apollonius proves his result characterizing the tangent to a parabola. We give the essentials of the argument using the modern algebraic representation of the parabola. In Figure 2.13 (see also Figure 2.6) let OCD be the parabola such that $y = \sqrt{x}$. We show that the tangent ACT at the point $C = (x_0, \sqrt{x_0})$ is characterized by the fact that the

[52] *Ibid.*, pp. 357–361. In his treatment of Cavalieri, Baron (1969) preserves this misconception by claiming that Cavalieri treated areas and volumes as sums of lines or planes.
[53] See Mahoney (1994), pp. 233–239.

subtangent $Ax_0 = 2x_0$. Suppose that there is a point (x, y) on ACT is such that $y < \sqrt{x}$. By similarity

$$\frac{y}{\sqrt{x_0}} = \frac{x_0 + x}{2x_0}.$$

Hence

$$\frac{\sqrt{x}}{\sqrt{x_0}} > \frac{x_0 + x}{2x_0},$$

so that

$$\sqrt{xx_0} > \frac{x_0 + x}{2}.$$

However, squaring both sides and transposing gives

$$0 > \frac{x_0^2}{4} - \frac{xx_0}{2} + \frac{x^2}{4} = \frac{(x_0 - x)^2}{4}$$

which is impossible.

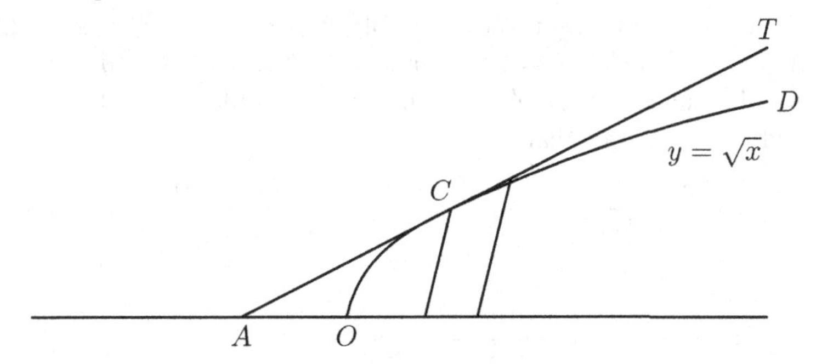

Figure-2.13

Notice that just as in the case of Descartes "double root" treatment of the tangent problem this demonstration does not explicitly involve any limit processes; it is a *reductio ad absurdum* proof. Similar derivations for tangent lines to the other conic sections are given by Apollonius. The difficulty, of course, lies in discovering how Apollonius arrived at his conclusion that the subtangent is $2x_0$. Such questions led early modern geometers to formulate various techniques some of which anticipate the modern definition that the slope of a tangent line to the curve with ordinate $y = f(x)$ at x_0 is given by

$$\lim_{x_1 \to x_0} \frac{f(x_1) - f(x_0)}{x_1 - x_0}.$$

Consider, for example, the arguments by which both Fermat and Descartes justified the Apollonian result concerning the tangent to the parabola expressed in modern terms by the equation $y = \sqrt{x}$.

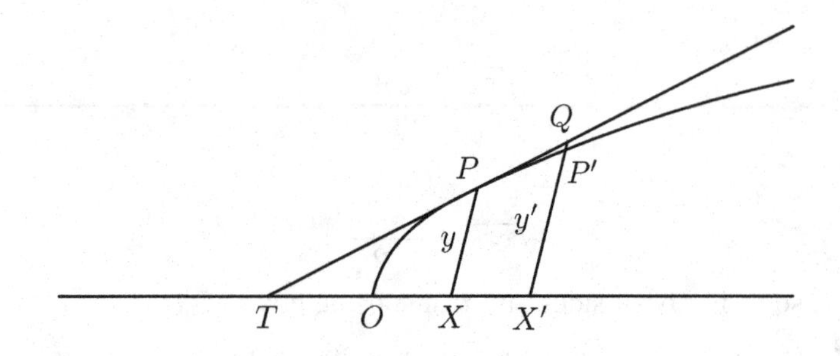

Figure 2.14

Fermat's use of limit argument is as follows: Set $y = XP$, $x = OX$, $y' = X'P'$, $x' = OX'$, $t' = TX'$, and $t = TX$. By the definition of the parabola, the fact that $y' < X'Q$, and the similarity of the triangles TXP and $TX'Q$ we get that

$$\frac{x'}{x} = \frac{y'^2}{y^2} \leq \frac{(X'Q)^2}{y^2} = \frac{t'^2}{t^2} = \frac{(t + (x' - x))^2}{t^2}.$$

Hence

$$\frac{x'}{x} \leq \frac{t^2 + 2t(x' - x) + (x' - x)^2}{t^2},$$

and so

$$\frac{(x' - x)}{x} \leq \frac{t^2 + 2t(x' - x) + (x' - x)^2 - t^2}{t^2} = \frac{2t(x' - x) + (x' - x)^2}{t^2}.$$

Canceling $(x' - x)$ and multiplying both sides by x and t^2 then gives

$$t^2 \leq 2tx + x(x' - x)$$

or

$$t \leq 2x + \frac{x(x' - x)}{t},$$

equality holding if, and only if, $x = x'$. Descartes was bitterly opposed to Fermat's view of tangents, and proposed a different argument. Here the chord $T'PP'$ is calculated and its "subchord" $T'X'$ is compared in the limit with the subtangent TX (Figure 2.15).

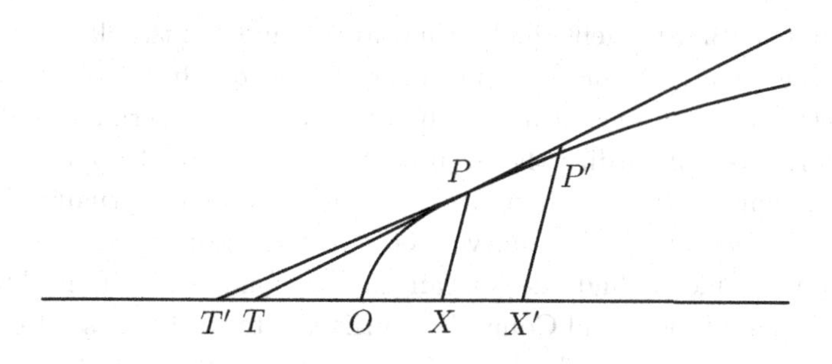

Figure 2.15

Using the same notation as in Figure 2.14 Descartes starts with the basic equality

$$\frac{x'}{x} = \frac{y'^2}{y^2} = \frac{t'^2}{t^2}$$

where $t' = T'X'$ and $t = T'X$. Therefore

$$\frac{(x' - x)}{x} = \frac{(t + x' - x)^2 - t^2}{t^2},$$

which yields the equality

$$t = 2x + \frac{x(x' - x)}{t}$$

as $x' \to x$ $t \to TX$ and in the limit becomes $2x$.[54] This treatment of tangents is contained in Descartes correspondence with Mersenne in 1632. But he never published it, preferring in his *Géométrie* his algebraic method involving double roots we have sketched above. However, his second method could be and soon was generalized so that the subtangent t for an arbitrary curve could be determined from the equation

$$\frac{y}{t} = \lim_{y' \to y} \frac{y' - y}{x' - x}.$$

We will see that Leibniz frequently uses this approach.

Still another "kinematic" view of tangents goes back to Roberval. He thought of a curve C as the locus of a moving point p, and then claimed the direction of the instantaneous velocity of p at a point (x, y) on C is the same as that of the tangent. This instantaneous velocity in turn could be determined by limit techniques by using the differential or characteristic triangle where the sides were interpreted as distances traveled by the point in an infinitesimal interval of time.

[54]The above arguments due to Fermat and Descartes are given in Whiteside (1961).

By mid-century tangents had been drawn to most of the classical curves such as the conics, cycloid, cissoid, limaçon, and quadratix, Archimedean spiral, etc. by variations on the techniques we have described; and there were efforts to standardize the arguments so they would apply to whole families of curves, thus replacing the need for often conceptually different methods for each individual curve. One of the most successful of these efforts was a rule to find the subtangent propounded by René Francois Sluse, "Canon of Liege and Counselor to His Electoral Highness of *Collen*," holding for a curve expressible by arbitrary polynomials The rule was stated without proof in the *Philosophical Transactions* of the Royal Society for 1672,[55] but Sluse claimed that it was easy enough for a young boy to learn. Sluse considers various subcases of a curve defined implicitly by the equation $f(x) = g(x,y)$ where both f and g are polynomials such that each term has the same dimension; i.e., $f(x) = \sum_{i+l=m} a_i^l x^i$ and $g(x,y) = \sum_{i+j+l=m} b_{ij}^l x^i y^j$. In the simplest situation g is a function of y alone. Then the subtangent t is given by

$$t = \frac{\sum_{j+l=m, j\geq 1} j b_j^l y^j}{\sum_{i+l=m, i\geq 1} a_i^l i x^{i-1}}.$$

For example, if

$$q^2 + bx - x^2 = y^2,$$

then

$$t = \frac{2y^2}{b - 2x}.$$

Suppose now that g involves x and y, let $h(x,y) = \sum_{i+j=m} c_{ij} x^i y^j$ be the sum of the terms of g containing x and consider the new equation $f(x) - h(x,y) = g(x,y)$.

$$t = \frac{\sum_{j+l=m, j\geq 1} j b_j^l y^j}{\sum_{i+l=m, i\geq 1} a_i^l i x^{i-1} - \sum_{i+j=m, i\geq 1} i c_{ij} x^{i-1} y^j}.$$

One of the examples Sluse gives is

$$x^5 + bx^4 = 2q^2y^3 - x^2y^3.$$

Setting

$$x^5 + bx^4 + x^2y^3 = 2q^2y^3 - x^2y^3,$$

[55] Sluse (1672). The quotation is from the article heading.

we find that

$$t = \frac{6q^2y^3 - 3x^2y^3}{5x^4 + 4bx^3 + 2xy^3}.$$

After contemplating how extremely awkward Sluse's procedure is, at least from our point of view, the reader may enjoy deriving Sluse's rule via implicit differentiation and, in fact, generalizing it so that f is a polynomial in x and y, or more generally replacing the equation in f and g by $h(x, y) = 0$ where h is a polynomial in two variables, which Sluse does not seem to have done.[56] A special case of Sluse's rule is that if $x^n = a^{n-1}y$ (where once again following Viète a is a line segment introduced to preserve dimensional homogeneity), then

$$t = \frac{a^{n-1}x^n/a^{n-1}}{nx^{n-1}} = \frac{x}{n}.$$

This was also discovered by Galileo's student Torricelli (1608–1647). A similar rule for which Sluse's version is a generalization was due to Jan Hudde (1628–1704), a mayor of Amsterdam and governor of the East India Company who had studied mathematics in private lessons under Schooten.[57]

As we have just seen in the generation or two before Leibniz infinitesimal problems concerning quadratures, volumes, and tangents were being hotly pursued by mathematicians in France, England, and Italy. To the individuals previously mentioned one could add Blaise Pascal, Christiaan Huygens, , John Wallis, Gregory St. Vincent (1584–1667), Isaac Barrow, Nicolaus Mercator (1620–1687), Galileo, and James Gregory (1638–1675).[58]

[56] A proof claiming to be based on Sluse's manuscripts is given on p. 360 of Whiteside (1961). Whiteside claims "to sketch his [Sluse's] rather lengthly treatment. The proof seems modern and holds for a general polynomial g in x and y, a case (probably because of its lack of homogeneity) Sluse did not consider in his published work. Also it leads directly to the equation $dy/dx = -(\partial g/\partial x)/(\partial g/\partial y)$. Although Sluse's method must be logically equivalent to Whiteside's proof, it is almost certain that Sluse could not have thought in this way. It would be nice to know what his original "lengthly treatment" actually was. Also see Baron (1969), p. 215f.

[57] Hudde also gave a method of finding extreme values of functions, essentially by solving the equation $f'(x) = 0$. Another of his results concerns double roots of polynomial equations. In modern notations if $p(x) = \sum_{i=0}^{n} a_i x^i = 0$ has a double root r and $\{b_i\}_{i=0}^{n}$ is a sequence in arithmetical progression then r is also a root of $p^*(x) = \sum_{i=0}^{n} p_i a_i x^i = 0$.

[58] Excellent accounts of this activity together with a detailed technical analysis of the achievements of most of the people we have mentioned may be found in the books mentioned at the beginning of this essay. The most thorough are Margaret Baron's *The Origins of the Infinitesimal Calculus* (1969), C. E. Edward's *The Historical Development of the Calculus* (1972), and Derick Whiteside's *Patterns of Mathematical Thought in the later Seventeenth Century* (1961). A valuable account of the seventeenth-century mathematical background may be found in Mahoney (1994). For mathematicians of sixteenth-century Italy, see Rose (1975).

Tremendous progress had been made in the sense that tangents, areas, centers of gravity, volumes of solids of revolution, etc. associated with quite exotic curves could be determined. Some problems considered in the period would still be challenging today even for a student armed with modern methods. All this preparation was certainly a necessary condition for Leibniz's achievement. Without it one cannot imagine the invention of any kind of mathematics we would be willing to call "calculus." On the other hand, the search by historians for premonitions of calculus obscures the fact—as should already be evident—that the mathematical style and presuppositions of the first half of the seventeenth century differ profoundly not only our own, but also as we shall soon show, from those of its final decades. On a technical level seventeenth century mathematicians were constrained by what we have called a "geometric paradigm" or "thought style." This meant that tangents were defined by their subtangents, not in terms of a derivative. They were line segments to be constructed. Our modern functional outlook was unknown. Algebra which had begun to be applied to geometric problems amounted mainly to an analysis of line segments or other geometric entities which helped to unravel the relations and ratios involved in a geometric problem. A curve might be represented by the equation $y = f(x)$, but f did not represent a function. Instead, the curve was a locus defined by segments y and x whose relation was given by the equation. Hence algebraic operations required dimensional homogeneity since there was no meaning to an expression like $a^2 + b$ (the addition of a square to a line). Just as finding a tangent was a type of construction, a successful quadrature meant finding a square equal to a region, not a number representing an area. In fact infinitesimal or other geometric problems were already free of numbers since a basic magnitude for geometers was the line segment. This implied in particular that ratios were relations as defined by Euclid, not quotients of numbers. These constraints represented a mathematical value system which was common to most early modern mathematicians. It certainly allowed some very profound mathematics to be produced, but left no room for key insights of calculus such as the inverse relation between integration and differentiation. There were some exceptions. Descartes, as we have seen, was a partial one, and John Wallis was another. In fact, as we shall show in the next two chapters not even Leibniz fully escaped the classical paradigm. His calculus was a geometric calculus, not a functional one as it has been since the mid-eighteenth century.

We can understand this fundamental conservatism present in a rapidly changing discipline by the fact that on a philosophical level most early seventeenth century mathematicians—like their "cousins" the Renaissance literary humanists—stood in the dispute between the Ancients and the Moderns resolutely with the Ancients. In their basic values they resemble those who wanted to restore the Latin of Cicero, or return medicine to the purity of Galen. Most did not think that they were transcending ancient Greek mathematics. Some wanted to develop it beyond the existing texts; others wanted to "restore" it by rediscovering the lost methods which they felt to lie behind the exquisite obscurity of the classical authors. Given the quotation we have already seen, this was probably the initial motive behind Descartes algebraic approach to geometry. Also Pierre Fermat, a classically trained provincial lawyer and councilor to the Parlement of Toulouse, in the his number theory was inspired by and wished to perfect the *Arithmetica* of Diophantus.[59] Another classically motivated project of Fermat was his attempted restoration of the *Plane Loci* of Apollonius which led him to invent a form of analytic geometry independently of Descartes. Even Cavalieri valued his technique of indivisibles as a tool to discover the methods by which Archimedes arrived at his more difficult quadratures whose motivation was concealed behind his polished *reductio ad absurdum* proofs.

One might argue, as the work of Descartes shows, that the received geometric paradigm was becoming less and less "pure" as the century progressed. Since the work of François Viète and William Oughtred (1574–1660), whose textbook of algebra was very popular, algebra had become an increasingly valuable tool for seventeenth century mathematicians even in geometric analysis. As Michael Mahoney has noted, this caused a certain tension. On the one hand, even in the middle of the seventeenth century the classical geometric tradition of Euclid, Apollonius, and Archimedes was still basically intact in spite of the inroads of algebra. We can still find a shared set of geometric values even in mathematicians of the stature of Newton or Descartes who from our vantage point transcended them. Algebra was certainly valued, but only as an adjunct to geometry not as an independent discipline which could replace it. In the minds of many mathematicians algebra offered the means for the rediscovery of methods of discovery used by the ancients. On the other hand, it is also the case that mathematicians were becoming gradually aware of the power of their new methods and the fact that they were making discoveries which were not present in the classical canon. Descartes, in particular, soon realizes that he has gone

[59]The best account of Fermat is Mahoney (1994).

far beyond the task of finding the method of the ancients underlying their geometry and congratulates himself on the discovery of methods unknown to the Greeks. Newton, despite his many achievements, is a excellent illustration of the almost schizoid aspect to this tension. Newton is an awesome mathematician, far deeper than most of his contemporaries (even Leibniz), and it is certainly true that Newton used the new algebra for many of his discoveries. His fluxional calculus which he developed more than a decade before Leibniz would have been impossible without it. He was certainly also aware that his achievements exceeded anything in the classical corpus. Yet at the same time he was a master of Greek geometrical techniques, and the *Principia* for all its profundity is deliberately written in the archaic style of Archimedes and Apollonius; its mathematical flavor in fact is so severely classical that we find it difficult to appreciate how Newton could naturally *think* in a way that is quite beyond the capacity not only of modern professional mathematicians but of their nineteenth century predecessors who were closer to this tradition. The judgement of the British philosopher of science and theologian William Whewell is illuminating:

> The ponderous instrument of synthesis, so effective in [his hands], has never since been grasped by one who could use it for such purposes; and we gaze upon it with admiring curiosity, as on some gigantic implement of war, which stands idle among memorials of ancient days, and makes us wonder what manner of man he was who could wield as a weapon what we can hardly lift as a burden.[60]

Even by the standards of the time Newton's mathematical value system was conservative. Although he would admit that algebraic analysis was a valuable tool of discovery, it had to perform this function in a concealed way. Newton felt that the algebraic scaffolding was unworthy of publication. The final result, no matter how obtained, should be justified by a formal synthetic construction or proof in the manner of the ancient geometers.[61] Somewhat ironically his mathematical bête noire was Descartes from whom as an undergraduate he had learned modern methods via Frans

[60]Quoted in Guicciardini (1999), p. 2.

[61]In other respects also, it has long been known that Newton was pre-modern, halfway between a Renaissance magus and a modern scientist. He was addicted to alchemy, Biblical chronology, and the Book of Revelations as much as to what we regard as science and left possibly more unpublished manuscripts on these subjects than on mathematics or physics. These features of Newton were first pointed out by the economist Lord Keynes who had collected many of Newton's unpublished papers. See his still valuable essay [Keynes, 1947)].

van Schooten's text. He recognized the revolutionary impact of Cartesian methods, Yet he disliked them as *ugly* mathematics:

> ... men of recent times, eager to add to the discoveries of the ancients, have united specious arithmetic with geometry. Benefiting from this, progress has been broad and far-reaching if your eye is on the profuseness of output but the advance is less than a blessing if you look at the complexity of its conclusions. For these computations, progressing by means of arithmetical operations alone, very often express in an intolerably roundabout way quantities which in geometry are designated by the drawing of a single line.[62]

In contrast to the present era, for the ancients Newton says:

> ... geometry was contrived as a means of escaping the tediousness of calculations by the ready drawing of lines. ... they never introduced arithmetical terms into geometry; while recent people, by confusing both, have lost the simplicity in which all elegance of geometry depends.[63]

It is exactly this separation of algebra and geometry that Descartes and his followers violated.

Newton's attitudes were not unique. The same can be found in many other contemporary mathematicians including Christiaan Huygens, Blaise Pascal, and Isaac Barrow. Several were geniuses, and all had genuine talent. But one should not forget that almost all of them had a pre-modern mathematical outlook and value system. There was a certain irony to this situation. As a consequence of Newton's own excellence the archaic mathematical values embedded in his revolutionary physics will survive for 150 years longer in England than they will on the continent.

Although there are elements of resemblance, it would be too strong to say that the unstable mixture of algebra and geometry in the mid-seventeenth century was the mathematical analogy of a scientific crisis in the Kuhnian sense. Mathematicians of the time may have felt a tension between the respective roles of algebra and geometry, but there was nothing in mathematics as serious as the anomalies in, for instance, the Ptolemaic system and mathematician were generally not (at least consciously) bothered by the cognitive dissonance implied by their sometimes self-contradictory views. Yet there was a definite "paradigm shift." The differences between

[62]Quoted in Guicciardini (2009), p. 77. Guicciardini has a valuable discussion of Newton's mathematical classicism.

[63]Quoted in *Ibid.*, p. 64, from Newton's *Lucasian Lectures on Algebra*.

early and late seventeenth century mathematics were not merely quantitative and did not arise solely because of the growth in mathematical knowledge. They are qualitative and conditioned the entire structure of mathematics. We agree with Michael Mahoney:

> Mathematicians and mechanicians at the end of the century did not understand their subjects better than had their predecessors of the early 1600s; they understood them differently.[64]

As we will hope to show, a fundamental architect of this change was Leibniz.

[64]Mahoney (1990b.)

Chapter 3

Isaac Barrow: A Foil to Leibniz

To get a more intimate familiarity with the geometric values we have been describing and to appreciate their technical impact on infinitesimal problems of interest prior to Leibniz, let us look in some detail at Isaac Barrow, Newton's mentor at Cambridge, who is a perfect mathematical representative of them and in many ways an antithesis to Leibniz. Barrow is not the only exemplar of the classical paradigm. We could have picked several of the mathematicians previously mentioned, including Leibniz's mentor Christiaan Huygens. But Barrow is a particularly illuminating example, both because of his chronological closeness to and purported influence upon Leibniz.

Barrow had an interesting if relatively short life. Several of his ancestors had been professionals, a great-grandfather being a physician and probable tutor to one of Elizabeth I's ministers, the Lord Treasurer Robert Cecil. His father Thomas Barrow, however, was lower in the social scale being a linen draper. Born in 1630, Isaac was either the only child of his father's first marriage or the only one to survive. His mother Ann died in 1634 when Isaac was four, and his father remarried around 1636. Thomas, intending that his only son should become a scholar rather than a merchant, sent him to Charterhouse, a prestigious grammar school, paying double the usual tuition in return for a promise of special attention to his son. However, it seems that young Isaac's education was neglected, and that he developed the reputation of a bully, playing the role of the school ruffian and fighting with other students. A second school Essex in Felstead, where he was sent in 1640, had a much better outcome. Here Isaac rapidly learned Latin, Greek, and Hebrew. When his father suffered grave financial reverses as a result of the rebelion in Ireland and could no longer pay the school fees, the headmaster Martin Holbeach, recognizing Barrow's intelligence, kept him on anyway, allowing him to live in his own house and to help pay for his education by tutoring an aristocratic student Thomas Fairfax who by inheritance had become an Irish Viscount. In 1643 Barrow received a scholarship to Peterhouse College, Cambridge where his uncle was a Fellow. Unfortunately the uncle was dismissed from Peterhouse due to his Royalist and high church views in January, 1644 by Parliamentary representatives, and Isaac's scholarship was given to another student. After a period of destitution in London as a companion to Fairfax,[1] Barrow was finally able to enter Trinity College, Cambridge in 1646 with the help of support from both a friend and his father. He graduated in 1649 and was elected a Fellow of Trinity the same year, receiving the M.A. degree in 1652. Although the bulk of his education was in the traditional form of the Greek and Latin classics and Aristotelian philosophy, Barrow seems to have begun to study mathematics seriously around 1648. He soon mastered Euclid, Archimedes, and Apollonius, as well as Descartes with whom he was particularly impressed. By the early 1650s he was probably familiar with all the latest mathematical trends.

Politically, the Barrow family were ardent royalists. Such opinions were not the safest in the years following the Civil War. Not only had his uncle lost his position and Isaac the Peterhouse scholarship because of them, but he narrowly escaped expulsion twice as an undergraduate at Trinity. Also the universities were under attack by various protestant religious extremists. To escape this atmosphere, in 1655 Barrow began an extended period of foreign travel, visiting Paris, Florence, and Turkey. On the voyage to Turkey, he helped fight off an attack by Algerian pirates. Returning to England in 1659, he witnessed and was delighted by the restoration of Charles II in 1660. Later that year he was appointed Regius Professor of Greek at Cambridge and three years later to the just established Lucasian Chair of Mathematics. He held that position for five years. Barrow was probably not Newton's teacher as once believed, but he was a mentor, treating Newton as a younger colleague, and helped to arrange Newton's appointment as his successor to the Lucasian Chair. Upon leaving mathematics in 1669, Barrow devoted the remainder of his life to religion, becoming a Doctor of Divinity in 1670, Master of Trinity College in 1675, and Vice-Chancellor of the University in 1675. He died in London at age 47 of "malignant fever" and is buried in Westminster Abby. According to those who knew him Barrow was a kind and good natured individual. This was unusual in the

[1] Fairfax had made an unwise marriage and as a result his guardian cut off his income.

seventeenth century when one considers the endless priority and personality disputes among mathematicians and scientists of the time.[2] A later epitaph stated that

> He was a Godlike, and truly great man, if Probity, Piety, Learning in the highest degree, and equal Modesty, most holy and sweet Manners, can confer that title ...[3]

Barrow's most significant original mathematical works, which were based on his Cambridge courses were the *Lectiones Geometricae* (1670) and *Lectiones Mathematicae* (1683). Other publications include a very popular edition of Euclid's *Elements*, his lectures on optics, and editions of the works of Archimedes, and Apollonius' four books on Conic sections. With respect to Divinity, his sermons were well regarded, and most of them were published after his death between 1678 and 1680.[4] His theological masterwork *A Treatise of the Pope's Supremacy* was published in 1680.

In his time Barrow was considered a mathematician of the highest talent and second only to Newton. His abilities were praised unanimously by contemporaries such as Huygens, Sluse, James Gregory (1638–1675), John Collins (1625–1683), and Gabriel Mouton (1618–1694), as well as Henry Pemberton (1694–1771), the editor of the third edition of the *Principia* who thought that Barrow "may be esteemed as having shewn a compass of invention equal, if not superior to any of the moderns," Newton excepted.[5]

Unfortunately, his reputation began to decline beginning with the outbreak of the priority quarrel between Newton and Leibniz. The partisans of each accused the other of having borrowed (or worse) from Barrow, while at the same time arguing that he had absolutely nothing to teach their own man. And because Barrow's mathematics was not the wave of the future after the achievements of Newton and Leibniz, Barrow has emerged particularly among modern historians as little more than a learned and competent but quite unoriginal codifier of existing ideas.[6] This writer agrees

[2]Such as, for example, the quarrels of Wallis and Hobbes, Newton and Leibniz, Descartes and Fermat, and Robert Hooke with virtually everyone.

[3]For information concerning Barrow's life I am indebted to Feingold (1990). For the epitaph by John Mapletoft see p. 90.

[4]According to the Eleventh Edition of the *Encyclopedia Britannica*: "His "*Sermons* have long enjoyed a high reputation; they are weighty pieces of reasoning, elaborate in construction and ponderous in style". [Anonymous (1910)].

[5]Feingold (1993), p. 331.

[6]*Ibid.* In this article Feingold assigns Barrow a greater degree of originality and merit than he is often given. He also suggests that Barrow had some real influence on both Newton and Leibniz:

with Feingold's rehabilitation of Barrow, but the opposite opinion may have been reinforced by the fact that Barrow's mathematical writings are *very* difficult for those trained in modern mathematics to understand. Hence there may be the temptation to dismiss them.

To experience the hermeneutic problems in one part of the seventeenth-century mathematical mainstream, we need only examine his *Lectiones Geometricae*. The thirteen lectures of this book comprise a beautiful and polished type of mathematics, but it is no longer ours. His outlook is totally geometrical. For him the new algebra developed by Viète and his successors is not even a branch of mathematics: it should be considered part of logic, and he avoided it whenever possible. Instead, he regarded geometry as the fundamental mathematical science. It includes arithmetic. Numbers simply represent geometric magnitudes.[7]

Like all his contemporaries, Barrow has a complete grasp of Apollonius, Archimedes, and Euclid as well as the "mechanical" curves constructed by ancient geometers, e.g., the cissoid, the quadratrix, the spiral of Archimedes, or more modern examples such as the cycloid investigated by Roberval, Huygens, and Pascal. He thinks of his work as a continuation of the glorious tradition of the Greek geometers. As in Newton's *Principia* the raw material of proof is clever citation or use of the classical canon especially the Apollonian theory of conic sections or the Euclidian doctrine of proportion. Since this training has almost vanished in the modern world, his arguments are just as difficult to grasp as Newton's. Even when we succeed, what is missing is the whole network of associations, concepts,

> ... it remains an inconvertible fact that contemporaries accorded prominence to Barrow's mathematical learning, particularly to his *Geometrical Lectures* (which was almost universally praised), and that many talented seventeenth-century mathematicians recognized a close affinity between Barrow's work and that of Leibniz and Newton.

(p. 337).

[7]These opinions are particularly strongly expressed in the *Lectiones Mathematicae*, Lecture II. See also Mahoney (1990a), pp. 185–190 and p. 200f. Barrow is reacting against John Wallis and other partisans of algebra who argued that geometry should be subordinate to algebra. As a sample of this attitude Barrow argues that the identity

$$\lim_{n \to \infty} \frac{\sum_{i=0}^{n} \sqrt{i}}{(n+1)\sqrt{n}} = \frac{2}{3}$$

"can never be demonstrated by any Method of Arithmetic itself" rather it is a geometric result. To prove it Barrow appeals to Archimedes quadrature of the parabola. The reader will note that this is the exact reverse of the modern argument. See Panza (2008), pp. 372–388.

aesthetic considerations, motivation, and mathematical "taste" that would enable modern mathematicians to *naturally* think as he does. The problem is similar to reading Aristotle's *Physics* or the works of some scholastic like Suarez (1548–1617) or Zabarella (1543–1580). The words are there, understanding can be teased out of the argument, but it is almost impossible to get "inside" the writer's mind and to think naturally using the same tools as he does about the problems which are important to him.

Consider Barrow's treatment of tangents. Some of his notions are familiar. Curves are the loci of moving points generated by concurrent "local motions," a view that Newton probably borrowed and which made possible his concept of a "fluxion" as a rate of change. In modern language essentially the same idea is expressed by representing the x and y coordinates of the moving point by the "parametric" equations $x = x(t)$ and $y = y(t)$ where t often represents time in a physical situation. But although in an introductory Lecture he has a discussion of uniformly accelerated motion similar to Galileo (or to Nicole Oresme (c. 1320–1382) in the fourteenth century), for Barrow tangents of curves in their polished geometric form have nothing to do with what we call "rates of change," a Newtonian fluxion, or any sort of limit process. Instead, as for most of his precursors and contemporaries (especially Fermat and Descartes) tangents are static geometrical objects to be *constructed* by finding the "subtangent" or normal to the curve at the point of tangency which are then characterized in terms of the locus properties of the curve.[8] That a line cutting a curve at a point is a tangent is usually demonstrated not by the employment of limit arguments but by clever inequalities showing that in a neighborhood of the point of tangency the line is on one side of the curve.

Barrow has several ingenious geometrical techniques to find tangents. One to prove that if one curve has a known tangent, we can use this information to find the tangent of a related curve. The following (Theorem 5 of Lecture X) is one of many examples of this method. In Figure 3.1 AEG and AFI are two curves, such that the line ET is a known tangent to AEG whose length is equal to the arc AE. Then, if the ordinate DF defining

[8]Recall that a subtangent is the line segment between the points of intersection of the tangent and a line (often what we would consider the x axis) and a perpendicular from the point of tangency to the line. But see Propositions 11 to 13 of Lecture IV of the *Lectiones* for a partial exception to our judgement where Barrow relates motion of the subtangent to the vertical component of velocity at the point of tangency. Soon afterwards he notes that the results of Galileo and Torricelli on parabolic trajectories can be derived using these ideas. See also the historically sensitive discussion in Mahoney (1990a), pp. 207–213.

AFI is also equal to the arc AE and TR is perpendicular to AP, Barrow proves that RF is tangent to AFI. This result can be derived by modern methods, but the proof is a computational mess; Barrow's argument is a short paragraph in length.[9]

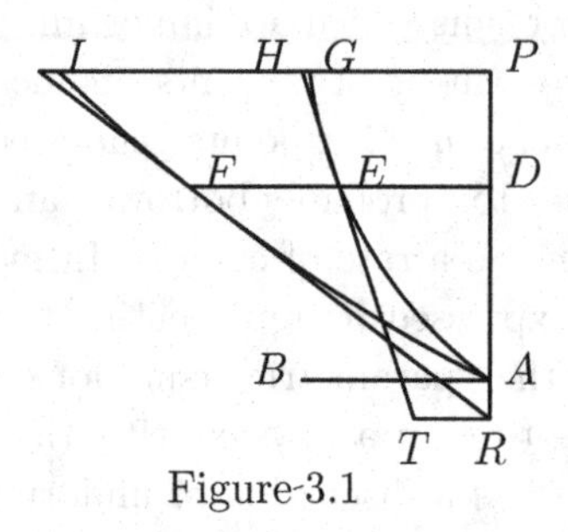

Figure-3.1

A second technique, often combined with the first, is to introduce an auxiliary curve whose tangent is known. Then Barrow proves that this curve is tangent to the given curve at the required point. The tangent to the first curve will then be tangent to the second curve. For instance, in Proposition 8 of Lecture VIII it is shown that a tangent to a curve at a point can be constructed by "fitting" a hyperbola to the curve, so that the tangent to the hyperbola at the point is tangent to the curve. Because this is such a good example of Barrow's reasoning we give some of the details. We have three curves XEM, YFN, and ZGO such that if a line $DEFG$ is drawn from a fixed point D intersecting the curves respectively at G, F, and E, then $EG : EF$ is always a constant ratio R. (Figure 3.2) Let ET and GT be tangents to XEM and ZGO at G and E intersecting at the point T. To find a tangent to YFN at F, proceed as follows: Let $T'FK$ be the locus of points such that if DL is any straight line passing through D, cutting the tangents TE and TG at the points I and L and YFN at the point N, then $IL : IK = R$. It is known that $T'FK$ is a hyperbola. Now $T'FK$ passes through the point F and since $|IK| > |IN|$, $T'FK$ does not cut YFN. Let FS be the tangent to $T'FK$ at F (which can be constructed by a theorem of Apollonius). Then FS will also be the required tangent to YFN at F.

[9] See Barrow (1916), p. 114f.

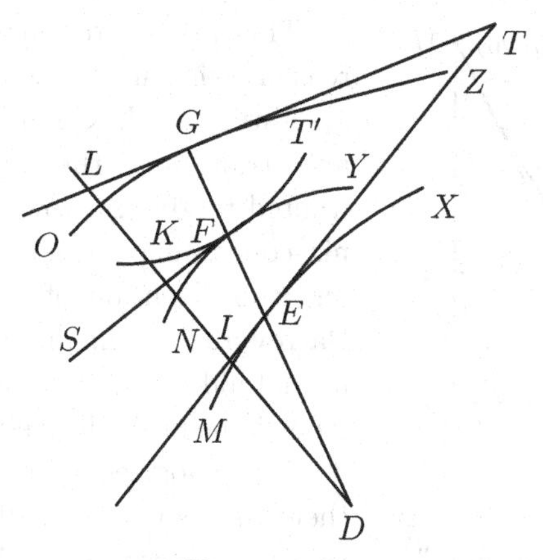

Figure 3.2

In this way a well known classical theory concerning tangents to conic sections can be employed for arbitrary curves. We leave it to the reader to show how the previous two examples can be proven using the language of modern calculus.

To be fair, Barrow also presents a casual computational method to determine subtangents. This is the well known "a, e" method in which the characteristic triangle is clearly drawn; formally it is suggestive of the modern process of "implicit differentiation". How does this method work? Given a curve with equation $f(x, y) = 0$, Barrow wishes to construct the tangent at point M by determining the length of the subtangent PT. (See Figure 3.3.) To this end, let the coordinates of M be (x_0, y_0), $MR = a$, and $NR = e$. Then Barrow expands the equation $f(x_0 - e, y_0 - a) = 0$ according to three rules which we quote:[10]

> RULE 1. In the calculation, I omit all terms containing a power of a or e, or products of these (for these terms have no value).
>
> RULE 2. After the equation has been formed, I reject all terms consisting of letters denoting known or determined quantities or terms which do not contain a or e.
>
> RULE 3. I substitute m (or MP) for a, and t (or PT) for e. Hence at length the quantity of PT is found.[11]

To illustrate the method Barrow applies it to five curves including the Folium of Descartes and the Quadratrix.

[10] *Ibid.*, Lecture X, p.120
[11] *Ibid.* Analogous methods were invented by Fermat and Sluse.

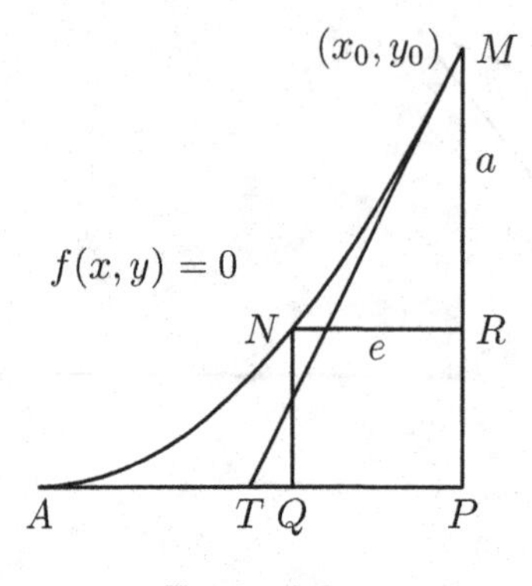

Figure 3.3

This procedure was one of the main reasons why his modern editor and commentator J. M. Child, who republished and translated the *Lectiones* in 1916, wanted to give Barrow priority over Leibniz concerning the characteristic triangle and the invention of calculus. However, Barrow regards his method as only an informal aid "by calculation" to precise geometric results. He sees little potential in it and apologizes for including it because there are "so many well-known and well-worn methods of the kind." As we have already pointed out, there is absolutely no sense here of a "derivative." For him as for Fermat, Descartes, Sluse, Huygens, et al., the tangent line problem was nothing more than an interesting and difficult problem in constructive geometry. Indeed, elsewhere he says that a central aim of the work is "an investigation of tangents, freed from the loathsome burden of calculation."[12]

In the case of quadratures one of Barrow's favorite methods is what Leibniz called "transmutation." Barrow transforms one figure into another one such that the areas encompassed by both figures are related. A typical result of this kind is Problem 24 in Lecture XI.

Referring to Figure 3.4 Barrow asserts that:

> If DOK is any curve, D a given point on it, and DK any chord; also if DZI is a curve such that when any point M is taken in the curve DOK, DM is joined, DS is drawn perpendicular to DM, MS is tangent to the curve, DP is taken along DK equal to DM, and PZ is drawn perpendicular to DK, so that PZ is equal to DS; in this case the space DZI is equal to twice the space $DKOD$.[13]

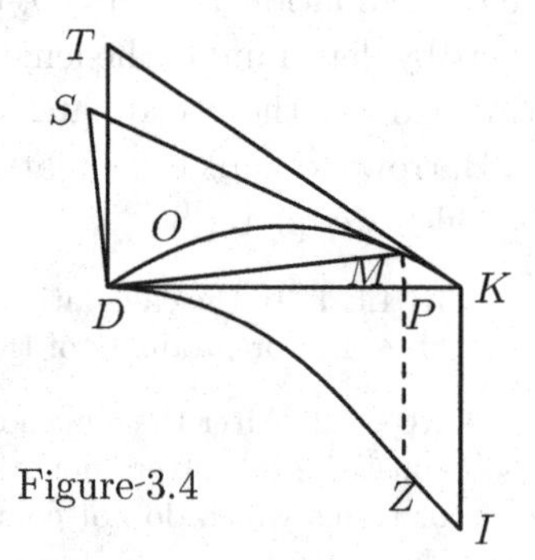

Figure 3.4

The idea behind the proof of this curious result is to observe that for M

[12] *Ibid.*, Lecture VI, p. 66.
[13] Barrow (1916). The diagram is given in Leibniz (1920), p. 173.

close to K, the circular arc $\widehat{MP}$ is nearly a line segment parallel to DS and the triangles MPK, SDK are nearly similar. Therefore

$$MP : PK \approx DS : DK,$$

so that $MP \cdot DK \approx DS \cdot PK$. Since $PZ = DS$, the area of $PZIK \approx$ twice the area of ΔMDK. If we continually repeat this argument taking M as the new K, the region $DKOD$ is partitioned into triangular sectors with common vertex D. Each of these sectors is equal to half the area of a trapezoidal region like $PZIK$. As we shall see in Chapter 5, this method of dividing an area into triangular sectors together with Figure 3.4 may have been the inspiration behind a fundamental transmutation theorem of Leibniz.[14] There are several other theorems of this general type, some of which may have influenced Leibniz. Indeed, Barrow concludes his proof with the statement "Should any one explore and investigate this mine, he will find very many things of this kind." We will describe in Chapter 6 another of Barrow's transmutation theorems involving subnormals which actually shows up in Leibniz's 1686 paper (without attribution) on integration in the *Acta Eruditorum*. As in the previous example, Barrow's results dealing with quadratures often also involve tangents. Here is one (Proposition 11 of Lecture X) that links the two concepts together in an unusually fascinating way. In Figure 3.5:

> Let ZGE be any curve of which the axis is AD; and let ordinates applied to this axis, AZ, PG, DE, continually increase from the initial ordinate AZ; also let AFI be a line such that, if any straight line EDF is drawn perpendicular to AD, cutting the curves in the points E, F, and AD in D, the rectangle contained by DF and a given length R is equal to the intercepted space $ADEZ$; also let $DE : DF = R : DT$, and join DT. Then TF will touch the curve AIF.

As a corollary which can be deduced "from the preceding," Barrow imagines that the curves ZGE and AFK (here K seems to replace I as a point on the curve) are so related such that the square on DF is equal to "twice the space $ADEZ$.

[14]See Figure 5.5. Child (*Ibid.*, p. 172) speaks of "deliberate plagiarism" on Leibniz's part. But this may be too strong a judgement.

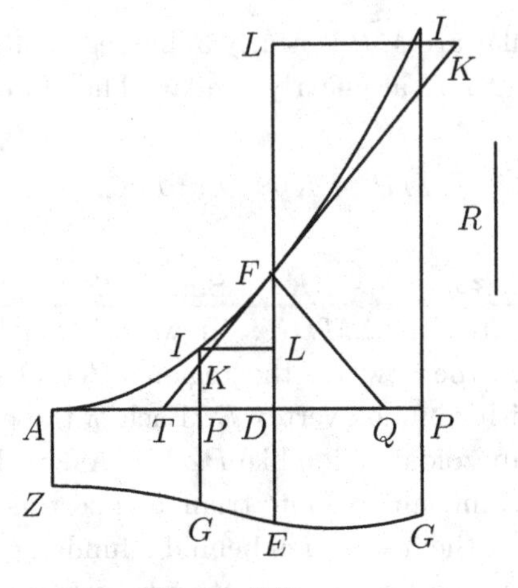

Figure 3.5

Then if DQ is constructed along AD (prolonged if necessary) such that $DQ = DE$, the line segment FQ is normal to the curve AFK.[15]

To further support of his thesis that Barrow is responsible for many of the central ideas of calculus, Child finds throughout the *Lectiones* analogues of the derivatives of products and quotients of functions, the chain rule, the differentiation of the trigonometric functions, etc. and describes at the beginning of each Lecture the calculus formulas said to be contained in it. In particular, he claims that the above result is actually the FTC, an opinion which has also been endorsed by Dirk J. Struik[16] Indeed, *if* we put the conclusions Barrow's geometric arguments into modern notation as Child constantly does, we *will* find analogues to almost every formula in calculus, including formulas (1.1)–(1.6) of Chapter I, and in the special case of the previous theorem it is not *too* difficult using modern methods to derive equation (1.3) where y is identified with DE. (The reader is invited to try.) Therefore in Child's opinion, Barrow should be given credit as a coequal inventor of calculus. But despite these "translations," the situation in the case of quadratures is similar to that concerning tangents. For Barrow and his contemporaries before Newton and Leibniz a "quadrature" is simply a problem of showing that some region equals a square or rectangle; there is no conception either of "area" as a numerical measure or of an "area function" of a curve whose tangent is the ordinate of the curve.

It is interesting also to recall that Child's thesis and the kinds of cor-

[15] *Ibid.*, p. 118.
[16] Struik (1969), pp. 255–263.

respondences observed by both him and Struik were fully anticipated by Leibniz's contemporaries, especially by his friend Ehrenfried Walter von Tschirnhaus in 1678–1679 and later by James Bernoulli (1654–1705) in a 1691 article published in the *Acta Eruditorum*.[17] Both were so used to thinking in the traditional geometric fashion that it was easy for them to use Barrow's methods to solve many of the problems considered by Leibniz. Therefore they judged Leibniz's calculus to be mainly a matter of notation and being very comfortable with existing methods failed to appreciate its immense power and revolutionary implications. Tschirnhaus, in fact, did not even consider the notation an improvement—finding Barrow "much more readily intelligible" than Leibniz.[18] As we have already mentioned, the assertion that Leibniz essentially appropriated the material of Barrow was a standard charge of Newton's partisans in the priority dispute between him and Newton. (Leibniz returned the accusation with interest.)

Shall we say then that Barrow discovered, the "characteristic triangle", "differentiation", or a form of the FTC? Or, despite the apparent similarities, would it not be more accurate to conclude that Barrow's mathematics is operationally incommensurable with our own or at least to use vocabulary of Ludwik Fleck representing a different and incompatible "thought style." There is indeed a one-to-one "logic preserving" map between some of Barrow's results and those of a modern textbook. But if we were able to make contact with Barrow (perhaps by employing one of Heinrich Cornelius Agrippa von Nettlesheim's (1486–1535) conjurations of angelic spirits) and make him understand this quasi-isomorphism by carefully explaining the Leibnizian notation, he might agree with the details, but fail to see their point—for he would claim that he was doing mathematics while Leibniz was doing some sort of ugly algebraic cabala which he has taken great pains to avoid. He would also probably judge that Leibniz was poorly trained in geometry since hardly any of his arguments (as we will see in Chapters 5 and 6) depend on more than simple properties of similar triangles. There

---

[17] Bernoulli writes in his article *Speculum calculi differentialis* in the January 1691 issue of *Acta Eruditorum*:

> Yet, to speak frankly, whoever has understood Barrow's calculus ... will hardly fail to know that the other discoveries of Mr. Leibniz, considering that they were based on that earlier discovery, and do not differ from them, except perhaps in the notation of the differentials and some abridgement of the operation of it.

Quoted in Feingold (1993), p. 337.

[18] *Ibid.*

is no sign in Leibniz's work of either the ingenious constructions of which Barrow was a master or of much knowledge of the classical canon.

We can multiply similar examples of results that seem to anticipate calculus not only from Barrow, but also from many of his contemporaries. Mathematicians who shared his geometric values were able to solve what they considered to be quite challenging problems. These included the areas under the "higher parabolas" having equation $y = x^n$ (Cavalieri, Wallis, Fermat, Roberval, Pascal), the area enclosed by the cycloid (Roberval, Fermat, Descartes), the logarithmic properties of the hyperbola (Gregory St. Vincent), rules for finding subtangents similar to Barrow's (Sluse, Fermat, Hudde), the area enclosed by the cycloid (Roberval, Fermat, Descartes), the logarithmic properties of the hyperbola (Gregory St. Vincent), rules for finding subtangents similar to Barrow's (Sluse, Fermat, Hudde), the rectification of arcs (Huygens, Wallis, Gregory), as well as various cubatures and the finding of centers of gravity. And Christiaan Huygens, Leibniz's initial mentor and critical correspondent, who is today mostly remembered as a physicist, gave a supremely elegant geometrical demonstrations to show the isochronism of the cycloidal pendulum as well as to rectify and find evolutes of the cycloid and parabola.[19] One might even conjecture that the integral formula (1.7) in Chapter 1 might not be beyond pre-Leibnizian infinitesimal techniques, the difference being that this problem might have been the subject of an entire treatise instead of being a second semester calculus exercise.

It is tempting (and partially correct), therefore, to conclude that our mathematics is more developed than that of Barrow or his contemporaries because we can solve using present techniques many of their problems in a "easier" and more "systematic" way. But it is perhaps more accurate to say that we can solve the "images" of their problems which are of interest to us under the interpretative map we have constructed. But there are other problems which they solve which we have forgotten, have no interest in, can no longer understand, or sometimes can solve only with difficulty using modern techniques. Whether or not Kuhn's view that knowledge is both gained and lost in scientific revolutions is true, this does seem in a sense to be the case in mathematical revolutions. Whatever the gains initiated by the Leibnizian revolution, the classical geometrical tradition and the knowledge it entailed represented by Barrow, Huygens, Newton, and others

[19]For an account which does justice to the classical flavor of Huygens's methods, see Yoder (1988). For reasons we discuss below, he had great difficulty in accepting the Leibnizian calculus.

has for all practical purposes been lost among practicing mathematicians.[20]

In Barrow's case (and also for several of his contemporaries), we can show that their work in infinitesimal analysis contains results anticipating or even identical to our own *when and only when* modern notation is supplied. This fact has stimulated a vast literature by historians of mathematics. But while the resulting literature can be valuable for its mastery of the technical details of early modern mathematics, it is in a profound way un-historical. There is an almost insoluble hermeneutic problem here. On the one hand, given the changes in the mathematical climate since the seventeenth century, such "translations" may be the only way we can gain access to old mathematics or summarize it in a concise way. But on the other hand, the notation, as we have already pointed out, carries with it a baggage of mathematical perceptions, associations, and values which were not present in the original work,[21] so that this procedure may be as much of a historical falsification as nineteenth-century historians arguing that a Renaissance magus like Giordano Bruno (1548–1600), who was soaked in magic and hermeticism, was a martyred protagonist of Modern Science because of his Copernicanism and belief in the plurality of worlds.[22] To give a parallel illustration in the history of physics, to argue that Barrow's or other pre-Leibnizian mathematics has been incorporated in any simple way into modern mathematics is similar to arguing that Newton has been incorporated into Einstein. Seventeenth-century formulas (suitably translated) do indeed remain in modern mathematics just as the Newtonian formula $F = ma$ remains in Einstein's theory. But:

> to make the transition Einstein's universe, the whole conceptual web whose stands are space, time, matter, force, and so on, had to be shifted and laid down again on nature whole.[23]

---

[20]This writer has colleagues who are successful research mathematicians who know no geometry at all, as they have never had even a high school course in it.

[21]To give another example in addition to those mentioned in the Preface, Baron (2003) states a rule found by Fermat, Sluse, and Hudde for finding the length of the subtangent t to the curve $f(x, y) = 0$ as

$$t = -y\frac{f_y(x, y)}{f_x(x, y}.$$

This is formally correct, but the modern notion of a partial derivative or implicit differentiation would have been meaningless to any of these mathematicians.

[22]The best modern study of Bruno is Yates (1964). She demolishes the nineteenth-century "martyr of science" point of view and fixes him solidly in the Renaissance hermetic tradition. See also Yates (1982).

[23]Kuhn, 1962, p. 148.

An analogous conceptual gulf separates Barrow and most other seventeenth-century mathematicians from our mathematical world. We should not disparage them. They were not trying to do modern calculus in a clumsy, limited way. Instead, they inhabited a mathematical world very different from our own. While, some (but not all) of their theorems can fairly easily be shown to be equivalent to results derivable from modern calculus, the mode of thinking behind them is foreign to us. It is tempting to think that Barrow, in particular, found his results by first using something akin to modern techniques and then casting them into the classical geometric form, but it is more likely that (unlike us) he could naturally *think* in the classical way. We may be able to prove his results (once they are stated) using our formalism, but this method gives no insight into their motivation. Just how and why he was able to state and prove the theorems in the form he did remains mysterious to us as we cannot really enter into his mathematical world. There is an obvious psychological analogy here with the tension and incommensurability between two radically different scientific paradigms or thought styles. But taken on its own terms, much of Barrow's mathematics is as profound, original, and difficult as anything which followed.

As has been pointed out by Molland (1993),[24] and Mahoney[25] the very conservatism of the mathematics Barrow and his contemporaries pursued in an ironic way provided a fertile seedbed for further development; for much of the classical corpus in mathematics as in literature was only partially extant and needed restoration. The effort of scholars to fill in the gaps in the ancient writers led inevitably to the production of new mathematics especially involving algebra (e.g., Fermat's effort to improve Diophantus). Also those works which had survived in a complete form (such as much of Archimedes) evoked vague discontent. since their motivation and method of discovery (as opposed to their elegantly polished proofs) were invisible. There was a feeling which naturally stimulated new efforts that the ancient methods had been hidden and therefore awaited recovery. It is perhaps just because of the luxuriant growth of mathematics between 1500 and 1650 that we tend to exaggerate its similarities to our own mathematics. But had the tradition we are attempting to describe continued to develop without Leibniz or Newton the future evolution of mathematics—and as a consequence the modern world—might have simply gone on a different trajectory (a sort of unrealized Leibnizian "possible world").

[24] p. 106f.
[25] 1990a.

Chapter 4

A Young Central European Polymath: Between the Scholastics and the Moderns

It is a familiar cliche that mathematical talent is exhibited early in life, in most cases evident in childhood or adolescence. If so, Leibniz is an exception. There were few clues in his early life that he would become a mathematician, let alone a mathematical revolutionary. Although his prodigious talents were evident at an early age, they seemed at first to lie in the direction of philosophy, theology, and law rather than mathematics and science.

He was fortunate in being born in Leipzig and in 1646, two years before the Treaty of Westphalia finally ended the Thirty Years War. Although this terrible war may have killed nearly a third of the German population, most of the destruction was over by 1646 and Leipzig together with the surrounding territory of Saxony was not devastated nearly as badly as other parts of Germany. Indeed, Saxony, although fervently Lutheran, had sided with the Austrian Hapsburgs and had even profited from the war, gaining territory as a result.[1] Had Leibniz's parents to be lived, say, in Magdeburg which had been sacked by the Imperialist general Tilly in 1631 with the massacre of most of its 36,000 residents, very probably the consequences for the subsequent intellectual history of Europe would have been profound. His ancestry can be traced back to the sixteenth century and both his grandfather and great-grandfather had been government officials. His father Friedrich (1597–1652) who died when Leibniz was six was a professor of Moral Philosophy at the local university. Friedrich's first two wives had died and Leibniz together with a sister Anna Catharina (1648–1672) were the products of a third marriage; he had a half-brother and half-sister by

[1] However, the two great battles of Breitenfeld (1631) and Lützen (1632) had been fought nearby. On both occasions the city was besieged and part of the University was destroyed by artillery fire. The city began to recover when it was occupied by the Swedes in 1642. [Antogonazza (2009), pp. 50f.]

the first marriage.[2] The family was reasonably well off and had good social and academic connections on both sides (Leibniz's mother was the daughter of a law professor and when orphaned at age thirteen was brought up in the homes of two other Leipzig professors). It was soon evident that Leibniz was a prodigy. At age six (when his father died) Leibniz had already learned to read German and would rather study and read than play. At age seven (1653) he began formal schooling at the Nikolaischule, a highly respected local grammar school where he remained until 1661. The Nikolaischule taught a wide spectrum of the population and specialized in preparing the most talented of its students for the university. Its rather arid curriculum was based on the medieval *trivium*, that is, grammar, rhetoric, and logic together with Lutheran theology. More than half of the instruction was devoted to Latin (which both students and faculty were required to speak at school) and Greek. Very little time was spent on mathematics, physics, or astronomy. Leibniz found the instruction there rather slow, and he supplemented the instruction in Latin by reading Livy and "a chronological thesaurus by Sethus Calvisius." He found Livy difficult but became able to read it by using the illustrations to puzzle out the text. One of his teachers, being annoyed by this behavior, complained to his father that Leibniz should only be allowed to read the school textbooks, and the more advanced books taken from him. However, a local nobleman, apparently a friend of the family, who witnessed the teacher's complaint, persuaded his father not to frustrate Leibniz's talent and to allow him access to his library. There at age eight he began to read Cicero, Seneca, Pliny, the early Christian fathers and other classical authors.[3] As a result by the age of twelve he was fluent in Latin and well on the way to mastering Greek. The next year he astounded his teachers by delivering at a school celebration three hundred hexameters of Latin verse he had composed in a few hours. A few years later, still before the age of fifteen when he left the Nikolaischule, he mastered the later scholastics such as the Paduan philosopher Zabarella and the Spanish Jesuit Francis Suarez, reporting in later life that he read and understood the latter as easily as other boys his age "read fairy tales and romance."[4] As a result when Aristotelian logic was introduced in his last two years of school he was able to challenge his teachers on various points, especially the adequacy of Aristotle's list of the categories.

In 1661 Leibniz left the Nikolaischule and began attending the Uni-

[2]Mackie (1845), p. 16.
[3]*Ibid.*, pp. 19–22.
[4]*Ibid.*, p. 24.

versity of Leipzig. As in other universities of the time, instruction was centered on Aristotelian philosophy. He was especially impressed by Jacob Thomasius (1622–1684) who taught moral philosophy, rhetoric, and dialectic. Thomasius remained Leibniz's friend and correspondent for life and urged him to read Aristotle directly instead of confining himself to scholastic commentators who Thomasius felt had distorted Aristotle. The kind of training he received under Thomasius' direction is revealed by his bachelor's dissertation *Disputatio metaphysica de principio individui*, published in 1663 when Leibniz was seventeen. This work was concerned with the "principle of individuation," i.e., what is it that defines an individual and makes it different from other individuals out of a species to which it belongs. Two individuals such as identical twins may share the same essence or form,[5] but are numerically different. How can this be explained? To express this a little differently, any collection of properties possessed by an individual can be possessed by other individuals, so that individuality must consist of the union of a common form with something else. What this "something else" is constituted the problem. Aristotle's solution was that it was the difference in the matter on which the same form may be imprinted that makes different individuals: "The whole thing, such and such a form in this flesh and these bones, is Callias or Socrates."[6] This problem, initially discussed by Boethius (ca. 480–524) and the Arabian commentators on Aristotle, Avicenna (c. 980–1037), and Averroes (1126–1198), was intensely debated by scholastics such as Aquinas, Duns Scotus (c. 1265–1108), and William of Ockham (c. 1288–c. 1348); and it was still of vital interest in seventeenth century German universities.[7] Leibniz begins by listing four different theses concerning the principle of individuation. These are:

- Negation: A thing is an individual because of the properties it does not share with other individuals.
- Existence: A thing may be defined by a common nature or form shared with other things, but it is an individual only because it exists.
- *Haecceity*: This is the position of Duns Scotus. A thing is an indi-

[5]The essence or form is the collection of properties that make a thing the kind of thing it is.

[6]*Metaphysics*, 1034a, 5–8. If this were true as St. Thomas Aquinas (1225–1274) argued each individual angel having no matter would have to constitute a species.

[7]See Gracia, (1994) for a collection of papers tracing the history of this dispute in great detail from 1150 to 1650.

vidual because it is the union of a common nature and "thisness"
- "The whole entity" or the union of *both* matter and form constitute the individual.[8]

Leibniz gives elaborate arguments against the first three criteria of individuation and defends the last. In the course of his analysis he shows himself a Nominalist. To Leibniz nothing exists independently of individuals. Common properties shared by individuals do not actually exist *in re*; they are purely creations of the mind. While the existence of an individual may be conceptually distinct from its essence, the two are not actually separable. In this position he was influenced by Suarez, and it may have led to the Identity of Indiscernibles which was an key element in Leibniz's later philosophy.[9]

Leibniz's thesis was a technically brilliantly done classroom exercise. But of greater probable intrinsic interest to him was his discovery just before he turned fifteen and entered the university of the "moderns" such as Bacon (1561–1626), Hobbes, Descartes, Gassendi (1592–1655), Kepler, and Galileo. These philosophers apparently had a different outlook than Aristotle as interpreted by the scholastics. Instead of a physics resting on substantial forms and qualities, they proposed quantitative and mechanical explanations. Nature was written in the language of mathematics and consisted of a law bound collection of material corpuscles in motion. All material phenomena should be explained using only the concepts of extension and motion. Leibniz recalled in 1714 spending a day wandering in a grove just outside of Leipzig trying to decide between these two approaches and said that "Mechanism finally prevailed and led me to apply myself to mathematics, the very sap of the new philosophy."[10] This, however, is only partially true. For much of the rest of his life he wrestled with the problem of combining mechanism with a properly understood Aristotle in a way that would both account for natural phenomena according to mechanical principles and yet explain the underlying reality in a way similar

[8] *Ibid.*, p. 58.

[9] This principle asserts that there cannot be two individuals that are identical in every respect but numerically different. Whether or not Leibniz's dissertation supported an early version of this principle is disputed. Mates (1986) and McCullough (1996) believe that it did and shows that important features of Leibniz's metaphysics were present at an early age. Garber (2009) disagrees. He feels that the dissertation was only an academic exercise supporting a position of his teacher Thomasius. Had Thomasius supported another of the four options, so would have Leibniz's dissertation. Leibniz's position on the problem of individuation is analyzed in great detail in McCullough (1996).

[10] Quoted in Antognazza (2009), p. 53. See also Mercer (2001), p. 24.

to Aristotle.[11] This was a daunting task. No two philosophies seemed more dissimilar to each other, and mechanism in particular was thought to have alarming theological consequences. In the case of the Eucharist, for instance, Catholic dogma (especially after the Council of Trent) taught that the Substance of Christ's Body and Blood after the consecration by the priest in the Mass replaced the substance of the bread and wine, only the "accidents" (visual and physical qualities) of the bread and wine being preserved. It was difficult to see how this "transubstantiation" could occur (or how Christ would be present at simultaneous Masses in different locations) if a physical body was characterized solely by an arrangement of atoms in motion. Anyone other than Leibniz confronting this problem might have become a Hobbesian materialist or rejected (as Thomasius did) mechanism in favor of some form of Aristotelianism either as revealed in the original texts or in some major scholastic interpretation. Leibniz, however, did neither. Instead he began to demonstrate a life-long eclectic effort to combine apparently contradictory doctrines into a harmonious whole. Leibniz took a novel counterintuitive approach by claiming that nothing in the new theories was incompatible with a properly understood Aristotle divested of scholastic misinterpretations. At the phenomenal level, he agreed that mechanical explanations were to be preferred. Yet the motion and continuing existence material bodies, he felt, demanded that they possess an incorporeal principle of activity and organization which constitutes their substance. Leibniz thinks of this as a mind and identifies it with the scholastic/Aristotelian substantial form. In an essay of 1668 *On transubstantiation* he applies these ideas to the Eucharist, arguing that in the Mass the mind of Christ replaces the "mind" determining the substance of the bread and wine. These then become the Substance of Christ's body and blood while preserving the physical properties that exist before consecration.[12]

By the time he graduated from Leipzig Leibniz was very likely thoroughly familiar with all the major ancient, scholastic, and modern philosophers, yet he had very little training in mathematics. His only actual exposure to it seems first to have been some obscure lectures on Euclid at Leipzig given by one Johann Kühn, followed by further elementary instruction during the summer session of 1663 at the University of Jena from the Professor of Mathematics Erhard Weigel Most students found Kühn's lectures unintelligible. Weigel was a improvement. He was a minor but

[11]Antognazza (2009), pp. 53–58.
[12]Mercer (2001), pp. 82–88 and Chapter 3.

reasonably well trained mathematician and seems to have taught Leibniz some combinatorial theory. Weigel also had interests in both Aristotle and the newer philosophers such as Descartes. He believed that the new mechanical philosophy was in large part actually an extension of Aristotle and may have even inspired Leibniz's desire to reconcile in some sort of synthesis Christianity, Aristotle, and the new mechanical philosophy.[13] Weigel's philosophical influence was probably of more value than his mathematics for in later life Leibniz was fond of stating how little mathematics he knew in this period. He confesses that he did not "except for the most ordinary practical rules, know anything about geometry; he had scarcely even considered Euclid with anything like proper attention being fully occupied with other studies."[14] Perhaps he knew a little of the first book of Euclid, arithmetic, and whatever Weigel taught him; but he was quite unfamiliar with current literature such as Descartes *Géométrie*. Training in mathematics, he said:

> ... was entirely neglected in the places where my mind received its early discipline; and had I, like Pascal, spent my youth in Paris, I should perhaps have sooner made original contributions to science.[15]

Returning to Leipzig in the Fall, Leibniz obtained a Master's degree in Philosophy in February 1664 with a thesis showing a growing interest in law and entitled *Specimen quaetionem philosophicarum de jure collectarum* and then, after formally turning to legal studies, a Bachelor's degree in law the following September. So far, although a precocious (and very young) student, there was little in his background and apparent interests to distinguish him from other talented students or to indicate his future mathematical brilliance. His education in late scholastic philosophy and law was a standard one, and it was probable that his teachers expected him to pursue an academic career in one or both of these fields. If he had done so and kept on a conventional contemporary intellectual track (broadly interpreted), he probably would have become either a successful law or philosophy professor at some German university or a bureaucrat in one of the German principalities and, like his teachers, forgotten today except by specialists.

[13] An interesting discussion of Weigel and his possible influence on Leibniz may be found in Mercer (2001).

[14] Leibniz (1920), p. 35. In a letter to James Bernoulli (April 1703) he also speaks of his "superb ignorance of mathematics." *Ibid.*, p. 12.

[15] Quoted in Mackie (1845), p. 28f.

One of the first indication of some mathematical interest in Leibniz's intellectual development was shown in his habilitation thesis entitled *Disputatio arithmetica de complexionibus*, defended in March 1666 before the Faculty of Philosophy, and which was published the same year in an enlarged and revised version with the title *Dissertatio de Arte Combinatoria.*[16] Its mathematical content consisted of a fairly elementary part of the theory of permutations and combinations. He calls a subset of a larger whole a "complexion." The number of members in a complexion is the "exponent," "after the example of a geometric progression,"[17] while the number of members in the whole is called the *number*. What we call the number of combinations of n objects taken, say, 3 at a time he writes as "con3nation," and so forth. He then gives a table of the number of combination of n things taken r at a time according to a rule which is equivalent to Pascal's triangle and for which he gives a verbal description: "Add the complexions of the number of the number preceding the given number, by the given exponent and by the exponent preceding it; the sum will be the desired complexions."[18] In modern notation, this is equivalent to the addition law

$$\binom{n}{r} = \binom{n-1}{r} + \binom{n-1}{r-1},$$

by which the table can be constructed. Leibniz is aware of the formula con2nation $= n(n-1)/2$ where n is the "number," but does not exhibit an algebraic representation of "con$r$nation" equivalent to $\binom{n}{r}$. Given a geometric progression with base 2, he states that the total number of "complexions" (i.e., subsets) of a collection of n members is "the number of terms of the progression whose exponent is the given number minus 1,"[19] and illustrates this by a table without giving a proof, since "It is difficult to understand the reason or demonstration for this or explain it if understood."[20] Furthermore, much of this as Leibniz admits is unoriginal. Of the four problems he considers he says "we owe the latter part of the first

[16]Another may be his thesis for the bachelor's degree in law. One topic in this work was an attempt develop a logic of uncertain judgements applied to the law. He assigns the numbers 0, 1/2, and 1 to laws which respectively depend on impossible, uncertain, or necessary conditions. [Aiton (1985), p. 17.]

[17]*Dissertatio*, quoted in Leibniz (1969), p. 78.

[18]*Ibid.*, p. 79f.

[19]*Ibid.*, p. 80.

[20]*Ibid.*, Note that Leibniz seems unaware of a very simple proof: Think of an urn and for each member of the set of either dropping it in or keeping it out (2 possibilities). Every time we do this the contents of the urn constitute a "complexion." Therefore there are 2^n possible complexions. Since one possibility is the "empty complexion" (nothing in the urn) which doesn't count, we subtract 1.

problem, and the second and fourth to others; the rest we ourselves have discovered."[21] Most of the results, in fact, do not go beyond contemporary books by Daniel Schwenter[22] and Philipp Harsdörffer with both of which he was familiar. He also knew of the commentary on an early medieval textbook on astronomy, the *The Sphere of Sacrobosco* by Christopher Clavius, a Jesuit mathematician, which in part dealt with combinations. But some had been overlooked in the literature, and Leibniz's arguments in places were his own.[23]

However, we should not think of the *Dissertatio* as even primarily mathematical. Much of it exhibits a quite alien pre-modern mental universe. There is an elaborate proof of the existence of God similar to Aquinas' cosmological argument. There are theses in metaphysics such as "God is substance; creature is accident."[24] Leibniz uses his doctrine of complexions to derive the four elements: earth, air, fire, and water from the four primary qualities: heat, cold, dryness, and moistness, thus clarifying an argument of the Pythagorean Ocellus Lucanus mentioned by Aristotle.[25] There are also applications to law such as the number of contracts in favor of a subset of contracting parties—the mandator, the mandatory, and a third person. Leibniz's claims correctly that there are seven possibilities. This, however, differs from the accepted answer of five. He is able to whittle down the number to six and wonders why the jurisconsults also eliminated the case of a contract in favor of all three parties.[26] Still other applications involve poetic meters, determination of the number of tones possible (4,095) in an organ with twelve pipes, and the calculation of the number of modes of the syllogism. Of perhaps greater interest, the dissertation shows the beginnings of his life-long efforts to symbolize logical argument and his belief that the *ars combinatoria* developed in the thesis was the "key to all the sciences" and to finding an art of discovery which would be the "mother of all inventions"[27] Since these projects are so important in Leibniz's subse-

[21] *Ibid.*, p. 78f.
[22] *Deliciae physico-mathematicae* Nuremberg, 1651–53.
[23] Hofmann, 1974b. One apparently new result discovered by Leibniz is the identity $2(n+1)! - n \cdot n! = (n+1)! + n!$.
[24] Leibniz (1969), p. 75.
[25] *Ibid.*, p. 81. The elements are determined by the four qualities taken two at a time, which would give six elements. However, heat and cold or dryness and wetness taken together cannot form elements since they are contrary qualities; so there are only four possibilities.
[26] Leibniz refers to an accepted legal practice of discarding the contract in favor of the "mandatory alone, because this would be advice rather than a mandate." (*Ibid.*, p. 81f.)
[27] Antognazza (2009), p. 63.

quent career, we will discuss them and the Llullist influences behind them in detail elsewhere.[28] To sum up, the *Dissertatio de Arte Combinatoria* seems like an almost indigestible casserole containing an elaborate combination of ideas derived from the contemporary intellectual climate and showing nearly incredible learning and considerable originality.

Leibniz was turned down for the Doctorate of Law at Leipzig in 1666. It seems that the older students backed by the faculty objected (in their own self-interest) to any extremely young student (Leibniz was twenty) being awarded this degree and competing with them. Consequently he left Leipzig in September, enrolled in the University of Altdorf near Nuremberg, and was awarded the degree in February 1667.[29] For the remainder of his life he returned to Leipzig only for a few short visits. At this point Leibniz seemed destined for a conventional legal career. He was offered a professorship in law at the University of Altdorf in 1667 but declined it. Since he was still very young, he may have felt that it was too early to settle down. As he still needed a paying job, he became secretary of a small quasi-Rosicrucian group based in Nuremberg and pursued the study of alchemy.[30] Later in 1667 he left Nuremberg, briefly visiting Mainz, enroute to Frankfurt as part of a projected grand tour. In Frankfurt he met and became friends with Johann Christian von Boineburg (1622–1672), the former chief minister of the Elector-Archbishop of Mainz, Johann Philipp von Schönborn (1605–1673). The Elector was a major figure in the Empire. He had been closely involved in the negotiations that had ended the Thirty Years War and was currently involved in two ambitious plans for reform: one of the legal system and the other an attempt to reconcile Protestantism and Catholicism in Germany. Excited by the Elector's projects, Leibniz put off his idea of a grand tour and submitted a quickly written plan for the reform of the teaching of law[31] to Johann Philippe at the end of 1667. This and Boyneburg's patronage allowed Leibniz to move to Mainz and enter the service of the Elector. One of his duties was to assist in an ambitious plan for the reform of the civil code. Here he remained for the next five years. The Elector was so impressed with his work that he appointed Leibniz as a judge of the High Court of Appeals, even though Leibniz was only twenty-four and a Lutheran while Mainz was a Catholic

[28] Chapters 7–9.
[29] Hofmann (1974a), p. 5.
[30] For an interesting account of the influence of Rosicrucian ideas generally and the Kabbalah in particular on Leibniz see Coudert (1995).
[31] *Nova Methodus Discendae Docendaeque Jurisprudentiae.*

territory. The years in Mainz were exceedingly productive. The Court at Mainz was a major intellectual center. Leibniz met a large number of intellectual and scientific figures including Heinrich Oldenburg (1615–1677), a fellow German who was Secretary of the English Royal Society, with whom he became friends and entered into correspondence with other scientists and intellectuals. During this period Leibniz, as he was to do for the rest of his life, worked on an immense array of projects. Besides his official work in legal reform, he wrote and published under an assumed name a memorandum in favor of a candidate for the Polish throne supported by the Elector. In theology he composed a defense of Trinitarianism for Boinburg[32] and began a vast never completed treatise on the essentials of the Christian faith common to both Protestants and Catholics called the *Demonstrationes Catholicae.*[33] A third essay was published in 1669 under the title *Confessio naturae contra atheistas.* All these writings were intended to confute the enemies of religion in general and to help reunite the churches. The latter project exhibited in the *Demonstrationes Catholicae* remained dear to him his entire life and again is an illustration of Leibniz's perpetual effort to mediate between seemingly unreconcilable positions.

There is evidence that Leibniz was beginning to become more interested in mathematics and physics at this time. He continued his meditations on the new mechanical philosophy of Hobbes, Gassendi, and Descartes. Again, as is shown in two letters to his old teacher Thomasius in 1668 and 1669, his philosophical aim in this period, possibly influenced by his teacher Weigel, was to somehow reconcile the new mechanical philosophy with Aristotle. These letters also show a particular interest in motion: how to explain it, and its relation to the continuum. In this period he also began efforts to construct an artificial language or "Universal Characteristic," whose possibility was already hinted at in *De Arte Combinatoria.* This language would be free of the ambiguities in natural languages and also would function as an aid to discovery and as a means to settle differences in an absolutely reliable manner resembling calculation calculations in arithmetic. Since the characteristic bears such a close relationship to Leibniz's mathematical style, we will defer further discussion of it until a later chapter.[34]

The major initial influence in Leibniz's turn in the direction of mathematics and physics seems to have been the work of Thomas Hobbes. This was perhaps was not the best choice since Hobbes, now elderly and possi-

[32] *Defensio Trinitatus per nova reperta logica.*
[33] See Antognazza (2009), p. 90.
[34] Chapter 8.

bly senile, had by the late 1660s made himself a mathematical laughing-stock. Beginning with his exposition of corpuscular philosophy *De corpore*, Hobbes had repeatedly attempted to square the circle. Each time his purported quadrature was shown to be mistaken. This in itself was no disgrace. The quadrature problem was not known to be impossible of solution, and other mathematicians such as Gregory St. Vincent had also produced mistaken quadrature constructions. Initially, Hobbes had realized his mistakes and simply tried again. Most recently, however, he refused to accept criticism and argued that his many constructions were correct. Moreover, he was publishing doubts concerning well known mathematical results such as the Pythagorean Theorem and the value of π (which he held to be around 3.2).[35] As early as 1663 when he was primarily involved in legal studies, Leibniz showed familiarity with Hobbes and admired the clarity and precision of Hobbes' philosophical writings although, of course, he strongly opposed Hobbes' materialism and reputed atheism.[36] Unlike many of his contemporaries, Leibniz did not recoil in horror from the "Monster of Mamesbury." Rather, Hobbes was an able foe whose arguments contained valuable insights and when pernicious had to be answered. This respect caused Leibniz in July, 1670 to write a flattering (but unanswered letter) to Hobbes claiming to have "read almost all your works, in part separately and in part in the collected edition."[37] As mentioned above Hobbes' egregious mathematical eccentricities were very well known among mathematicians by 1670, and Leibniz was certainly aware of some of them. We find marginal notes by him various parts of Hobbes' *Opera philosophica* (which Boineburg had purchased at a book fair probably in 1669) expressing astonishment at his denial of the Pythagorean theorem and refusal to accept the square root of the area of a square as measuring the length of its side.[38] However, Leib-

[35]Even though Hobbes was mathematically incompetent, he was a firm believer in the geometric paradigm sketched in Chapter 2. In response to his critics he rejected even the possibility of applying algebra to geometry, arguing that it gave incorrect dimensional results. See the discussion in Jesseph (1999).

[36]See e.g., Leibniz's *Confessio Naturae contra Atheistas* (1669) in Leibniz (1969), p. 110 and Goldenbaum (2009), p. 63.

[37]Leibniz (1969), p. 105. Leibniz had sent this letter to Hobbes via Heinrich Oldenburg, a fellow German and secretary of the Royal Society with whom he had also initiated correspondence in the summer of 1670, probably with the encouragement of Boineburg. The reason why it was unanswered may have been because Oldenburg may not have forwarded it because he was aware of Hobbes reputation and felt that the letter would harm Leibniz's own scientific reputation in England (Philip Beeley, personal communication).

[38]*Ibid.*, p. 61. As we have said, the failed proofs of the quadrature of the circle in earlier editions of *De Corpore* or the duplication of the cube (published in *La Duplication du Cube par V.A.Q.R.* (Paris, 1661) are one thing, but the incredible additional mathe-

niz still had very little technical command of mathematics and may have lacked the training to have been able to independently evaluate Hobbes' incredibly bad mathematics. At any rate, Leibniz was mainly interested in Hobbes' political theories and in his dynamics. Whatever his opinion of Hobbes' mathematics, Leibniz's letter expressed a diplomatic politeness towards an old man whose general philosophy and dynamical and political theories he respected. He also seems to have felt that Hobbes' corpuscular philosophy was essential to physical explanation, but he believed that it needed be integrated on a metaphysical level with both Aristotle and his own idealism. Leibniz became especially intrigued with the concept of *conatus* developed by Hobbes in his *De corpore.* Hobbes begins with an idiosyncratic conception of a point. For Euclid a point was "that which has no parts." But according to Hobbes whatever exists in nature is a material body. For him, therefore,

> ... by a point is not understood that which has no quantity, or which can by no means be divided (for nothing of this sort is in the nature of things), but that whose quantity is not considered, i.e., neither its quantity nor any of its parts are computed in demonstrations, so that a point is not taken for indivisible, but for undivided. And as also an instant is to be taken as an undivided time, not an indivisible time.[39]

Points then are apparently extended and instants are durations, but their magnitudes are so small that they are not taken into consideration in geometric or physical arguments. This "materiality" of spatial points, furthermore, means that they can differ in "size." It follows also that lines, being composed of points and contrary to Euclid, have breadth.[40] Time, however, is constant in the sense that moments have equal duration *Conatus* is then defined as motion through a spatial point in a moment. But here Hobbes adds a qualification that seems to contradict his definition of point:

> *conatus* is a motion through a space and time less than any given, that is, less than any determined whether by exposition or assigned by number, that is, through a point.[41]

matical mistakes with made in the 1668 edition of *De Corpore*, the refusal of Hobbes to admit *any* mistakes, and his disagreement with standard mathematical results such as the Pythagorean Theorem, or the accepted values of π or $\tan 30°$ in other writings of the late 1660s are another. They totally destroyed any respect for Hobbes among the mathematically literate. See Jesseph (2008), Chapter 6.

[39]Hobbes, *De Corpore*, 3.15.2, also quoted in Jesseph (1999), p. 103.

[40]On occasion Hobbes will try to answer various critics such as Huygens who pointed out his many geometric mistakes by asserting that they fail to take into account these properties of points and lines. See e.g., Jesseph (1999), p. 286ff.

[41]Hobbes, *De Corpore*, 3.15.2, also quoted in *Ibid.*, p. 102.

There is a vague limit concept here; but we can ask how the limiting point can be extended and divisible if its magnitude is less than any assigned magnitude?[42] *Conatus*, although it is motion at a point, is not what we think of as instantaneous velocity. This Hobbes calls *impetus* which is defined as the quantity of *conatus*, that is the distance traveled in an indefinitely small time divided by the time.[43] The influence of these ideas upon Leibniz is most concretely shown by one of his earliest writings on natural philosophy, the *Theoria Motus Abstracti* published in 1671 and dedicated to the French Academy of Sciences. Much of it is almost a paraphrase of relevant parts of *De corpore*.[44] Like Hobbes, Leibniz argues that points have "parts," can be divided, and can be of different sizes. He accepts Hobbes' conception of time and his definitions of *conatus* and *impetus*. Both also use the idea that points may differ in magnitude to explain how a *conatus* can vary in strength: a stronger *conatus* moves through a "longer" point in the same moment of time than a weaker one, or in his words: *Punctum puncto, conatus conato major est, instans vero instanti aequale.*[45] But, there is one important difference: Hobbes' points are extended while Leibniz is (easily!) able to show that they are unextended.[46] How then can they have parts? Leibniz's answer is that the parts have no distance from each other; they are "superimposed" on each other.[47]

What are we to make of these early speculations? Douglas Jesseph has suggested that Hobbes' concept of *impetus* may have played a role in Leibniz's development of the calculus. Given a curve representing distance D versus time t, Hobbes definition of the *impetus* at a specific t amounts to finding the slope of the tangent line to the curve at t, since we must calculate the "average" impetus over a small distance $D(t+\delta t)-D(t)$ where δt is a time interval ultimately "smaller than any that can be assigned." What we call the characteristic triangle and the derivative follow immediately from this reasoning. The problem with such an interpretation, however, is that it is not clear that Hobbes ever actually drew these *graphical* conclusions from his definitions. Secondly, similar notions were prevalent among many other mathematicians, notably Barrow and Pascal. Since they ac-

[42] Possibly Hobbes meant that given any magnitude, we can consider a point as having a "length" less than that magnitude.

[43] Jesseph (2008), p. 219.

[44] Bernstein (1980), pp. 25–37.

[45] Quoted in Dugas (1958), p. 455.

[46] Leibniz (1969), p. 139.

[47] I am indebted to Philip Beeley (personal communication) for his comments on these matters.

tually dealt with characteristic triangles they may have a better claim to have influenced Leibniz than Hobbes. The real problem, however, is that around 1670 Leibniz had hardly any real knowledge of advanced mathematics and we suspect that even Hobbes' few decent mathematical ideas (as distinguished from his dynamical philosophy, concept of conatus, etc.) could not have had much *technical* mathematical impact on Leibniz at this time. However, even if Hobbes contributed nothing of value to Leibniz's mathematical development, he probably provided a catalyst. For, as Ursula Goldenbaum has suggested, in order to understand them he would have had to study geometry intensively. In this sense, if in no other, Hobbes then may have been indirectly responsible for Leibniz's gradually increasing technical competence.[48]

Although as we have said the *Theoria motus abstracti* was strongly influenced by Hobbes' ideas, the actual inspiration for it was Leibniz's discovery of work on the mathematical laws of collision by John Wallis, Christopher Wren (1632–1723), and Christiaan Huygens, recently published in the *Philosophical Transactions* of the Royal Society, in the late summer of 1669 when he accompanied Boineburg to the spa town of Bad Schwalbach and was shown the journal by a law professor friend of Boineburg. These authors, reacting to some mistakes of Descartes had derived the correct laws of elastic and in some cases (Wallis) of inelastic collision.[49] In contemplating this work, Leibniz criticized it for being limited to the phenomenal realm. These "laws" may be supported by our experience, but they are not the true laws which are derivable simply from the rational analysis of the concept of motion. In fact, experiments

> ... must be eliminated from the science of the abstract reasons for motion, just as they should be eliminated from geometrical reasonings. For they are demonstrated not from fact and sense, but from the definition of the terms.[50]

The *Theoria Motus Abstracti* represented Leibniz's attempt to find these laws by applying and extending Hobbes' ideas. Like Hobbes, he argues that a body at rest cannot be the cause of its own or any other motion. This implies that a body at rest, no matter how large, offers no resistance and

[48]Goldenbaum (2008), p. 67f.

[49]Wallis and Wren used the law of the conservation of momentum to obtain their results, while Huygens also introduced the conservation of kinetic energy in the case of elastic collisions.

[50]Quoted in Garber (2009), p. 15. The quotation is from Leibniz's letter to Hobbes of July 1670 which may be found in Leibniz (1969), p. 105ff.

will be moved upon impact with an arbitrarily small body with the speed of that body. In general if two moving bodies collide, the two together will move with a speed which is the difference between their speeds and in the direction of the faster body. Size or mass plays no role in collisions; only the speed is a factor.[51] The "laws" of collision promulgated in *Theoria Motus Abstracti*, however, are not as irrational as they seem at first glance. In his later *Specimen Dynamicum* (1695) Leibniz indicates that at the time he agreed with the basic ideas of Descartes concerning matter. Like Descartes he assumed body to be "understood in mathematical terms only–magnitude, figure, and their change" and made no use "of metaphysical notions such as active power in form or of passive power and resistance to motion in matter."[52] Leibniz's laws are then reasonable conclusions given these assumptions.

What explains then the disagreement between the apparent "laws" of motion discovered by Wren, Wallis, and Huygens and their true form, derivable from the foundations laid by Hobbes and Descartes, and expounded in the *Theoria motus abstracti*? Leibniz's answer is contained in his *Hypotheses physica nova* or *Theoria motus concreti* which is complementary to his *Theoria motus abstracti*, begun soon after his visit to Bad Schwabach in 1669, and presented to the Royal Society (to which it was also dedicated) in 1671.[53] The explanation Leibniz gives is truly ingenious. Suppose a ball B hits a row of balls $a_1, \ldots, a_n$ moving with a common speed s_1 in the opposite direction to B and that the speed s_2 of B is greater than s_1. After hitting a_1 by the true laws of motion the speed of $B + a_1$ will be $s_2 - s_1$. This process is repeated with a_2, a_3, etc. The speed of B diminishes and may even be reversed. Since physical bodies are "discontinuous" or made of smaller disjoint parts, we can see how the size of bodies can affect their behavior upon collision. But Leibniz has greater ambitions for the *Hypotheses physica nova* than to merely explain the mathematical laws of collision. He gives an elaborate account of an aether filling all space and uses it to explain light, magnetism, gravity, sound, heat, chemical phenomena, the nature of liquids, hardness, elasticity, and it generates the elements earth, air and water in the form of bubbles which are the foundation of all earthly

[51]See the discussion and references in Garber (2009), p. 17. Here again Leibniz's views of the laws of motion and collision are very similar to those of Hobbes. See Jesseph (2006), pp. 129–139. For a discussion of Leibniz's belief that the true laws of nature must be discovered by abstract reason alone and transcend the phenomenal level see Beeley (2009).

[52]Leibniz (1969), p. 440.

[53]The two essays were printed together as Leibniz (1671).

beings. Gravity in particular is an effect of the circulation of the aether and has the property that the heaviness of objects decreases as the square of the distance from the center of the earth.[54] It is interesting that *Hypotheses physica nova* was well received by the Royal Society. Leibniz had sent an incomplete copy of the work to Oldenburg in March 1671 who gave it to a committee composed of Robert Hooke (1635–1703), John Pell (1611–1685), Wren, and Wallis for evaluation. The opinion, if any, of Pell is unknown. Wren may not have seen the work. Hooke was (as usual) rather unflattering, but Wallis liked it a great deal and wrote a long positive report. Wallis also was sent *Theoria motus abstracti.* Fortunately, Wallis claimed that because of a lack of time, he did not look over it in depth, and confined himself to a few ambivalent remarks. Leibniz was lucky. Wallis and Hobbes hated each other. Had Wallis closely studied *Theoria motus abstracti* he would have been aware of Hobbes influence on Leibniz with probably deplorable results for Leibniz's future with the Royal Society. But because the positive reception of *Hypotheses physica nova* and Oldenburg's support he gained election to membership in the Royal Society in April 1673, just after his first visit to London.[55]

By 1672 Leibniz had already shown interest and written in a bewildering number of fields: among them law, theology, scholastic and modern philosophy, physics, the universal characteristic, logic, and elementary combinatorics. He had made, moreover, an excellent impression on a number of important people, among them Boineburg, Wallis, and Oldenburg. A letter of Boineburg to Hermann Conring (1606–81), April 22, 1670, registers what was probably a typical impression:

> He is a young man of twenty-four years from Leipzig, a doctor of law, and indeed more learned than can be seen or believed. He understands all philosophy and happily excogitates on the old and the new. He is equipped with an ability to write of the highest level. He is mathematically very inclined and well versed in physics, medicine, and mechanics.[56]

Boineburg is perhaps exaggerating a bit when he speaks of Leibniz's mathematical inclinations and knowledge of physics or mechanics. With the exception of the combinatorial parts of his Master's thesis, Leibniz had demonstrated little technical command of contemporary of mathematics. He was grossly familiar with Cavalieri and some other mathematicians,

[54]Dugas (1958), pp. 456–458; Garber (2009), pp. 19–21.
[55]See Beeley (2004).
[56]Quoted in Beeley (2004), p. 54.

probably from the discussions in Hobbes' *De Corpore*, but almost certainly had not yet read them in the original. He probably would had difficulty following Archimedes, Apollonius, Descartes *Geometrie*, or even some of the later books of Euclid. It was the same with physics or mechanics. He could discourse on the qualitative, "metaphysical" aspects of these disciplines as they were understood in the seventeenth century, but could he have understood Huygens or for that matter the mathematical arguments underlying the laws of collision derived in the *Philosophical Transactions* papers of Wren, Wallis, and Huygens? A similar situation is true in physics. However interesting *Theoria motus abstracti* and *Hypotheses physica nova* are as registers of contemporary mechanical philosophy applied to physics, as a discoverer of real physical laws Leibniz is simply not yet in the same league as Huygens or even Wren and Wallis. Although extremely bright, ambitious, well educated, and lucky in the powerful support of his countryman Oldenburg, he had not yet really separated himself from what must have been a fair number of other young contemporary German university graduates, well versed in philosophy, who went on to successful careers as professors, lawyers, or functionaries in German principalities and who today are mostly forgotten.

Chapter 5

First Steps in Mathematics

The opportunity that changed the history of western mathematics was a consequence of a memorandum, the *Consilium Aegyptiacum* (1671–72) drawn up by Leibniz recommending that the European states undertake a new crusade against the Turks. Anticipating Napoleon, he proposes that France should take the lead by an invasion of Egypt. Leibniz's proposal was part of an effort to divert Louis XIV's destructive ambitions from Holland (and potentially Germany). Boineburg had already vaguely communicated the plan in correspondence with the French foreign minister, Simon Arnauld de Pomponne (1618–1699). Since there seemed to be some interest Leibniz was sent to Paris in March 1672 by Boineburg with the mission of formally presenting the Egyptian project at the French Court. A secondary goal was to secure a considerable amount of money in the form of property rents and pensions which the French owed Boineburg.[1] Evidently Leibniz's memorandum was not taken seriously, since Louis allied with Charles II of England, Sweden, and the Bishopric of Münster had declared war on April 6 and duely launched an invasion of on Holland in May, barely two months after Leibniz's arrival in Paris (as late as September Leibniz was even unable to meet Pomponne).[2] A further effort is made in November 1672 when Leibniz is joined by Boineburg's son-in-law and nephew of the Archbishop-Elector Melchior Friedrich von Schönborn (1644–1717) who was carrying a proposal for a Peace Conference in Cologne. This too was ignored by the French and the war continued until 1678.[3] But at least these diplomatic disappointments allowed Leibniz to spend the next four years in Paris, then

[1] Antognazza (2009), pp. 116–118, 124.
[2] *Ibid.*, pp. 118, 141.
[3] *Ibid.*, p. 145f. The war was ended by the Treaty of Nijmegen. The Dutch had repulsed the invasion of their country, but the French were granted some territory in the Spanish Netherlands (now Belgium).

the culturally most developed city in Europe, with relatively few demands put on his time by the authorities in Mainz.[4]

Leibniz's exposure to the wider world of seventeenth-century science and mathematics really begins in the Paris period. In the fall of 1672 he meets and becomes friends with Christiaan Huygens, probably the premier European mathematician of the day. Huygens was a son of a distinguished poet diplomat Constantijn Huygens (1596–1687) and his family was of great political importance in Holland. When he and Leibniz met Huygens was employed by the French government as Director of the Royal Library and organizer of the *Académie Royale des Sciences.* Huygens acts as Leibniz's informal mathematical tutor and advisor, introducing him to the contemporary mathematical scene and pointing out the important literature that Leibniz should read. Leibniz writes of himself that on Huygens' advice:

> He began to work at Cartesian analysis which aforetime had been beyond him, and in order to obtain an insight into the geometry of quadratures he consulted the *Synopis Geometricae* of Honoratus Fabri, and a little book by Dettonville [i.e., Pascal].[5]

Additionally, he meets the mathematicians Jacques Ozanam (1640–1718) and Ehrenfried Walther von Tschirnhaus[6] and the philosophers Antoine Arnauld (1612–1694) and Nicolas Malebranche (1638–1715). Ozanam will acquaint him with number theory, proposing such problems as the finding of three integers x, y, z such that $x - y$, $x - z$, $x^2 - y^2$, $x^2 - z^2$, and $y^2 - z^2$ are all squares,[7] and he will engage in a number of joint projects with Tschirnhaus; a friendship and correspondence between the two lasts until Tschirnhaus' death in 1708.

Almost incredibly it is also in this period, very soon after he begins the study of mathematics, that he completes his earliest original results. Leibniz's first discovery occurs soon after his initial meeting with Huygens. He remarked to Huygens that it might be possible to sum even infinitely many terms of a series provided the terms could be written by some rule. This is especially evident, Leibniz said, if the terms can be represented as differences, that is, if we can write $b_n = a_{n+1} - a_n$, then

$$b_1 + b_2 + \cdots + b_n = (a_0 - a_1) + (a_1 - a_2) + \cdots + (a_n - a_{n+1}) = a_0 - a_{n+1}.$$

[4] Antognazza (2009), pp. 116–118, 124.

[5] Leibniz (1920), pp. 37–38.

[6] Besides being a first rate mathematician who focused mainly on algebra, Tschirnhaus is considered to be the inventor of European porcelain which hitherto had only been available as an import from China or Japan.

[7] Aiton (1985), p. 49. This was one of few problems which Leibniz could not solve.

It is interesting that this simple idea did not occur to Leibniz directly, but rather as a consequence of philosophical and logical meditations on the Euclidean axiom that "the whole is greater than the part." Intrigued by these remarks, Huygens asked him as a test of his ability if he could sum the series $1 + 1/3 + 1/6 + 1/10 + \ldots$, whose terms are the reciprocals of the triangular numbers, a problem which Huygens had already solved and which had occurred to him in 1665 in connection with a question in probability theory. Today this and similar problems are routine second semester calculus exercises when the topic of infinite series is introduced, for the terms of this series can be written as

$$\frac{2}{n(n+1)} \equiv \frac{2}{n} - \frac{2}{n+1}.$$

Hence

$$\sum_{i=1}^{i=n} \frac{2}{i(i+1)} = 2 - \frac{2}{n+1}.$$

Since $\frac{2}{n+1} \to 0$ as $n \to \infty$, the sum of infinitely many terms ought to be 2. Although we would think that this would have been an immediately obvious result to Leibniz, he was led to it by a rather indirect route. In his *Opus geometricum* Gregory St. Vincent (1584–1667) had given an exceeding awkward treatment of the problem of summing the lengths of a series of line-segments $\{[b_k b_{k+1}]\}_{k=1}^{k=\infty}$ such that $|b_{k+1}b_{k+2}|/|b_k b_{k+1}| = r < 1$ for all k (See Figure 5.1). Saint-Vincent proves that the sum S of this series is $|b_1 b_\infty|$ where b_∞ is a point such that $|b_2, b_\infty|/|b_1 b_\infty| = r$.

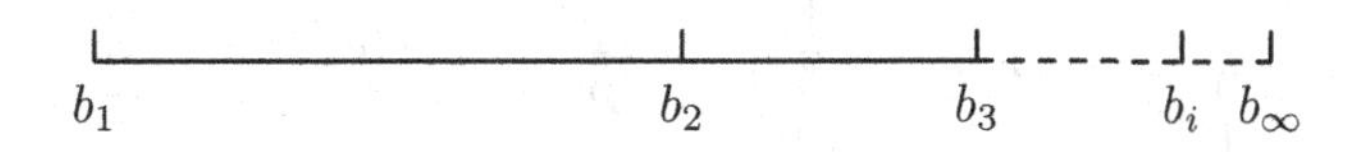

Figure 5.1

But instead of lining the segments up end-to-end as Saint Vincent did, Leibniz realized that the problem is much easier if all the segments share the same left endpoint b_1 If, for example, in Figure 5.2 we take $r = 1/3$, and write $|b_1 b_n| = r^{n-2}$, $n \geq 2$,

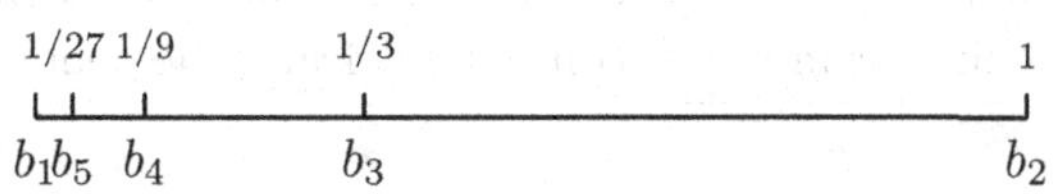

1/27 1/9 1/3 1

$b_1 b_5$ b_4 b_3 b_2

Figure 5.2

Leibniz notices that the sum of the differences $b_n - b_{n+1}$, $n \geq 2$, of the successive terms b_n of this series is 1. On the other hand, if

$$S = b_1 + b_2 + b_3 + b_4 \cdots \equiv 1 + 1/3 + 1/9 + \ldots$$

and since $b_n - b_{n+1} = 2/3^{n-1}$, $n \geq 2$, we have that

$$1 = 2(1/3 + 1/9 + 1/27 + \ldots) = 2(S - 1).$$

Hence $S = 3/2$, and for general r the same analysis shows that $S = 1/(1-r)$. He then applies a similar idea to Huygen's problem. He considers line-segments of length $1/n$ and lets $[b_1 b_\infty] = [0,1]$, $[b_1, b_2] = [0, 1/2], \ldots [b_1, b_n] = [0, 1/n]$. Then the lengths of the segments $[b_2, b_\infty], [b_3, b_2]$
$\ldots [b_{n+1}, b_n] \ldots$ clearly add up to 1 and

$$|b_{n+1}, b_n| = \frac{1}{n} - \frac{1}{n+1} = \frac{1}{n(n+1)}.$$

Leibniz does not stop with the reciprocals of the triangular numbers. He uses it to sum several other series including the reciprocals of various generalizations of triangular numbers such as "pyramidal" and "pyramidal trigono-trigonal" numbers. Here is a table of such series dating from around 1676.[8]

TRIANGULUM HARMONICUM

$\frac{1}{1}$	$\frac{1}{1}$	$\frac{1}{1}$	$\frac{1}{1}$	$\frac{1}{1}$	$\frac{1}{1}$	etc.
$\frac{1}{2}$	$\frac{1}{3}$	$\frac{1}{4}$	$\frac{1}{5}$	$\frac{1}{6}$	etc.	
$\frac{1}{3}$	$\frac{1}{6}$	$\frac{1}{10}$	$\frac{1}{15}$	etc.		
$\frac{1}{4}$	$\frac{1}{10}$	$\frac{1}{20}$	etc.			
$\frac{1}{5}$	$\frac{1}{15}$	etc.				
$\frac{1}{6}$	etc.					

Note that the denominator of the fraction in row i and column j is the sum of the denominator in row i, column $j-1$ and the denominator of the

[8]From Leibniz (1993), p. 85.

fraction in row $i-1$ and column j. The first column is the harmonic series and Leibniz writes its sum as $\frac{1}{0}$. In general if the sum of the $j-1$ column is $\frac{a}{b}$ the sum of the j column is $\frac{a+1}{b+1}$. For example, the sum of the third column of "pyramidal numbers" or

$$1+\frac{1}{4}+\frac{1}{10}+\frac{1}{20}+\frac{1}{35}\cdots$$

is $=\frac{3}{2}$ and the sum of the fourth column or

$$1+\frac{1}{5}+\frac{1}{15}+\frac{1}{35}+\frac{1}{70}=\frac{4}{3}.$$ [9]

The reader may enjoy finding the sums of the other columns.

In a letter to James Bernoulli Leibniz indicates another early discovery. Pascal in his *Lettres de A. Dettonville* (1663) in the process of giving a proof "of the mensuration of a sphere by Archimedes" had found the moment of a quarter circle with respect to its diameter. In Figure 5.3

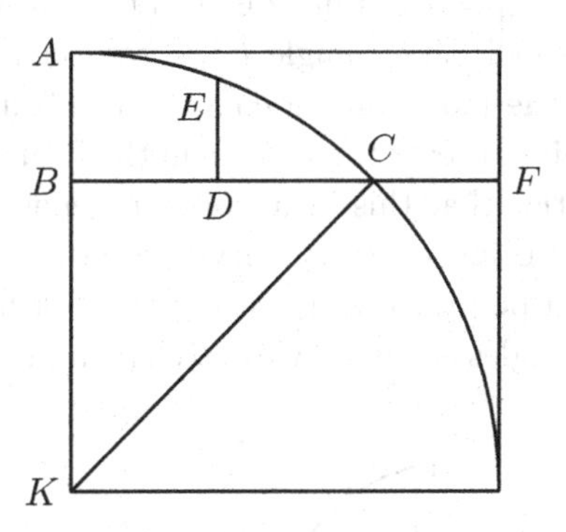

Figure 5.3

[9] *Ibid.* and Hofmann (1974a), pp. 15–19. It is interesting that Newton had a different method for summing this type of series. In the case of the reciprocals of the triangular number, start with the harmonic series $1+1/2+1/3+\cdots+1/n+\ldots$ and subtract from it all the terms but the first, obtaining

$$\begin{aligned}1&=(1-1/2)+(1/2-1/3)+(1/3-1/4)+\cdots+(1/n-1/(n+1)+\ldots\\&\equiv 1/2+1/6+1/12+\cdots+1/n(n+1)+\ldots\end{aligned}$$

Therefore multiplying by 2, we find that 2 is the desired sum. If in the last series we again subtract all the terms but the first, we get that

$$1/2=(1/2-1/6)+(1/6-1/12)+(1/20-1/30)+\cdots=1/3+1/12+1/60+\ldots$$

This argument is given in Newton's *Account of the Commercium Epistolicum* in Hall (1980), p. 273.

we assume that the triangle EDC is so small that the arc $\widehat{EC}$ may be considered a line segment. Then the triangle EDC is similar to the triangle CKB and so

$$EC : CK = ED : BC.$$

Hence $EC \cdot BC = CK \cdot ED$. Now the left side is the moment of the arc $\widehat{EC}$ about the diameter AK. Since CK is a constant radius, adding up these moments gives a total moment of the arc $\widehat{AEC}$ equal to the rectangle ABF. In modern notation setting $\widehat{EC} = ds$, $ED = dx$, $AB = x$, $BC = y$, and $CK = r$, Pascal has shown that

$$\int_0^x y\,ds = r\int_0^x dx = rx, \tag{5.1}$$

so that in particular the total moment of the quarter circle will be r^2. Further, Pascal observed that this total moment is proportional to the surface area of the hemisphere obtained by rotating the quarter circle about the axis AK. Leibniz calls the triangle EDC the "characteristic triangle." Of course, it will be the most critical component of his new calculus, but its systematic usage lies at least three years in the future. Leibniz's original contribution is to notice that this idea works in general. In Figure 5.4 the normals C_1P_1, C_2P_2 to an arbitrary curve are no longer radii of a circle meeting at a common point. Nevertheless there is still similarity between the characteristic triangles and triangles $B_1C_1P_1$ and $B_2C_2P_2$, etc.

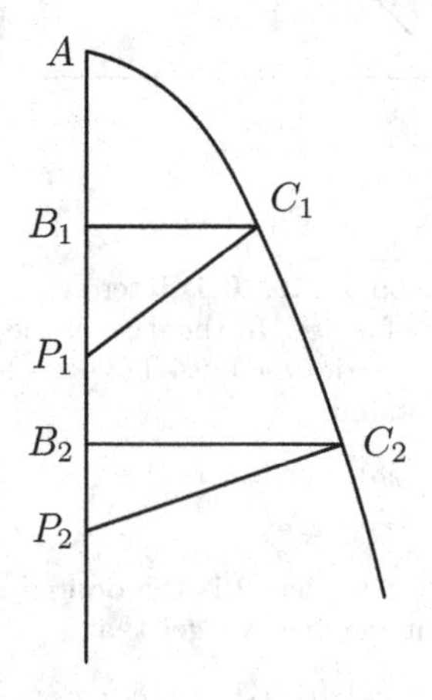

Figure 5.4

And so using modern notation and setting $C_1P_1 = n(x)$, we get that

$$\int_0^x y\,ds = \int_0^x n(x)\,dx. \tag{5.2}$$

As Leibniz remarked in a 1686 publication in the *Acta Eruditorum* he realized that the area defined by the curve $y = n(x)$ and the x axis is proportional to the surface area of a solid of revolution.[10] Both these determinations of the surface area of a hemisphere or a solid of rotation follow at once from the fact that the left-hand integrals in (5.1) and (5.2) multiplied by 2π are the modern formulas for surface area of a solid rotated around the y axis. When Leibniz showed this generalization of Pascal's result to Huygens:

> He was struck with wonder and said 'Now that is the very theorem upon which depend my constructions for finding the area of the surfaces of parabolic, elliptic and hyperbolic conoids; and how they were discovered neither Roberval or Bullialdus were ever able to understand.[11]

Leibniz's third major discovery in toward the end of 1673 or early in 1674 was much more profound. This is the beautiful result that

$$\frac{\pi}{4} = 1 - \frac{1}{3} + \frac{1}{5} - \frac{1}{7} + \cdots + (-1)^n \frac{1}{2n+1} + \ldots \tag{5.3}$$

Again, this is now a simple calculus exercise. If $\theta = \arctan(x)$, then using the chain rule

$$\sec^2(\theta)\frac{d\theta}{dx} = 1,$$

so that

$$\frac{d\theta}{dx} = \frac{1}{1+\tan^2(\theta)} = \frac{1}{1+x^2}.$$

The last term on the right in this equation is the sum of the geometric series

$$1 - x^2 + x^4 + \cdots + (-1)^n x^{2n} + \ldots$$

provided $|x| < 1$. Hence integrating term by term we have for $\theta < \pi/4$ that

$$\theta = x - \frac{x^3}{3} + \frac{x^5}{5} + \cdots + (-1)^n \frac{x^{2n+1}}{2n+1} + \ldots$$

[10] *De geometria recondita et analysi indivisibilium atque infinitorum* in Leibniz (1989), p. 140.

[11] This and the entire letter to Bernoulli may be found in Leibniz (1920), pp. 15–18.

from which the result follows using the fact that the series represents a continuous function and converges if $|x| \leq 1$.

Leibniz's method however is quite different. He outlines it in his unpublished *Historia et origo calculi differentialis* written in 1714.[12] The first step is what he called his "transmutation" theorem.

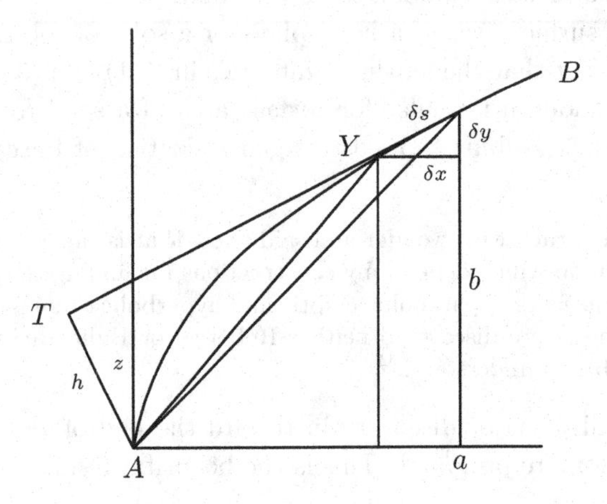

Figure 5.5

In Figure 5.5 let AYB be a smooth convex curve and TY tangent to ABY at Y. Consider the characteristic triangle at Y with sides δx, δy and δs, the triangle being so small that δs may be considered both part of the tangent and part of the curve. By similar triangles $h/z = \delta x/\delta s$, so that $z\delta x = h\delta s$. Now the thin triangular sector with vertex A and base δs has area $(h/2)\delta s$. Therefore the sum of all these sectors from $x = 0$ to $x = a$ plus the area of the triangle with sides a and b should be the area under the curve for $0 \leq x \leq a$. In our notation this is

$$\int_0^a y\,dx = (1/2)\left[ab + \int_0^a z(x)\,dx\right] \equiv (1/2)\left[ab + \int_0^{\bar{s}} h(s)\,ds\right] \tag{5.4}$$

where in the last expression $\bar{s}$ is the length of the curve between $x = 0$ and $x = a$. In other words, Leibniz has shown that the area under the curve $y = f(x)$ can be expressed in terms of the areas under the curves defined by $h = h(s)$ or $z = z(x)$. Both h and z can be computed since the tangents

[12] As a reply to the accusations from the partisans of Newton that he had plagiarized the calculus.

(or more precisely the subtangents) to a great many curves could be found by methods due to Fermat, Descartes, Sluse or Hudde. But since the use of h demands rectification of the curve, in most cases the curve defined by z is the most convenient to use.

If the curve in Figure 5.5 is a semicircle with radius 1 we have more information. In Figure 5.6 the triangles CUA and BYX are similar. To see this, note that the triangles CUA and CUY are congruent. Also BYC is isosceles and $\angle ACY$ is the sum of the two base angles of BCY. It follows that $\theta = \phi$.

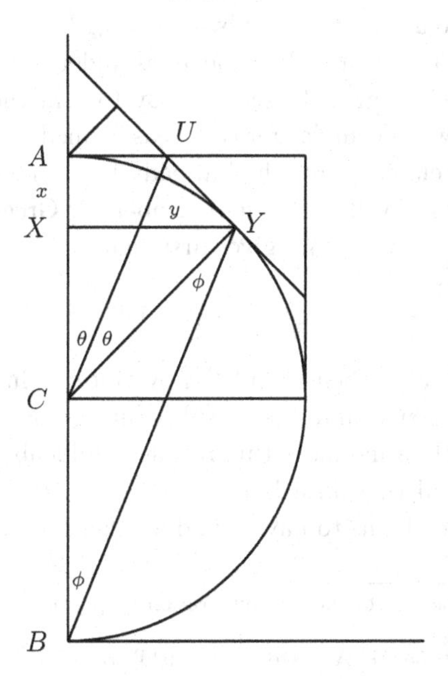

Figure 5.6

Then we have from similarity

$$\frac{x}{2-x} = \frac{x(2-x)}{(2-x)^2} = \frac{y^2}{(2-x)^2} = \frac{z^2}{1^2}.$$

Solving for x, we get that

$$x = \frac{2z^2}{1+z^2}.$$

As in the general case the area $\pi/4$ of the quarter circle is $1/2(1 + \int_0^1 z\,dx)$. But by applying a method of reducing one quadrature to a different but equivalent one using moment arguments (see below)[13] and the previous expression for x, it follows that

$$\int_0^1 z\,dx = 1 - \int_0^1 x\,dz = 1 - 2\int_0^1 \frac{z^2\,dz}{1+z^2}.$$

Hence

$$\frac{\pi}{4} = 1 - \int_0^1 \frac{z^2\,dz}{1+z^2}. \tag{5.5}$$

Leibniz had read Nicolaus Mercator's (1620–1687) *Logarithmotechnia* which squared the hyperbola $y = b^2/(b+x)$ by first using long division to express y as an infinite series and then finding the areas under the "higher parabolas" which constitute each term of the series.[14] By 1674 methods to find all such areas were well known. Formulas for the areas defined by $y = x^r$, $r > 0$, and rational or closely related results had already been proven in various ways by Fermat, Torricelli, Wallis, Huygens, Roberval, Gregory, and of course Newton.[15] Mercator's procedure gives first that

$$\frac{z^2}{1+z^2} = z^2 - z^4 + \cdots + (-1)^{n-1}z^{2n} + \ldots$$

The derivation of (5.3) is then completed by "integrating" the above series term by term between 0 and 1 and substituting the result into (5.5).[16] Leibniz triumphantly announces this result to Oldenburg first in a letter of July 15 (N.S.) and then in a letter of October 16, 1674 (N.S.). In the second letter he also claims to have found a series for the arcsine, but does not exhibit it.[17]

[13]This will be equivalent to the modern integration by parts.

[14]Hofmann (1974a), p. 60.

[15]Hofmann (1974a), pp. 55–57. Also see Boyer (1949), p. 121 and Baron (1969), pp. 151, 162, 191, 210.

[16]Note that the series to be integrated does not converge for $z = 1$. Leibniz, however, bypasses the questions of whether in this case term by term integration is valid.

[17]Newton (1959), I, pp. 315, 324. In April of the following year Collins forwarded via Oldenburgh a letter giving a series for the half circumference, zone, and segment of a circle together with series for the sine and arcsine, and Gregory's series for $\tan x$ and $\arctan x$. Leibniz replies on May 20 thanking Oldenburg for the series, but saying that he has not had the time to compare them with his own. Oddly, a year later Leibniz seems to have forgotten about both Collin's letter and his own statement of 1674. He states in a letter to Oldenburg on May 12 (N.S.), 1676 that Georg Mohr, "a native of Denmark, who is very skilled in geometry and analysis" has brought from England series for both the sine and arcsin, and asks Oldenburg for the proof. In return Leibniz promises to send the proof of his series quadrature of the circle. [Newton (1959), II, p. 4.]. This request motivated Newton to send Leibniz via Oldenburg his "First Letter" or *Epistola Prior* dated June 13 (O.S). (and forwarded by Oldenburg July 26 (O.S.)), concerning his series

We briefly mention two other applications of the transmutation theorem. Leibniz related in a letter to a certain La Roque in late 1675, that it gave a simple way to find the areas under a class of the higher parabolas $y = x^r$ even when r is a rational number p/q provided $q > p > 0$. In this case $z = (q - p)y/q$. Combining this with (5.4) where $a = x$ and solving for $\int_0^x y\,dx$ (using modern notation[18] gives that

$$\int_0^x y\,dx = \frac{qxy}{p+q} = \frac{x^{p/q+1}}{p/q+1}.$$

Another application was the determination of the area of certain segments of the cycloid, a result he published in the *Journal des Sçavans* in 1678. As he writes Oldenburgh in the July 15 letter:

> In geometry I have made some discoveries by rare luck rather than by much study. Out of many I shall recount to you one very neat theorem, hitherto, as far as I know, unknown to the very greatest geometers, at least with whom I have spoken.[19]

Specifically, (see Figure 5.7) he announces that the area of the region $AEGA$ is the area of the triangle AFB or half the square of the radius of the generating circle whose center is F. This complements a result of Huygens proven by a different method that the area of the segment AEI where AI is half the radius is a quarter of the area of the hexagon inscribed in the generating circle.[20]

---

methods (a summary will be given below). Leibniz replies on August 27 (N.S.), claiming to have received the letter only the previous day. He lays claim to the co-discovery of series for the $\sin x$, $\cos x$, and what we now call $\exp(x)$ This apparent forgetfulness on Leibniz's part combined with his claims to independent discovery was later to be material for accusations of plagiarism against Leibniz. See Newton's *Account* in Hall (1980), pp. 272–279.

[18]Hofmann (1974a), p. 55. As we shall see, Leibniz did not introduce the integral sign until 1675, and initially it meant an aggregate of Cavalierian indivisibles equivalent to an area. The introduction of dx and the concept of the integral as a sum of strips with height y and width dx does not occur until 1676.

[19]Newton (1959), I, p. 315.

[20]Hofmann (1974a), p. 59.

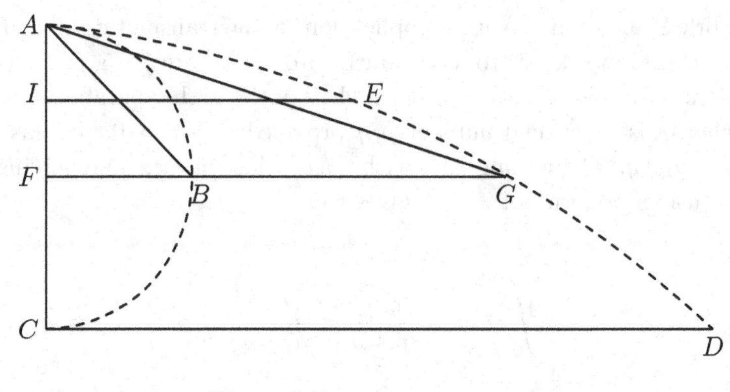

Figure-5.7

Leibniz was excited by his new discoveries. Somewhat earlier, in 1673 he writes:

> From a simple diagram depicting nothing but a short arc and some straight lines intersecting one another, I have deduced over thirty wonderful propositions, through which many curves are squared or transformed into other curves by means of a method as simple as that used to treat rectilinear figures in Euclid's *Elements*. The whole thing depends on a right angled triangle with infinitely small sides, which I am accustomed to call 'characteristic' in similitude to which other triangles are constructed with assignable sides according to the properties of the figure.[21]

It is certainly true that these achievements were spectacular for young man just beginning a serious study of mathematics, and that his work on series summation via differences together with the generalization of Pascal's moment theorem had impressed Huygens. They revealed Leibniz's extraordinary talent and his ability to quickly get to the frontier of current mathematical research. But Leibniz had a habit of enthusiastic exaggeration which others would note. There is an impression of grandiose promise which is never *quite* fulfilled. If the exuberant paragraph quoted above refers to transmutation, there are in the extant manuscripts the three applications we have mentioned (or possibly a few more) but no sign of "thirty wonderful theorems." If it is also referring to his generalization of Pascal's moment theorem for the circle, it does not seem—at least in the short term—that Leibniz derived any important consequences from it. Moreover, no matter

[21]Quoted in Antognazza (2009), p. 158.

how impressive his early work was, much of it was already known. Leibniz was so new to mathematics that he simply did not know the existing literature. As he will soon learn, most or all of his work on series using differences was either known or already in print. Also Leibniz's characterization of the areas under certain higher parabolas simply re-derived results, more general versions of which were known, as we have pointed out, to several contemporary mathematicians, albeit in a new and ingenious way. Although in the case of the cycloid Leibniz's specific quadrature was new, many results concerning the cycloid had been found by Pascal, Roberval, and Honoré Fabri (1607–1688).[22] Finally, several British mathematicians who were Leibniz's contemporaries had recently discovered characterizations of π of elegance and difficulty equivalent to Leibniz's result for $\pi/4$. John Wallis, for instance, in his *Arithmetica Infinitorum* (1656) had arrived at the infinite product

$$\frac{4}{\pi}=\frac{3\cdot 3\cdot 5\cdot 5\cdot 7\cdot 7\ldots(2n+1)\cdot(2n+1)\ldots}{2\cdot 4\cdot 4\cdot 4\cdot 6\cdot 6\ldots 2n\cdot 2n}$$

Lord Brouncker (1620–1684) another early member (and second President) of the Royal Society had expressed $\pi/4$ as the continued fraction

$$\frac{4}{\pi}=1+\cfrac{1^2}{2+\cfrac{3^2}{2+\cfrac{5^2}{2+\cfrac{7^2}{2+\cdots}}}}.$$

And as far back as the 1590s Viète had derived the infinite product representation

$$\frac{2}{\pi}=\sqrt{1/2}\cdot\sqrt{(1/2)(1+\sqrt{1/2})}\cdot\sqrt{(1/2)(1+\sqrt{(1/2)(1+\sqrt{1/2})})}\ldots$$

Leibniz, as shown in his letter to Oldenburg announcing his circle quadrature, is aware of Wallis' and Brouncker's work, but he was very unpleasantly surprised to find that even his series for $\pi/4$ (like his other results in series) had been previously discovered.[23] Five years earlier (1668) in his *Geometricae pars universalis* James Gregory had derived a series expressing both the tangent in term of the angle and the angle in terms of the tangent. In the latter case Gregory's series is

$$\arctan x=x-\frac{x^3}{3}+\frac{x^5}{5}+\cdots+(-1)^n\frac{x^{2n+1}}{2n+1}+\ldots$$

[22]Hall (1980), p. 76; Hofmann (1974a), p. 58.

[23]To the end of his life Leibniz could hardly accept or come to terms with the fact that his circle quadrature series had been discovered earlier. See Hall (1976), p. 142.

which obviously includes Leibniz's result as a special case. By the early 1670s there had in fact been a great deal of work done on series related to the circle. Gregory had also derived the series for both $\tan x$ and $\sec x$ in terms of x. Finally, Newton had in 1669 composed a short paper which he showed to Barrow giving the series for $\arcsin x$, $\sin x$ and $\cos x$.[24] This unfortunate beginning of both rediscovering the work of others—no matter how brilliantly done—and also his habit of bragging about it will dog Leibniz for the remainder of his life and may have been an important psychological factor in the much later charges of plagiarism from Newton.

We come now to an account of Leibniz's further progress towards calculus during his stay in Paris. His efforts in this direction are contained in a collection of manuscripts and notes written from the fall of 1675 to the summer of 1676, first collected and published by Gerhardt in the 1870s and later translated and published with an extensive commentary by J. M. Child in 1920. In 2008 the Leibniz Research Center in Hanover and the Göttingen Academy of Sciences published a new edition of his *Mathematische Schriften* containing 98 manuscripts relating to infinitesimal analysis and totaling 626 pages, written between the summer of 1674 and December 1676.[25] It is still difficult to make complete sense of this enormous amount of material. Some of it consists of commentaries on the work of other mathematicians or fairly complete papers, but much of it comprises rough notes and calculations. There are also a number of mistakes, often in algebra, or possibly errors of transcription. Many of the ideas are difficult to interpret and seem unrelated to any progress toward calculus or any other recognizable goal. When compared with Newton's initial efforts to construct the fluxional calculus in 1665 and 1666,[26] the latter is by contrast a model of clarity. It is an ironic fact that the man who transformed seventeenth century mathematics sometimes committed elementary errors that would have been impossible from Newton, Barrow, or any other well-trained contemporary mathematician. Yet there are also many brilliant flashes of insight. Of necessity we will focus on them; this does some violence to the actual historical record as it makes Leibniz's road toward calculus seem smoother than it really is. But since a complete analysis of all of his known mathematical writings in this period would require an entire volume, we really

---

[24]See the article "Circle" in the eleventh edition of the Encyclopedia Britannica, Vol 6, p. 386. See also Newton's *Account* in Hall (1980), pp. 272f, 275 and Hall (1976).
[25]Leibniz (2008).
[26]For example, *De Analysi* in Newton (1967), II, pp. 206–247 or the Portsmouth collection of manuscripts in Newton (1962).

have no other choice.[27]

The most prominent psychological characteristic of Leibniz's writings between 1674 and 1676 is his conviction that he is creating a *calculus* to analyze tangent and quadrature problems even before he has arrived at many concrete results. By this he means a tool or approach that will make such problems subject to a mechanical procedure that will churn out solutions in an near effortless manner by applying certain rules. In particular, there will be no need for the detailed geometrical analysis concentrating on what is specific to each problem in infinitesimal analysis (and hence different for each one) characteristic of Barrow, Huygens, and most other writers of the period.[28] To this end, we find almost from the beginning that Leibniz invents new notation and symbols–without as it sometimes seems initially *quite* understanding their exact meaning and implications. He will then "play" with and improve the notation until its significance becomes clearer. In this way it will frequently happen that the symbolism itself will suggest either new directions or that the solution of one problem is equivalent to the solution of another. This facility for the creation and sometimes "blind" manipulation of notation "by which the imagination is freed from a perpetual reference to diagrams"[29] is what is decisive for Leibniz and something new and unanticipated in seventeenth century mathematics except to a limited extent in algebra. The divorce between symbolism and geometry which Leibniz was able to implement by 1677 eventually led in the hands of eighteenth century mathematicians—particularly Leohard Euler (1707–1783)—to a new functional and algorithmic way of looking at tangent and quadrature problems. No longer would a tangent a geometrical object to be constructed by finding its subtangent or subnormal; instead it will be characterized by its *slope* which is simply one function among others to which the curve is related. In this way, although Leibniz never quite reaches this stage, his new mathematical approach, to use a metaphor introduced by Kuhn, will represent the beginning of a true "gestalt shift."

Leibniz's route to calculus in the Paris period appears to have developed along two paths, each using quite different ideas. The first owes a great deal to Cavalieri's method of indivisibles. In two notes of October

[27]The most complete analysis of Leibniz's mathematics during the Paris period may be found in Hofmann (1974a). This book is well worth study, but its conclusions differ considerably from ours.

[28]Leibniz's goal was, as we shall see in Chapter 8, part of a lifelong general program to formulate what he called the "Universal Characteristic" which would be a calculus applying to all human thought.

[29]*Historia et Origo* in Leibniz (1920), p. 25.

25 and 26, 1675 Leibniz uses a mixture of Cavalieri-like ideas and mechanics to arrive at quadrature formulas which from the modern point of view represent instances of integration by parts.[30] The first manuscript entitled *Analysis Tetragonista Ex Centrobarycis*[31] considers (Figure 5.8) the two complementary variable areas $AECB$ and $AECD$.

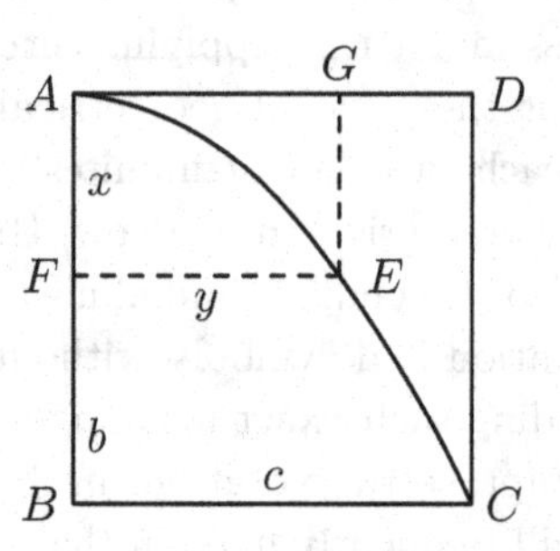

Figure 5.8

Let A denote the origin, $GE = AF$ and $AG = FE$ have variable lengths x and y respectively, with the "last" abscissa AB and ordinate BC being denoted by b and c. Then the moment of $y = FE$ about AD is yx. So in Cavalieri's notation the total moment of $AECB$ about the axis AD should be omn.yx, $0 \leq x \leq b$, $0 \leq y \leq c$. But the moment of GE about AD is $x \cdot x/2 = x^2/2$, so the total moment of $AECD$ about AD is omn. $x^2/2$, $0 \leq x \leq b$. But the moment of the rectangle $ABCD$ about AD is $bc \cdot b/2 = b^2c/2$. And since the sum of the moments of $AECB$ and $AECD$ is the moment $ABCD$, we get in Leibniz's notation that

$$\overline{\text{omn.}yx \text{ to } x} = \frac{b^2c}{2} - \overline{\text{omn.}\frac{x^2}{2} \text{ to } y}.^{32}$$

Using our notation, this is equivalent to the integration by parts formula

$$\int_0^b xy\,dx = \frac{b^2c}{2} - \left(\frac{1}{2}\right)\int_0^c x^2\,dy.$$

The next manuscript written a day later, which is a continuation of the first, uses moments of differences to express one area in terms of another. Referring to Figure 5.9,

[30]Translations and commentary may be found in Leibniz (1920), pp. 62–72.

[31]"Analysis of Quadrature using Centers of Gravity."

[32]The vinculum here and elsewhere indicates the scope of an operation such as "omn." Later we see it applied to "$\int$" and "d."

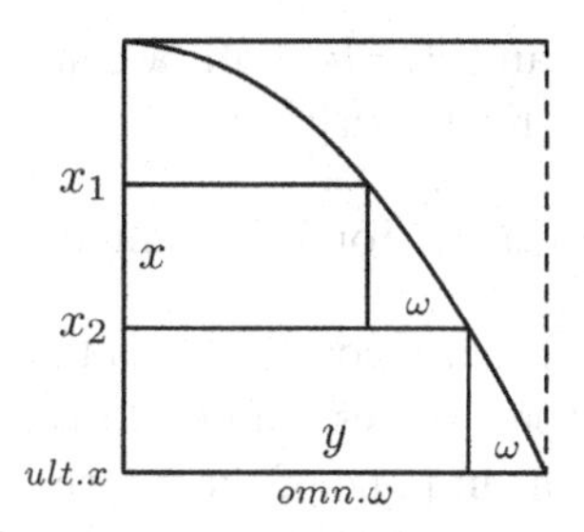

Figure-5.9

Leibniz derives the formula

$$\text{omn.}\overline{xw} = \text{ult.}x \cdot \overline{\text{omn.}w} - \overline{\text{omn.omn.}w}^{33}$$

which he must have arrived at by first thinking of all the moments xw as the sum of Roberval-like strips making up the area above the curve and omn.omn.w as the union of indivisibles making up the area below the curve. Since the sum of the two areas is the area of the rectangle, the result follows. Again if we set $\omega = dy$, $a = \text{ult.}x$, and $b = \text{omn.}\omega$, Leibniz has obtained the integration by parts formula

$$\int_0^b x\,dy = ab - \int_0^a y\,dx.$$

Next Leibniz does something really remarkable. He states (without proof) that the previous formula should hold *independently* of any diagram or the fact that w is a tiny difference so that if, for example, $w = a/x^2$ or a/x we should have

$$\text{omn.}\frac{a}{x} = \text{ult.}x \cdot \text{omn.}\frac{a}{x^2} - \text{omn.omn.}\frac{a}{x^2}$$
$$\text{omn.}a = \text{ult.}x \cdot \text{omn.}\frac{a}{x} - \text{omn.omn.}\frac{a}{x},$$

"the last theorem expresses the sum of logarithms in terms of the known quadrature of the hyperbola."[34] Both formulas are equivalent to

$$\int xy\,dx = x\int y\,dx - \int\left(\int y\,dx\right)dx$$

[33] I have altered this formula slightly by replacing a comma by "." and omitting two commas before the minus sign. The ult.x means the final value of x or in modern terms the upper limit of the integration, the lower being 0.

[34] i.e., $ax = ax\ln x - a\int_1^x \ln x\,dx$.

for the choices $y = a/x^2$ and $y = a/x$. In a continuation written October 29 he further generalizes this by writing

$$\text{omn.}xl = x\,\text{omn.}l - \text{omn.omn.}l \tag{5.6}$$

where "l is the term of a progression, and x is the number which expresses the position or order of the l corresponding to it ..."

It is almost certain that in none of these formulas was Leibniz thinking of anything like modern integration by parts—which is the inverse of the product rule for differentiation—since he not yet formulated either concept. A hint of how Leibniz may have come upon (5.6) and the previous formulas is given by a note in the October 29th manuscript. He says:

> ... for, if omn. is prefixed to a number or ratio, or to something indefinitely small, then a line is produced, also if to a line, then a surface, or if to a surface, then a solid; and so on to infinity for higher dimensions.[35]

If we think of l as a function value $f(x)$ for $0 \leq x \leq a$ with $f(0) = 0$, then omn.xl represents the volume of a pyramidal prism (call this P_1) resting on its flat side with vertical rectangular cross-sections having horizontal side $f(x)$ and vertical side x and bottom bounded by the x and y axes, the graph of f, and the line $x = a$ (or as Leibniz often calls this x either "ult x" or just x). The cross sections are "indivisibles" and their aggregate omn.xl (according to an informal interpretation of Cavalieri) constitutes the volume of P_1. $x\,$omn.l should then be the volume of a solid P_3 whose height is the "final" $x = a$ and whose bottom is the same as that of P_1.[36] The inner omn. in the second term on the right is the area of a horizontal cross-section of the first solid up to some intermediate value of x. The aggregate of these cross-sections "omn.omn.l is then the volume of a solid P_2 such that $P_1 \cup P_2 = P_3$. The formula then expresses the fact that the sum of the volumes of P_1 and P_2 is the volume of P_3.[37] In the same manuscript of October 29, he also begins to abandon the "omn." notation and writes

[35] Leibniz (1920), p. 80.

[36] Had Leibniz been careful he would have probably written something like "ult. x omn. $\overline{l\text{ to ult. }x}$."

[37] The reader may find all this easier to visualize by taking $l = 1$ and $a = 1$. Then P_1 is the wedge with base the unit square having vertices $(0, 0, 0)$, $(1, 0, 0)$, $(1, 1, 0)$ and $(0, 1, 0)$, and height $z = x$. P_3 is the unit cube with vertices $(0, 0, 0)$, $(1, 0, 0)$, $(1, 1, 0)$, $(0, 1, 0)$, $(0, 0, 1)$, $(1, 0, 1)$, $(1, 1, 1)$, and $(0, 1, 1)$, and P_2 is the wedge congruent to P_1 which is the complement of P_1 in P_3. P_1 and P_3 are analogous to the congruent triangles bounded by the diagonal of a square.

$\int l$ for omn.l, so that (5.6) becomes

$$\int \overline{xl} = x \int \overline{l} - \int \int l, \tag{5.7}$$

a much more "modern" looking formula.[38] The analysis given above is only a conjecture. Leibniz may have just felt that (5.6) and its precursors just *should* hold in general. But whether or not Leibniz is actually thinking of $\int xl$ as the volume of the solid P_1, he also realizes that it represents the area under a curve with ordinate xl. This fact allows him to *easily* derive formulas for areas defined by the higher parabolas $y = x^n$. If, for example, $l = x$ we have that

$$\int x^2 = x \int x - \int \int x.$$

Now Leibniz has already shown that $\int x = x^2/2$, probably by taking $l = 1$ in (5.6) or (5.7). Hence

$$\int x^2 = x \cdot x^2/2 - \int x^2/2,$$

so that $\int x^2 = x^3/3$. Obviously we can extend this argument to arbitrary n. Thus what Cavalieri and others did using long and complicated arguments (and in Cavalieri's case only up to $n = 9$), Leibniz is able to do in a few lines.[39] At the same time it is clear that Leibniz does not have firm grip on just how to use his notation or perhaps even its precise meaning should be, for one also finds here the curious formulas

$$\int \frac{\overline{l^2}}{2} = \int \overline{\int \overline{l} \frac{l}{a}}$$

and

$$\int \overline{c \int \overline{l^2}} = \frac{cx^3}{3} \text{ that is } = \frac{c \int \overline{l^3}}{3a^3}$$

where in both equations a is a unit scaling factor. In the first equation if (using the difference notation to be introduced in November) $l = dx$ the right-hand side seems to represent $\int x/a\, dx$ and the left side $x^2/2$. Even after Descartes had demonstrated that the product of two line segments

[38]But we should realize that $\int l$, etc. is not yet an antiderivative or "indefinite" integral nor is it the sum of infinitely thin rectangles. Although Leibniz will soon write expressions like $\int f\, dx$, at present we have only an aggregate of indivisibles expressing a particular area or volume, corresponding to modern definite integrals.

[39]Leibniz's procedure is even easier than that employed by a contemporary Calculus II class to evaluate definite integrals via Riemann sums. For he does not need formulas for the sums of powers of integers.

can be represented by a line segment, as we have seen, he still introduced powers of a unit a to preserve dimensional homogeneity. Leibniz seems to be doing the same. But there is some confusion. From our view point $\int x\,dx$ is the sum of rectangles, therefore two dimensional as are any of the strips $x\,dx$. But Leibniz thought that in general the $\int$ operator raised the dimension. While the inner $\int$ applied to an infinitesimal l produces a line, the product of this with l is two dimensional so that the final application of $\int$ produces a three dimensional solid, and so the factor a is necessary to match the dimension of the left-hand side. However if this interpretation is correct the presence of a^3 in the last formula is incorrect. All we can say is that matters are somewhat confused for Leibniz and that the tradition of Viète died slowly!

Leibniz's second path to the calculus is derived from his initial work on series, especially his realization that the sum of the difference of terms of a series is the first term minus the last which implies that series can be summed if we can express its terms as differences. We have seen how Leibniz used this idea to conclude that

$$\frac{1}{1}+\frac{1}{3}+\frac{1}{6}+\frac{1}{10}+\cdots+\frac{2}{n(n+1)}+\cdots=2.$$

According to Leibniz's much later (1714) unpublished account of the origins of the calculus *Historia et origo calculi differentialis*[40] he initially used the "d" and "$\int$" notation to express differences and sums of series of numbers. Thus dy is the difference of successive terms of a series whose general term is denoted by y while $\int y$ is a series whose terms are partial sums of the series associated with y. Higher differences such as ddy, d^3y or repeated sums like $\int\int y$, etc. can also be considered. In this notation, since $d\int y = y$ and $\int dy = y$ (assuming the first term is 0), Leibniz concludes that $\int$ and d are inverse operations. If we are to believe his *Historia et Origo* Leibniz used these finite sum and difference ideas to derive a great many formulas concerning series and differences involving their terms. Suppose, for instance, we have a series with terms a_k that are constantly deceasing toward a limiting value ω. If the successive differences written in modern notation are

$$\Delta a_j = a_j - a_{j+1}\ldots,\Delta^k a_j = \Delta(\Delta^{k-1}a_j),$$

[40]As we have already noted, this was written to answer the charges that Leibniz had plagiarized the calculus from Newton published by the Royal Society in its 1712 report *Commercium Epistolicum D. Johannis Collins et aliorum.* See Chapter 10 for a brief discussion of this famous controversy.

Leibniz gives tables stating that

$$a_1 - \omega = \Delta a_1 + (\Delta a_1 - \Delta^2 a_1) + \cdots + \sum_{j=1}^{k} (-1)^{j-1} \binom{k-1}{j-1} \Delta^j a_1 + \ldots$$

and that

$$\begin{aligned} a_1 - \omega &= 1 \cdot \Delta^2 a_1 + 2 \cdot \Delta^2 a_2 + \ldots j \cdot \Delta^2 a_j + \ldots \\ &= 1 \cdot \Delta^3 a_1 + (1+2) \cdot \Delta^3 a_2 + \cdots + (1+2+\cdots+j) \cdot \Delta^3 a_j + \ldots \\ &= \sum_{j=1}^{\infty} \beta_{jk} \cdot \Delta^k a_j \end{aligned}$$

where $\beta_{jk} = \sum_{i=1}^{j} \beta_{i(k-1)}$. In terms of the "$d$" and "$\int$" notation with y standing for a term of the series, x the natural numbers, and $\int^k$ repeated summation for $k \geq 3$ we have also the formulas

$$y - \omega = d(xy) - dd(y \int x) + d^3(y \int \int x) - d^4(y \int^3 x) + \ldots$$

$$\int y = yx - dy \cdot \int x + ddy \cdot \int \int x - d^3 \cdot \int^3 x + \ldots,$$

the latter formula holding when $\omega = 0$.

On November 11, 1675[41] less than two weeks after his introduction of the integral sign he begins to talk of differences which may be infinitesimal and manages to actually solve a "differential" equation but without having *any* notion of a "derivative." The problem is to find the curve with ordinate $y(x) = B_1C_1 = AD_1$ such that $B_1P_1 = b/y$ where referring to Figure 5.10, P_1 is the intersection of the normal to the curve at the point $(x, y(x))$ (where $x = AB_1 = D_1C_1$) with the x axis.[42]

---

[41]In the manuscript Leibniz subsequently altered the year to 1673. This has caused some controversy over Leibniz's honesty, etc. For a discussion of some of the issues see Leibniz (1920), pp. 90–93.

[42]I have slightly altered the letter labels in Figure 5.10. For Leibniz what I call B_1, C_1, or D_1 is written $(B), (C)$, or (D). Also there is a slip or misprint either in the original manuscript or in Child's translation, since Leibniz calls AD_1 both x and y.

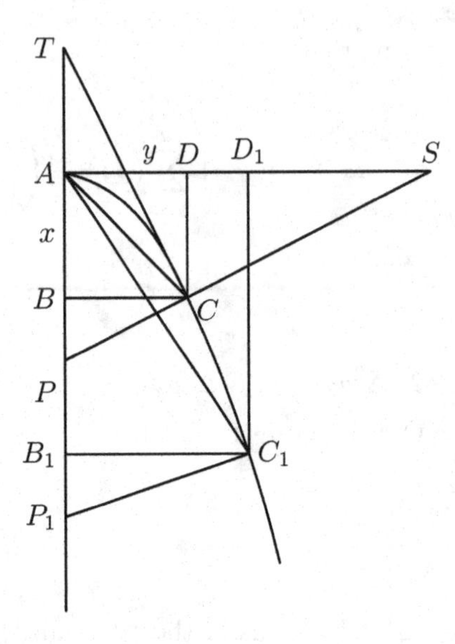

Figure-5.-10

Calling BP w and a small increment on the x axis at B z, he says that he has proved elsewhere that

$$\int wz = \frac{y^2}{2}, \quad \text{or} \quad wz = \frac{y^2}{2d}. \tag{5.8}$$

Here Leibniz denotes the differences by the notation $/d$ so that the difference of $y^2/2$ is expressed as $y^2/2d$ (which he will soon write as $d(y^2/2)$).[43] But this in turn is y "from the quadrature of a triangle."[44] Hence

$$\frac{b}{y}z = wz = y,$$

so that $z = y^2/b$. Applying $\int$ to both sides we get that $x = \int y^2/b$. Here once again the "$\int$" does not denote an integral in our sense of the term, but replaces and signifies the same thing as the earlier "omn." On the right hand side therefore we have an aggregate of the lines y^2/b which Leibniz associates with the area under the parabola $x = y^2/b$. But this area, as we have seen, was well known, and Leibniz had already known of it in 1673 as an application of his transmutation theorem and also in his manuscript

[43] In Leibniz's later notation $w = ydy/dx$. So $\int wz = \int y\,dy = y^2/2$.

[44] *Ibid.*, p. 93. This mysterious remark may anticipate the FTC and follows by considering $(y + \delta)^2 - y^2$ for small (infinitesimal) δ.

of October 29. Leibniz writes it as $y^3/3ba$. Hence the curve is the cubical parabola

$$x = \frac{y^3}{3ba}. \tag{5.9}$$

As before a is a unit scaling factor put in the formula to make the dimension of the right side the same as the left. He then verifies that the curve is correct by starting with (5.9) and using Sluse's method of tangents verifies that w is proportional to y^{-1}. This style of argument is characteristic of Leibniz in this period. It leads to a correct result but is rather confused. What, for instance, are we to make of the identity $z = y^2/b$, z being indefinitely small. One cannot imagine such reasoning from a Barrow, Huygens, or Newton.

As another application of the above technique Leibniz tries to solve a harder question. Suppose the abscissa x plus the the subnormal BP (or AP) is required to be proportional to y^{-1}. As before this leads him to the expression

$$x + w = \frac{a^2}{y}.$$

(Note the homogeneity again!) This in turn gives that

$$x + \left(\frac{y^2}{2d}\right) \cup \left(\frac{x}{d}\right) = \frac{a^2}{y}$$

where the $\cup$ denotes division by the difference x/d. Hence the previous equation may be rewritten as

$$x + y\frac{dy}{dx} = \frac{a^2}{y}.$$

Unfortunately for Leibniz this is a truly nasty differential equation having no solution in the ordinary functions or even an easily expressible power series solution.[45] Since Leibniz is still thinking in terms of differences in *series*, he thinks of both x and y as arithmetical progressions (which is only true if y is a linear function) so that dx and dy are "constant." By reasoning whose details we shall charitably omit, he arrives at equations worthy of

[45] According to the testimony of Professor A.R. Forsyth (1858–1942) to the editor of Leibniz's early mathematical manuscripts J. M. Child [Leibniz (1920), p. 91]. Forsyth was Sadleirian Professor of Pure Mathematics, a F.R.S., and a well known specialist in differential equations and author of a competent if ponderous text on the subject. He is nearly forgotten today.

beginning calculus students at the average second tier public university, namely that:

$$dx = \frac{x^2 + y^2}{2 \log y}$$
$$= \frac{y^2}{a^2 - xy},$$

so that eliminating dx the curve is

$$(y^2 + x^2)(a^2 - yx) = 2y^2 \log y.$$[46]

At this point Leibniz begins to replace x/d by the notation dx which he was to use permanently, and one of the most significant feature of this manuscript (especially in the the first demonstration) is Leibniz's realization, founded on his work on series, that when z is infinitesimal the "$\int$" and "d" operations are inverse to each other. This amounts to an application of the FTC. Using Leibniz's newly introduced notation, it then becomes very easy to derive (5.8). For since $w = ydy/dx$ and $z = dx$, we have first that

$$\int wz \equiv \int y\,dy = y^2/2,$$

and also

$$wz \equiv ydy = d(y^2/2).$$

It is doubtful, however, that Leibniz as of November 1675 realized the full universality of this idea. In fact his approach to (5.8) is still through moments. In the same manuscript of November 11, ydy is interpreted (Figure 5.11) as the moment of the difference dy about the x axis.

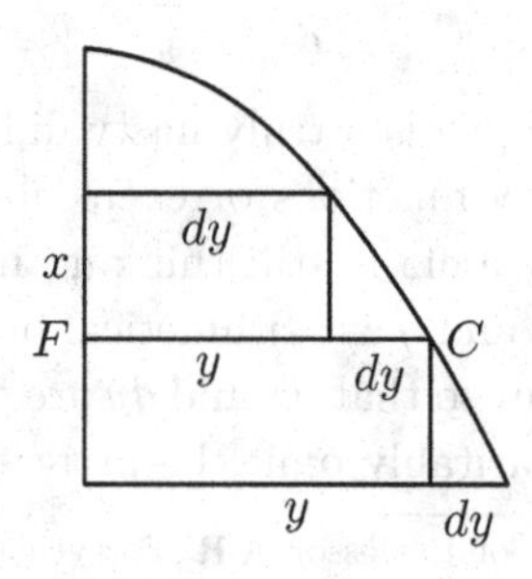

Figure-5.11

[46] *Ibid.*, p. 97.

Now, as Leibniz noted, this is not quite correct. The moment is actually $(y - dy/2)dy$ which can be regarded as ydy because of the last term "being infinitely small compared with the former." Since the sum of these moments is the moment of the sum of the differences dy,

$$\int y\,dy = y^2/2.$$

Moreover, Leibniz's hold on all these notions (or at least on the notation for them) is still rather uncertain. On the same page as the evaluation of $\int y\,dy$ he writes

$$yd\frac{y^2}{2} = a^2 - xy,$$

concluding that

$$ydy = \frac{a^2 - xy}{y},$$

"and therefore

$$\int \overline{\overline{y\,dy}} = \left(\int \frac{a^2}{y}\right) - \frac{x^2}{2}."$$

Since $\int \overline{\overline{y\,dy}} = y^2/2$, by taking differences of both sides Leibniz ends up with the rather strange expression

$$d\overline{x^2 + y^2} = \frac{2a^2}{y}.$$

He is obviously thinking here of $d\left(\int \frac{a^2}{y}\right)$ being a particular "indivisible" ordinate a^2/y.

Nearly at the end of the manuscript, Leibniz begins to wrestle some algebraic properties of the d and $\int$ signs. He asks first if $dxy = dx\,dy$ and on the basis of an example and bad mistakes in algebra (such as one cannot conceive of Barrow or Newton making!) that it is indeed the case that this relation holds. But then looking at the same question for $\int$ signifying a sum, he judges that $\int xy$ is not the same as $\left(\int x\right)\left(\int y\right)$. He also realizes that the quotient of the sums $\int dv \int d\psi$ is not the same as $\int(dv/d\psi)$. After some confused remarks trying to reconcile the difference between the behavior of the products of sums with those of differences, he finally concludes that he was mistaken a few lines earlier to conclude that $dxy = dxdy$. "This is a difficult point. But now I see how this is to be settled."[47] At this

[47] *Ibid.*, p. 101.

early stage, however, he has not yet developed formulas for differences of products or quotients.

The mathematics we have described in these manuscripts is certainly impressive, but it was not his only or perhaps even main interest during the Paris period. Throughout his life, Leibniz was always engaged simultaneously in an astonishing variety of projects—one wonders, in fact, how he found time to eat and sleep.[48] His four years in Paris were certainly no exception. Despite his achievements in this field, mathematics was far from his only interest. He also studied or showed interest in such things as:

> ... the trajectory of a projectile and of a water jet, isochronism of a pendulum, the work of the Archimedean cochlea, waves on the water surface spreading from a thrown stone, the passage of light through the border of two adjacent media, the origin of the rainbow ... methods of making maps, crystallography; magnetic properties of solids ... laths and grinding, ... the measuring of Egyptian pyramids, engraving, producing typographic images ..., cards and chess ... the invention of the pen and methods of ciphering ...[49]

In mechanics he produced two ingenious designs for a perpetual motion machine, one of which involved magnets.[50] A more substantial device was a mechanical calculating machine which he began to develop in 1671 even before leaving for Paris, a partially working model of which was ready by 1673. Pascal had produced an adding machine in 1642, but Leibniz's was superior in its ability to handle multiplication, division, the extraction of square and cube roots. It involved a stepped cylinder and anticipated some key mechanical features of nineteenth-century "Brunsviga" machines. A working and improved version was exhibited at the Paris Academy of Sciences in January 1675. One of the machines was ordered by Louis XIV's Minister of Finance Cobert (1619–1683) for the Royal Observatory and others were requested for use in various finance offices.[51] Several later copies still exist in Munich and an original prototype is on display in Hanover. In 1674 Leibniz also appears to have designed a machine that would perform enough alge-

[48] In later life his secretary Eckhart said that he would hardly ever leave his desk, sleeping in his chair and eating take-out meals from local taverns at it. [Antogonazza (2009), p. 558.]
[49] Kirsanov, (2008), p. 138.
[50] *Ibid.*, pp. 139–146. In later life, however when Leibniz understood conservation laws, he rejected the possibility of such machines.
[51] Antognazza (2009), p. 144, 160; Henrici (1910); Aition (1985), p. 53.

bra to solve equations.[52] Also around the same time, he presented to the Academy a chronometer, probably inspired by Huygens theoretical work on clocks in his *Horologium oscillatorium* employing two springs in a new arrangement; it seems, however, that this device was viewed as impractical and attracted little interest.[53] Nor was religion and philosophy neglected. In 1673 he wrote a short essay opposing a theological thesis that although God is responsible for whatever is real in sin, he is not the author of sin because that is a "pure privation of reality." This view Leibniz argued actually made God "the author of sin" while claiming not to. In the same year Leibniz composed an the essay, the *Confessio Philosophi*, which anticipated his famous view expressed in the *Theodicy* completed in 1708 that the present world is necessarily the best of all possible candidates God could have created.[54] Beginning in 1675 he made a careful study with many notes of Descartes *Principia Philosophia* and while visiting Claude Clerselier (1614-84), Descartes' literary executor, went through and made extracts from Descartes' (now lost) unpublished papers.[55]. Through the young Saxon nobleman and talented algebraist Ehrenfried Walther von Tschirnhaus whom he had met in 1675 and with whom he remained lifelong friends, he made an initial indirect contact with Baruch Spinoza (1632–1677). Leibniz is reported to have thought well of Spinoza's *Tractatus Theologico-Politicus*. But as a whole, although he respected his intellect, he disagreed with most of Spinoza's philosophy—especially his pantheism and determinism which to Leibniz denied freedom to either man or God.[56] Other works studied by Leibniz include Samuel Pufendorf's (1632–1694) huge treatise *De jure naturae et gentium* published in 1672, the doctrines of the French Abbé Foucher (1644–1696) who anticipated Berkeley by arguing that our sensations can give no proof of the existence of a mind-independent external world,[57] the *Essay de Logique* of Edme Mariotte (1620–1684), the works of the ancient Stoic Epicetus (c. 55–135 A.D.), the Cambridge Platonist Henry More (1614–1687), and Nicolas Malebranche's *Recherche de la vérité*.

[52] Couturat (1961), p. 115.

[53] Aiton (1985), p. 53f., Antognazza (2009), p. 160.

[54] *Ibid.*, pp. 144f, 482.

[55] *Ibid.*, p. 167.

[56] *Ibid.*, p. 168, Aiton (1985), p. 69. Stewart (2006) makes Leibniz's encounter with Spinoza the turning point of his life, and the desire to refute Spinoza becomes the engine driving his later philosophy. While there is some truth in this, we think it is an exaggeration. It is true that Leibniz wants to confute atheism and materialism, but Spinoza was not his unique bête noire. Hobbes also offended and cartesianism had dangerous implications.

[57] Leibniz corresponded with Foucher over this issue. See Garber (2009), pp. 268–279.

This period also sees the beginning of some of the characteristic doctrines of his mature philosophy including the thesis that every mind perceives however indistinctly everything in the universe and a kind of animistic atomism. In February and March of 1676 he reasons that "every body which is an aggregate can be destroyed." But at the same time "there seem to be elements, i.e., indestructible bodies." How is this possible? Leibniz's answer is that they possess minds which are not aggregates and are unextended and therefore indestructible. Elsewhere he speculates that the ultimate reality consists of tiny vortices endowed with minds. We see here ideas which possibly presage Leibniz's mature philosophy of monads developed decades later.[58] The mysteries of the continuum were also subject to detailed study and he found time to summarize in Latin three of Plato's dialogues[59] We cannot begin to justice here to the incredible variety and sheer bulk of Leibniz's efforts in these four years.

Leibniz was fortunate to have enjoyed the friendship of his fellow countrymen Oldenburg, Tschirnhaus, and Boineburg, as well as that of Christiaan Huygens in Paris. The favor and patronage of these three men was particularly helpful in the early part of his career. But a combination of bad luck and poor judgements on Leibniz's part soon proved to have damaging consequences. Early in his stay (November, 1672) in Paris his mentor Baron Boineburg sent his only son to Paris along with his son-in-law and diplomatic envoy Melchior Friedrich von Schönborn. Boineburg asked Leibniz to supervise the young man's education. Unfortunately, the following month (December 15) Boineburg died. This was followed two months later (February 1673) by the death of his employer the Archbishop and Elector Johann Philipp von Schönborn who had hired him and given permission to go to Paris. These events left Leibniz's situation very uncertain. His salary, although promised, had actually not been paid for the previously two years. Leibniz hoped that the new Elector Lothar Friedrich von Metternich (1617–1675) would allow him to continue to serve as a kind of cultural attache for Mainz in Paris, representing the interests of the Duchy and reporting on French progress in science. Permission was duely granted. Leibniz could remain in Paris indefinitely, but no salary or formal position was given or promised. Thus, as a result of the deaths of his two supporters, Leibniz no longer had access to income and patronage from Mainz. His only paying job was now confined to the tutoring of Boineburg's sixteen year old son, a

[58] Antognazza (2009), pp. 167–170, 173; Garber (2009), pp. 62–70, and Mercer (2001).
[59] Antognazza (2009), pp. 170–174. For Leibniz's work on the continuum in this period see Leibniz (2001).

duty for which it was soon apparent that he was unsuited. Leibniz began by drawing up an ambitious plan of study for the son which would begin at 5:30 a.m. and continue to 10 p.m. with three hours off for lunch, dinner, and attendance at mass. Naturally, as Leibniz ought to have realized, this was too much to expect of a teenager who wanted to enjoy the pleasures of Paris. The tutoring, consequently, did not go at all well and Leibniz complained about his student's disinterest and lack of application. When in 1674 Leibniz wrote Boineburg's widow to ask for payment of his accumulated salary as tutor and the recompense which Boineburg had promised him in writing for his (so far unsuccessful) efforts to recover Boineburg's rents and pension, he was coldly dismissed from the family service.[60] It is a reasonable guess that this was as a result of the boy's complaints, and it caused Leibniz severe financial inconvenience. He had already spent a small fortune on the construction of his calculating machine, and he now desperately began to search for a paid position. He fervently hoped to be able to stay in Paris. At first this seemed a possibility. In the fall of 1675 he was proposed to Colbert as the successor to Roberval who had died October 27, 1675. This was a chair in mathematics at the College Royal which the incumbent had to re-compete for every three years in a contest in problem solving (but which he was allowed to set). Unfortunately the nomination fell through. Probably the main reason was the fact that Leibniz was a foreigner and the French government was already employing two other foreign scientists, Huygens and Giovanni Cassini (1625–1712). But also because of a bad cold, he had to cancel a crucial interview with a key supporter. He then tried to be employed in Paris by the new archbishop of Mainz or to enter the service of the Emperor. These quests were also unsuccessful. For want of other options he then had to reluctantly accept an offer made in 1673 by the Duke of Hanover, Johann Frederick of Brunswick-Lünberg (1625–1679) to enter his service as court counselor which required that he leave Paris and reside in Hanover.[61]

There were also two visits to London at the beginning and end of Leibniz's time in Paris, both having mixed results. In January 1673, soon after his arrival in Paris, Leibniz joined a diplomatic delegation to London from Mainz. Since efforts to persuade the Sun King to cease his war with Holland and convene a peace conference or to take seriously the Egyptian project had failed, the only recourse was to see what could be done with Louis' ally

[60] Antognazza (2009), pp. 145–153; Hofmann (1974a), p. 46.
[61] Antognazza (2009), p. 174.

Charles II.[62] Naturally as a junior member of the delegation Leibniz played no role in the negotiations which were as fruitless as they had been in Paris. But he did have an opportunity to become acquainted with some mathematical and scientific members of the Royal Society. Oldenburg invited him to demonstrate a partially completed model of his calculating machine at a meeting February 1 of the Royal Society. Although most members were impressed by the device, even though because of the difficulties in its fabrication it could not yet perform multiplication and division as it had been designed to do, Leibniz had the bad luck to have Robert Hooke (1635–1703) in attendance. Hooke was an extremely irascible person and had the habit of quarreling with nearly everyone, either by dismissing their work, claiming that he had already done it, or by accusing rivals of plagiarism.[63] He had already expressed a bad opinion of Leibniz's 1670 *Hypothesis physica nova.* After taking Leibniz's machine apart he bitterly criticized it at a Royal Society meeting two weeks later at which Leibniz was absent, and a month later produced his own machine which Leibniz felt was a copy of his own. Several days after the February 1 meeting Leibniz met Robert Boyle at a party at Boyle's sister Lady Ranelagh's house where Boyle was living. This meeting went well, but the mathematician John Pell was also present. Pell, who seems somewhat misanthropic and was in ill-health, showed Leibniz little sympathy. When Leibniz stated some of his results on series, particularly relating to taking successive differences of the terms and their application to interpolation. Pell rather curtly told him that all these results had already been discovered and were in a book concerning the apparent diameters of the sun and moon published by Gabriel Mouton (1618–1694) in Lyons in 1670. Leibniz felt that Pell was making an implicit accusation of plagiarism, and was deeply embarrassed when the next day he looked at Mouton's book and found that Pell was correct. He then immediately wrote an agonized letter of explanation to Oldenburg rejecting any charge of deliberate plagiarism. After returning to Paris in early March he learned from a letter of Oldenburg written in early April, which contained a survey of recent mathematical progress by Oldenburg's mathematical advisor John Collins (1625–1683), that a great many of his

[62]Charles had allied Britain with France against Holland for a secret subsidy from Louis XIV for 100,000 pounds a year.

[63]Two of many of Hooke's charges against colleagues were (1) his claim of priority for Newton's planetary theory and (2) his accusation that Huygens had stolen ideas concerning the construction of clocks from him. In the course of the latter dispute he charged Oldenburgh first of giving away to Huygens his ideas and then of being a French spy and informer. [Aition (1985), p. 53.]

results on the summation of various series were also known and had been discovered by Mercator, Pietro Mengoli (1626–1686), François Regnauld, and others.[64] The second visit, which was for only a week, occurred in mid-October 1676 when Leibniz had left Paris for Hanover. He took a very slow indirect route, first traveling to England,and then to Holland where he met Jan Swammerdam (1637–1680), a biologist interested in insects, the mathematician Jan Hudde, Antoni van Leeuwenhoek (1632–1723), and then Spinoza with whom he had extensive philosophical discussions, one topic being the ontological proof that God's existence must necessarily follow from his definition.[65] Having left Paris on October 4, he did not arrive in Hanover until close to the end of December (perhaps a subconscious way of expressing a lack of enthusiasm for the new position).

This second visit to London greatly complicated Leibniz's future claim to the calculus. He had continued to maintain a correspondence with Oldenburg since the 1673 visit and had known at least late 1674 or early 1675 that Newton had some kind of general method to find quadratures, rectification of curves, centers of gravity and volumes of solids of revolution.[66] During the second visit he was allowed to see and copy from Newton's unpublished *De Analysi.* Also shortly before leaving Paris (August 24, 1676), Leibniz had received a letter from Newton—the so-called *Epistola Prior* concerning his work on series. A second much longer letter —the *Epistola Posterior* was written October 24, (O.S.) 1676. Both were written to Oldenburg who forwarded them to Leibniz. Neither seems to have played any role in Leibniz's development both because of their subject matter and in the case of the *Epistola Posterior* because of late delivery—Leibniz did not receive it until June 1677, well after he had formulated his own ideas. The *Epistola Prior* contains a statement of Newton's version of the binomial theorem for fractional powers and gives applications to determining infinite series representing such functions as $(c^2+x^2)^{1/2}$, $N/(y^3-a^2y)^{1/3}$, etc. Series for $\arcsin x$, $\sin x$, and the "versed" version of these functions[67] together with applications to certain geometric problems are also given. While interesting, the letter does not touch on infinitesimal problems. What Leibniz could read in the *Epistola Posterior* was again a great deal more of Newton's work on series and its applications. Some of this is very complicated material as is shown by the following example. Let the ordinate $y = dz^\theta(e+fz^\eta)^\lambda$

---

[64] Hofmann (1974a), pp. 31–34 and Chapter 4; Antognazza (2009), pp 149–151.
[65] Stewart (2006), Chapter 12 and Aiton (1985), p. 69f.
[66] Letter of Oldenburgh, December 8, 1674. See Newton (1959), I, p. 331.
[67] The versed sine of x is $1-\cos x$.

where z is the abscissa, the other parameters being constants. Suppose

$$\frac{\theta+1}{\eta}=r, \qquad \lambda+r=s, \qquad (d/\eta f)(e+fz^{\eta})^{\lambda+1}=Q,$$

and $\eta(r-1)=\pi$. Then the area of the curve can be expressed by the series

$$Q\times\left\{\frac{z^{\pi}}{s}-\frac{r-1}{s-1}\times\frac{eA}{fz^{q}}+\frac{r-2}{s-2}\times\frac{eB}{fz^{q}}-\frac{e-3}{s-3}\times\frac{eC}{fz^{q}}+\frac{z-4}{s-4}\times\frac{eD}{fz^{q}}-\dots\right\}$$

where A, B, C, D denote the immediately preceding term. Thus

$$A=\frac{z^{\pi}}{s}$$
$$B=\frac{r-1}{s-1}\times\frac{eA}{fz^{q}},$$

and so forth. When r is a fraction Newton says the series is infinite, and when r is an integer, it is finite. Newton gives several examples, the simplest of which $z=\sqrt{ax}$. Here

$$d=1, \quad \theta=0, \quad e=0, \quad f=a, \quad \eta=1,$$
$$r=1, \quad s=3/2, \quad Q=\sqrt{a}z^{3/2}, \quad \eta=0,$$

and the area is $(2/3)\sqrt{a}z^{3/2}$. In the same way if $z=cz^{\eta}$, the area will be $\frac{cz^{\eta+1}}{\eta+1}$.[68]

The only connection in the *Epistola Posterior* with Newton's fluxions and fluents was a cryptic Latin anagram[69] which meant "*Data aequatione fluentes quotcumque quantitates involvente fluxiones invenire, & vice versa*".[70] Assuming that the anagram was indecipherable, Leibniz could be excused if he thought that Newton's methods for finding areas and volumes simply involve term by term quadrature of series using the well known area formula for x^n. The case may be different, however, with *De Analysi*. As in the two letters, there is no mention of tangents, and the terms "fluent" and "fluxion" are not explicitly used. Yet this manuscript contains some material which could been of great assistance to Leibniz. However, it is found only on a few of the forty odd pages in the document and could be easily overlooked. Much of *De Analysi* in tone and subject matter resembles Newton's letters, especially the *Epistola Posterior*. It begins with rules of

[68]The letter may be found in Newton (1959), II, pp. 130–149. For this result see p. 134f.
[69]The anagram was "6accdae13eff7i3l9n4o4qrr4s8t12vx."
[70]*Account* in Hall (1980), p. 283. The translation is "Given any equation whatever involving fluents, to find the fluxions and vice versa."

quadrature for curves defined by $y = ax^{m/n}$ or linear combinations thereof. Quadratures for more complicated expressions such as $y = a^2/(b+x)$ or $y = \sqrt{a^2+x^2}$ are derived by first using division or "root extraction" to find infinite series representing the function and then integrating term by term. In the case of a circle where $y = \sqrt{a^2-x^2}$ the procedure yields an infinite series giving the area of the circle. There is also a long section where Newton shows how to find infinite series representing the roots of higher order equations. The area formulas for curves of the form $y = x^{m/n}$ may have been new to Leibniz, who as we saw had derived them in 1673 only for $m < n$. But such isolated statements would have had no impact on the methods he was to develop. However, there are three short sections of *De Analysi* in the last third of the manuscript which do embody and illustrate Newton's ideas of the inverse relationship between fluxions and fluents. Consider Figure 5.12.

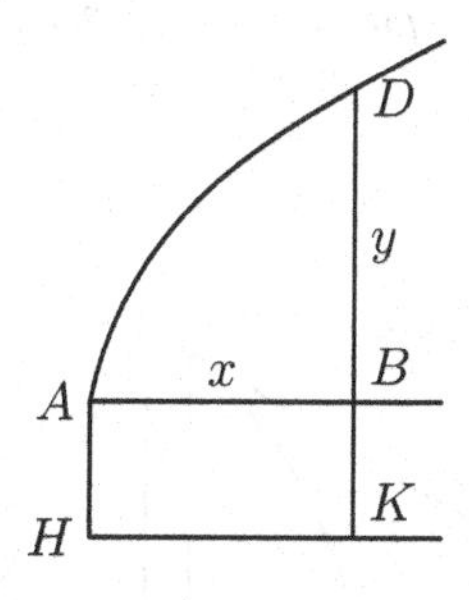

Figure 5.12

Let AD be an arbitrary curve, BD its ordinate y, AB the abscissa x, and $AH = BK$ a segment of unit length. Newton supposes that DBK moves uniformly to the right. Then he thinks of BD as the "moment" of the changing area ABD determined by the curve and BK as the moment of the changing rectangular area AK. Then "when given continuously the moment of BD, you can by the foregoing rules investigate the area ABD described by it or compare it with $AK(x)$ described by a unit moment."[71] As it stands this is rather unclear. Referring to this part of *De Analysi* in his 1713 commentary on the *Commercium Epistolicum*," Newton says that $BD = y$ and BK represent infinitely narrow strips. To use Leibniz's later

[71] Newton (1967), II, p. 232.

notation their areas are $y\,dx$ and $1 \cdot dx$ and are the "moments" by which the areas ABD and AK increase in an infinitesimal unit of time. It follows (calling $\mathcal{A}$ the area ABD) that $d\mathcal{A}/dx = y$, so that if y is a power series A can then be determined by applying term by term the rule of quadrature for x^n. Actually, however, Newton's reasoning would probably been different. The fluxion of A with respect to time or $\dot{\mathcal{A}} = y\,\dot{x}$ and the fluxion of area $AHKB$ with respect to time is $\dot{x}$ because $BK = 1$. Hence $\dot{\mathcal{A}}/\dot{x} = y$ and we have seen above that as early as 1665-1666 Newton could determine $\mathcal{A}$ for quite complicated y. As an application in Figure 5.13 we are given a circle with radius $AC = 1/2$ where A is the origin. The circle then has equation $y = \sqrt{x - x^2}$. Let $B = x$. The problem is to find the rectify the arc $\widehat{AD}$. Newton has previously remarked that any such problem can be reduced to finding "the quantity of a plane surface bounded by a curve line."[72] In the characteristic triangle DHG the "moment" $\widehat{DH}$ of the arc $\widehat{AD}$ satisfies the ratio

$$BK : \widehat{DH} = BD : DC = \sqrt{x - x^2} : 1/2 = 2\sqrt{x - x^2},$$

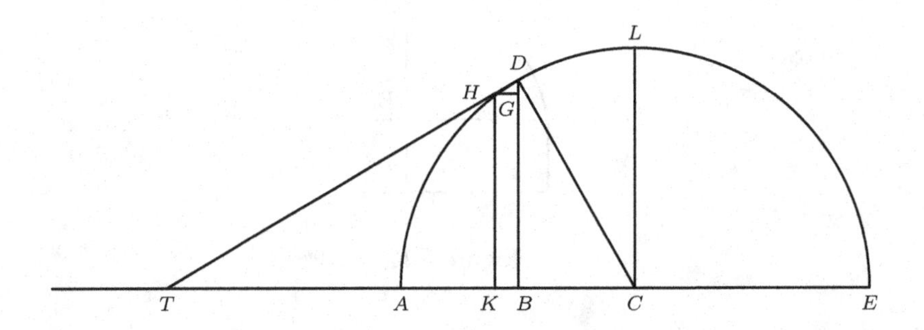

Figure 5.13

since $\widehat{DH}$ can be taken to be DH if the rectangle $HGBK$ is "indefinitely small." Therefore $\widehat{DH}$ the moment of the arclength should be the moment of area of a curve with ordinate $1/2\sqrt{x - x^2}$. The sum of the latter moments should then give the arclength. Newton's next step is to express the ordinate as a series and "integrate" term by term, using the rule that the

[72] *Ibid.*, p. 233.

area determined by a series is the sum of the areas determined by each term. A few pages later there is a still clearer example of the relation between the area under a curve considered as a fluent and the ordinate or fluxion of the area function. Newton wishes to prove his rule given at the beginning of *De Analysi* for the quadrature of the curves defined by $y = ax^{m/n}$. This argument is much clearer than the previous ones and is virtually identical to the modern proof that the derivative of the area function is the ordinate y. Let $AD\delta$ be any curve with abscissa $AB = x$ and ordinate $BD = y$ (Figure 5.13)

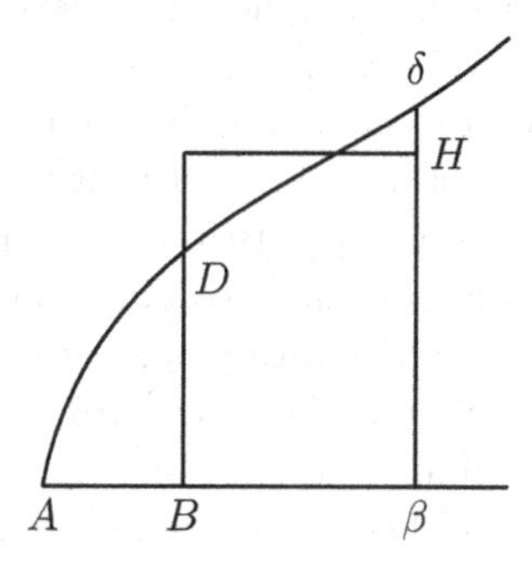

Figure 5.13

Call z the area ABD. Assuming we know the relation between z and x, we want to find $BD = y$. Let $o = B\beta$ be a slight increase in x so that $A\beta = x + o$. If $BH = v$ we have that the area $A\beta\delta \approx z + ov$. Suppose $z = (2/3)x^{3/2}$. We substitute $x + o$ for x and $z + ov$ for z, to obtain

$$(z+ov)^2 = z^2 + 2ov + (ov)^2 = (4/9)(x+o)^3 = (4/9)(x^3 + 3x^2o + 3xo^2 + o^3).$$

Now take away $z^2 \equiv (4/9)x^3$ from both sides and divide by o, getting

$$\frac{(z+ov)^2 - z^2}{o} = (4/3)x^2 + (4/3)xo + (4/9)o^2 = 2zv + ov^2 = (4/3)x^{3/2}v + ov^2.$$

If "$B\beta$ is infinitely small, that is o to be zero," the result is that we get that $(4/3)x^2 = (4/3)x^{3/2}y$ or $y = x^{1/2}$. "*Conversely therefore if* $x^{1/2} = y$, *then will* $\frac{2}{3}x^{\frac{3}{2}} = z$" [emphasis added]. We see in these arguments the essence of the FTC, or in Newton's language the fluxion of the area is the ordinate and the fluent of the ordinate is the area.

These contacts with Newton's thought together with the facts that many of Leibniz's results on series including his series for $\pi/4$ were already known in Britain were to be critical in the accusations of plagiarism made under the auspices of Royal Society forty years later. At this time, however, the relations between Leibniz and Newton were reasonably friendly. As late as the publication of the *Principia* in 1687, moreover, Newton was willing in a scholium or explanatory remark to grant Leibniz credit for an independent (although later) invention of calculus.[73] Of course, the modern consensus of historians is that Leibniz had arrived at the key ideas of his calculus independently of (although a decade later than) Newton, as early as the fall of 1675 before his second visit to London and communication with Newton. We shall discuss some of these issues at greater depth in Chapters 6 and 10; but this writer's opinion the exact nature of the psychological relation of Leibniz's work to Newton's may never be known.[74]

Although Leibniz had been unanimously elected a member of the Royal Society on April 19, 1673 just after his first visit to London, he had in fact made an uneven impression on some British scientists he had met. Probably the overall impression was still favorable and the attitude of Hooke may be discounted as his intemperate personality was well known in the Royal Society. But for some time since the 1673 visit Oldenburg, a once enthusiastic partisan of Leibniz, was showing a slight coolness in their correspondence. Oldenburg had been disappointed that Leibniz had failed to complete his calculating machine in a timely matter. Being no mathematician himself, Oldenburg had to rely on the opinions of others, especially his advisor on mathematical advisor John Collins. Collins was not a creative mathematician, but he knew the literature and was a nationalist unwilling to give much mathematical credit to foreigners. It was soon apparent to Collins, Pell, and others that Leibniz did not know the latest mathematical developments or even much mathematics at all. His ideas on series (including the arithmetical quadrature of the circle) were mostly known, and one detects a feeling that he promised more than he actually had achieved, since his announcements of grandiose discoveries often evaporated upon inspection. There is certainly sometimes a tone of boastfulness in Leibniz's correspondence that contrasts unfavorably with the businesslike directness of Newton. One of many samples of it is in the following passage contained in an August 17 (N.S.), 1676 letter to Oldenburg discussing Leibniz's achievements in transmutations, series, and algebra:

---

[73] Hall (1980), pp. 33, 65–68.

[74] For further discussion see Hofmann (1974a) and Hall (1980). The *De Analysi* itself and a reproduction of Leibniz's notes on it may be found in Newton (1967), II, pp. 206–259.

> But if anyone would not shun the labor I could teach him a general infallible analytical through which he could find the general roots of all equations. Indeed I might set out a better programme for those who would enjoy such a calculation, for I have under consideration analytical tables which might be of no less use in analysis than are tables of sines in practical geometry.... I am well aware that there is nothing of more importance in the whole of analysis, for with these tables most problems would at once be solved, or could be deduced from them without much trouble. The business depends upon a much greater matter, namely upon a combinatorial art that is indeed both general and true, and of a strength and power to which I am not aware that anyone has hitherto attained. In fact it does not differ from that supreme analysis, to the heart of which, so far as I can judge, Descartes did not penetrate; for in order to set it up we need an alphabet of human thoughts, and for finding this alphabet we need an analysis of axioms. But I am not surprised that no one has considered these matters adequately ...[75]

Similar passages abound in Leibniz's correspondence with Oldenburg and others. The last part of the quotation refers to his project of developing a calculus of reasoning, or the Universal Characteristic. We will look at this project, for which Leibniz is famous, in Chapter 8. The interesting thing is that the quotation suggests that he is in possession of it. In actual fact, however, its general construction remained little more than a dream.

It is perhaps too strong a judgement on Hofmann's part that British mathematicians regarded Leibniz as a "dilettanting beginner without adequate capacity for self-criticism and knowledge of the literature."[76] But this attitude does seem true of Pell and Collins and probably was communicated to Oldenburg. It did not help either that Oldenburg's own lack of mathematical knowledge sometimes led him to criticize Leibniz unfairly. For instance, when in 1674 Leibniz had communicated his series quadrature of the circle, Oldenburg replied that Gregory was at the point of proving that such a thing was impossible. Gregory's proof was in fact mistaken and beside the point, since Leibniz was talking about a *series* approximation to π while Gregory was considering the possibility of determining π by a finite sequence of algebraic operations.[77] A reasonable conclusion concerning the consequences of Leibniz's visits to London is that had he not been a

[75]Newton (1959), II, p. 70.
[76](1974a), p. 56.
[77]A proof that a circle could not be squared using ruler and compass or any finite algebraic process had to await the nineteenth century.

co-inventor of calculus but instead had done, say, a different type of mathematics, then any vaguely negative impressions he aroused would have been probably forgotten. But once the priority dispute with Newton flared, they could be resurrected and exaggerated.[78]

[78]There are thorough discussions of Leibniz's relations with the English in Hofmann (1974a) and in Hall (1976). In the writer's opinion Hofmann exaggerates the negativity of Leibniz's reception in London even on the part of Newton. A better balanced account may be found in Hall.

Chapter 6

The Creation of Calculus

Leibniz himself in a 1703 letter to James Bernoulli and in his unpublished *Historia et origo calculi differentialis*, written in 1713 as an answer to the *Commercium Epistolicum* report issued by the Royal Society accusing him of plagiarizing Newton, dates his discovery of the calculus to the 1673-4 period when he first extended Pascal's method of computing the surface area of a sphere to general solids of revolution and then arrived at his method of transmutation. He also mentions work in this period using differences as a tool to sum numerical series as well as a new notation "invented by him at a later date" in which $\int x$ and dx respectively represent the partial sums of a series (whose general term is x) and the difference of successive terms.[1] Still later (no precise date is given), he says that he realized that dx and dy represent infinitesimal differences of the abscissa and ordinate, since "the infinitely small lines occurring in diagrams were nothing else but the momentaneous differences of the variable lines."[2] The difference notation, originally applied to series, could then be used to denote the horizontal and vertical sides of the "characteristic triangle" which Leibniz says he discovered and named in 1673.[3]

A standard judgement by subsequent historians of mathematics, going back to C. I. Gerhardt,[4] essentially agrees with Leibniz's judgement, but puts the date about a year later than implied by the *Historia*, specifically to somewhere in the period 1674–76, but before the second trip to London. The verdict of A.R. Hall, for example, is that "we now know that calculus

---

[1] *Historia et origo* in Leibniz (1920), p. 33.

[2] *Ibid.*, p. 53f.

[3] *Ibid.*, p. 38.

[4] The nineteenth century editor of Leibniz's manuscripts and letters. See Gerhardt's essay "Leibniz in London" reprinted and translated in Leibniz (1920), especially p. 170.

was born in October 1675."[5] A similar opinion may found in Hofmann's classic *Leibniz in Paris*, Boyer's *The Development of Calculus*, Edwards *The Historical Development of Calculus*, and in Aiton's biography.[6] Margaret Baron speaks of "supremely important papers" developed by Leibniz between 1674 and 1676. She is especially impressed by his alternating series representation of $\pi/4$, his quadrature of the cycloid, his discovery of "omn." formulas equivalent to integration by parts and his introduction of the integration symbol $\int$ a month later.[7]

We have discussed many of the results cited both by Leibniz and these authors in the previous chapter. In our opinion their judgements (including Leibniz's own) seem a bit too generous and are not supported by the manuscripts we have consulted—particularly those written between October and November of 1675 and the summer of 1676.[8] It is quite true that by October 1676 Leibniz had put in an immense amount of effort into research relating to problems in infinitesimal mathematics. The manuscripts already described certainly do contain some suggestive "calculus-like" results including what we now interpret as integration by parts, the notation "$\int$" representing a collection of indivisibles in the sense of Cavalieri (which Leibniz had previously denoted by "omn."), x/d signifying a difference, and use of the characteristic triangle in his transmutation theorem and extension of Pascal in 1673. To get another sample of the flavor of Leibniz's mathematics prior to the fall of 1676, let us look again at one of his favorite topics, the method of transmutations whereby some complicated region is transformed into a simpler one of equal area whose quadrature is easier. In a reply to Oldenburg of August 17, 1676 (N.S.) concerning Newton's *Epistola Prior*, besides claiming independent credit for what we would now call exponential and cosine series,[9] Leibniz emphasizes the possibilities of the method and uses it derive series analogous to the quadrature of a quarter

[5]Hall (1980), p. 72.

[6]Hofmann (1974a), Chapter 13; Boyer (1959), Chapter 5; Edwards (1979), pp. 252–254; Aiton (1985), p. 57.

[7]Baron (1969), pp. 276–290.

[8]In the *Historia et origo* Leibniz was obviously concerned with putting the discovery of calculus as early as possible, certainly before the second trip to London in October 1676 and his receipt of the *Epistola Prior* and *Epistola Posterior* from Newton written in June and November of that year which became centerpieces of the accusation of plagiarism against him made by the Royal Society. This also may be a reason for later writers to focus on 1675 instead of 1666 or 1667.

[9]Leibniz calls l the "hyperbolic logarithm" of $1-m$ or $1+n$ and states that if $l-m<1$ then $m=l-l^2/2!+l^3/3!-l^4/4!+\ldots$ while if $l+n>1$ then $n=l+l^2/2!+l^3/3!+l^4/4!+\ldots$ [Newton (1959), II, p. 68].

circle for certain regions of general conics. In Figure 6.1

Figure 6.1

suppose that the curve AFQ is a circle, ellipse, or hyperbola having the general equation

$$\frac{(x-a)^2}{a^2} \pm \frac{y^2}{b^2} = 1,$$

and that "the rectangle under the semi-latus rectum and semi-latus transversum [is] unity." Then Leibniz states that the area of the sector QFA is given by the series

$$\frac{t}{1} \pm \frac{t^3}{3} + \frac{t^5}{5} \pm \frac{t^7}{7} + \ldots$$

where $\pm$ is $+$ for the hyperbola and $-$ for a circle or ellipse. This result probably dates from 1674 and its proof is similar to that given in Chapter 5 for the series representing $\pi/4$. The reader may enjoy trying reconstruct it himself.[10]

All this is very nice mathematics, but it is yet not the algorithmic calculus for which Leibniz is remembered. It also suggests that the central ideas involved in his calculus were perfected later than this letter. Since Leibniz regularly informed Oldenburg of his latest discoveries, a reasonable inference is that his methods of calculating tangents and quadratures were not yet in a state he wished to communicate in August 1676. Examination of the other manuscripts in the *Mathematische Schriften* further supports this impression. There are a great many special results concerning conics, especially hyperbolas, and the determination of centers of gravity, often involving some kind of transmutation. In several manuscripts Leibniz is fascinated by what he calls the "inverse tangent" problem which was also of interest to his contemporaries. This, however, was not the modern problem of determining the ordinate of a curve from a knowledge of the slope of the tangent, although the two problems are related. Instead, some (often

[10]The proof (correcting a small mistake in Leibniz's statement) is given in Appendix B.

complicated) geometric relationship is given involving the subtangent or normal and the problem is to find the curve. We have already seen some examples of this in Chapter 5 and will consider others below. The problem will usually lead to a "differential" equation involving the ratio dy/dx, and Leibniz will devise some *ad hoc* solution which may or may not be successful. From our vantage point inverse tangent problems as they were conceived in the mid-seventeenth century seem more difficult than finding a curve from the slope of its tangent; their very complexity may have made it difficult to see clearly the inverse relation between the tangent and ordinate or area and ordinate which is the essence of the modern FTC.[11]

The characteristic triangle will also frequently appear in Leibniz's work in this period, and use is made of its similarity with various triangles associated with the curve such as the triangles with side y, the subtangent, and tangent or y, the normal, and subnormal.[12] But the method is not used to systematically calculate tangents for classes of curves. As late as November 22, 1675 Leibniz is still, in fact, trying to find tangents by generalizing Descartes method of equal roots. But instead of a circle intersecting the curve he thinks of some other auxiliary curve for which the tangent is known intersecting the first curve at two points P_1 and P_2. The abscissae or ordinates of these points will be the roots of certain algebraic equations. If conditions can be determined so that $P_1 = P_2$ the curves will be tangent to each other. He suggests that a straight line can often be substituted for Descartes' circles, although for a conic where only y^2 appears a circle would be the most convenient choice.[13] In the spring and summer of 1676 there are also four manuscripts where tangents to various curves are determined by elaborate calculations. It is true that one manuscript of June 26, *Nova methodus tangentium* states that "the true method for general tangents is through differentials" and this is superior to the cartesian method of equal roots, but in none of the manuscripts is there any sign of the algorithm he developed a few months later after the second visit to London.[14]

---

[11] Several of the manuscripts featuring inverse tangent problems are briefly described in Rivaud (1914).

[12] Recall from Chapter 5, that this second triangle can be used for quadrature using a result of Barrow explained at the end of Chapter 3.

[13] *Methodi tangentium directae compendium calculi, dum jam inventis aliarum curvarum tangentibus utimur. Quaedam et de inversa methodo*, Leibniz (2008), pp. 341–344; translation in Leibniz (1920) pp. 111–113. Also see the October 1674 manuscript *De methodo tangentium inversa per aequ. duarum radicum aequalium* in Leibniz (2008), pp. 96–98.

[14] *Ibid.*, pp. 559–561. The other three manuscripts are *De Tangentibus et speciatim figurarum simplicium* (February—June), De tangentibus paraboloeidum (February—

Concerning quadratures, although we have seen the symbol $\int$ used to replace "omn" in denoting the "sum" or aggregate of Cavalieri's indivisibles and to arrive at results like $\overline{\int l} = l^2/2$, Leibniz remarks on a number of occasions that if we know the moment of a region R around three different lines we can find its area, but does little with this observation.[15] His interest in moments, which we have already seen in his "omn." results, probably derives from Huygens who was interested in mechanics.[16]

We should not think of these early efforts as imperfect attempts to find calculus. Rather, they represent Leibniz's exploration of infinitesimal problems in a perfectly valid and rational way, but one which is now remote from us. Then too, Leibniz was simultaneously occupied with other forms of mathematics. We have noted his interest and interaction with the British concerning series. There are also many letters and fragments dealing with algebra, number theory, higher order equations, their roots, etc.[17]

It is not until a manuscript of July 1676 that Leibniz begins to express integrals in their modern form as sums of Roberval-like rectangular strips having width dx.[18] Here we see correct integral forms like $\int \overline{dy} = y$, $\overline{\int x\,\overline{dx}} = x^2/2$, or $\int x^2\,dx = x^3/3$.[19] No proofs, however, are given. There is no evidence yet that Leibniz calculated that the difference of, say, $y = x^3$ for ordinates an infinitely small dx apart is $3x^2\,dx$, so that in analogy to his work on series the sum of these differences should be the difference between the cubes of the last and first ordinate or y^3 since $y(0) = 0$. As we

June), and *Calculus differentialis transcendens* (July) in *Ibid.*, pp. 449–480, pp. 481–495, and p. 607f. Note that the 31 page manuscript dealing with tangents to higher order parabolas would reduce to a few lines using his later method.

[15] e.g *Analysis Tetrgonistica Ex Centrobarycis*, Leibniz (2008), pp. 263–269; translation in Leibniz (1920) pp. 65–68. The result follows because the moment of R to any one of the lines is its area A times the distance of its center of gravity $(\bar{x}, \bar{y})$ to the given line. Therefore for three known lines we have three equations in unknowns $\bar{x}$, $\bar{y}$, A, and can solve for A.

[16] In his 1703 letter to James Bernoulli Leibniz indicates that Huygens explained the center of gravity to him, correcting a mistake Leibniz had made using the concept. See Leibniz (1920), p. 14.

[17] For instance, Jacques Ozanam (1640–1717), a French mathematician, proposed the problem of finding three numbers whose sum is a square and the sum of the squares is a fourth power. For the problem of finding three integers such that the difference between any two of them is a square, see, e.g., Rivaud (1914)., p. 72. Many problems in algebra and number theory which interested Leibniz are treated in Hofmann (1974a), Chapter 11.

[18] *Methodus tangentium inversa*, Leibniz (2008), pp. 598–602; translation in Leibniz (1920), pp. 118–124.

[19] The last integral is in Leibniz (2008) version, but does not appear in Child's 1920 edition of Leibniz's early mathematical manuscripts.

have seen in Chapter 5 these results can be obtained in several other ways and considered as area formulas they were well-known. The rest of the manuscript is devoted to two inverse tangent problems, the first of which Descartes claimed to have solved and the second proposed by Florimond de Beaune (1601–1652) to Descartes. Both are similar to the kind of inverse tangent problems he was doing in the Fall of 1675, and he is not able to solve either one completely. We briefly sketch the first problem. In Figure 6.2

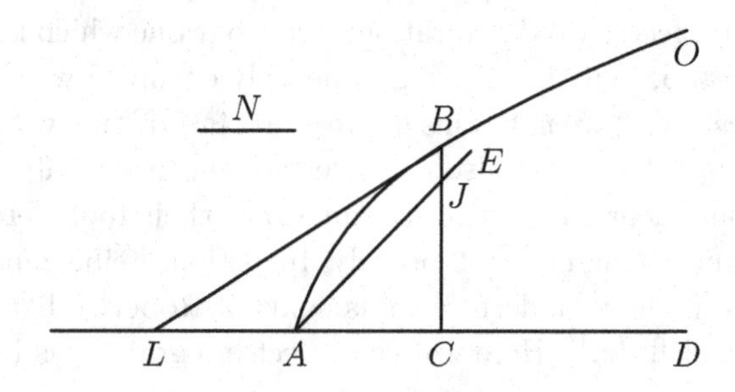

Figure 6.2

we want to find the curve ABO such that if AE is a line inclined at 45° and BL is a tangent then $(BC = y) : CL$ is the same ratio as $N : BJ$. Leibniz is led fairly quickly by exploiting ratios between sides of the characteristic triangle to the (linear) differential equation

$$\frac{dx}{dy} = \frac{N}{y - x}$$

where $x = AC$, which in fact is erroneous (the left side should be dy/dx). He does not solve it, but concludes that the curve has the property that the area $AJBA = BC \cdot N$, whence it follows that $\int y\, dx = (x^2/2) + Ny$.

In spite of this change in the meaning of the integral sign and Leibniz's ability to derive (and occasionally solve) differential equations appropriate to some geometric problem, one has the impression that Leibniz's ideas have not quite yet "jelled" into their final form. There is as yet no real algorithm to find either tangent lines or areas, and despite the fact that he has long recognized that summation of the terms of a series is inverse to taking their differences, there are no systematic applications of this fact to either quadratures or tangents. The only relation between the two concepts he had discovered by is the transmutation theorem of 1673/4, which is not yet the FTC (although we can think of it as a "cousin"). There are also (as we have already seen) in these early years mistakes and some confusion in Leibniz's manuscripts. In short, as late as the summer of 1676 Leibniz's *concrete* results, as we have repeatedly admitted, exhibited talent and ingenuity, but they did not separate him from other excellent mathematicians who were his contemporaries. Either his results like those concerning series were not new, or they were comparable in quality to results found by Wallis, James Gregory, Pascal, Fermat, Barrow, or Christiaan Huygens; they certainly could have been obtained by traditional methods, especially by mathematicians of the stature of Barrow or Huygens. And although Leibniz could absorb material extremely quickly, his knowledge of the literature and mastery of classical techniques were still probably inferior to all of these mathematicians. One wonders, in fact, if he could have really read the more complicated arguments of Apollonius. His work, although ingenious, rests on relatively simple similar triangle arguments and show none of the extraordinary technical ability of, for example, Barrow.

Leibniz's development of the calculus (and as we shall see his version in perfected form still differed profoundly from ours) took several years after 1675 and perhaps was not yet complete by the mid-1680s. It also did not attain a recognizably modern form centered on functions and their derivatives until the work of Euler in the mid-eighteenth century.[20] As late as October 1676 Leibniz's only real novelty was his desire to (and partial success in) creating a *calculus* for infinitesimal problems which promised to solve them by manipulating the notation according to formal rules, an idea which as we have already pointed out dates almost from his first exposure to mathematics. However promising, his actual accomplishments were not yet comparable to those of Newton a decade earlier. As early as 1665 Newton had powerful algorithms for both differentiation and integration and had clearly recognized the inverse relationship between the two processes For instance, in a manuscript of October, 1666 entitled "To resolve problems by motion" found in the Portsmouth collection, Newton had produced a table of some fairly difficult "integrals," expressed according to the following conventions: Let p and q respectively represent the velocity of a particle along the x and y axes (i.e., in modern notation $p = dx/dt$ and $q = dy/dt$). Then if

$$\frac{q}{p} = \frac{cx^{5n}}{x\sqrt{a + bx^n}}, \tag{6.1}$$

[20]See Bos (1974).

he states that

$$y = \left(\frac{1}{945nb^5}\right)(210b^4cx^{4n} - 240ab^3cx^{3n} + 288a^2b^2cx^{2n} - 384a^3bcx^n + 768a^4c)\sqrt{a + bx^n},$$

which, of course, is equivalent to finding an antiderivative for the right-hand side of (6.1). There are many other formulas of equal complexity in this manuscript, as well the simpler cases, e.g., if

$$\frac{q}{p} = ax^{\frac{m}{n}},$$

then

$$y = \frac{na}{m+n}x^{\frac{m+n}{n}}.$$ [21]

One wonders when Leibniz could have solved a problem of the difficulty of (6.1). It is likely that by the 1680s he could have done so, but probably not in 1675 or 76.

The situation, however, begins to rapidly change after Leibniz's second visit to London in October 1676. Something happened in this short period (we will look at the possibilities in Chapter 10) that enabled him to quickly produce a more "modern" version of his calculus. It is only when Leibniz systematically extrapolates his difference reasoning concerning series to the infinitesimal level, especially when in this setting he realizes that d and $\int$ are mutually inverse operations just as in the finite case and explores the consequences of this insight relating to tangents and quadratures that his calculus begins to develop into its mature form. The first substantial step in this direction comes a month after leaving London. The manuscript *Calculus Tangentium differentialis* was written in November 1676 in Amsterdam where he stayed from October to December 1676 during his trip from Paris to Hanover. In it for the first time he begins to think of dx and dy as infinitesimal (not necessarily constant) differences. He formulates correct rules for calculating these differences (but his notation initially confuses

[21] Newton (1962), pp. 19–26. These examples show that Newton very early in his career was a master of substitution of variable techniques in problems equivalent to integration.

differences with the ratio dy/dx).[22] He writes

$$\overline{dx} = 1, \qquad \overline{dx^2} = 2x, \qquad \overline{dx^3} = 3x^2, \quad \text{etc.}$$
$$\overline{d\frac{1}{x}} = -\frac{1}{x^2}, \qquad \overline{d\frac{1}{x^2}} = -\frac{2}{x^3}, \qquad \overline{d\frac{1}{x^3}} = \frac{3}{x^4}, \quad \text{etc.}$$
$$\overline{d\sqrt{x}} = -\frac{1}{\sqrt{x}}, \quad \text{etc.}$$

A few lines later there occur the correct formulas

$$\frac{\overline{dx^e}}{dx} = ex^{e-1}, \quad \text{and} \quad \overline{\int x\,\overline{dx}} = \frac{x^{e+1}}{e+1}.$$

Leibniz then uses the first formula on the general conic

$$ay^2 + byx + cx^2 + f^2x + g^2y + h^3 = 0$$

(note again Leibniz's unnecessary need for homogeneity) to obtain

$$a2\overline{dy}y + by\overline{dx} + 2cx\overline{dx} + f^2\overline{dx} + g^2\overline{dy} + bx\overline{dy} = 0.$$

Evidently Leibniz now knows the Product Rule. This is immediately followed by the equation

$$a\overline{dy}^2 + b\overline{dxdy} + c\overline{dx}^2 = 0,$$

obtained by setting the second order infinitesimal terms obtained in the difference process equal to 0, and the remark "This is the origin of the rule published by Sluse."[23] This is shortly followed by an only slightly incorrect application of the chain rule to "$\sqrt[2]{\overline{a + bz + cz^2}}$" (one of Leibniz's notation for the square root). He introduces the substitution $x = a + bz + cz^2$. Using either d or dx/dz to signify the slope of a tangent line, he notes that $\overline{d\sqrt[2]{x}} = -1/2\sqrt[2]{x}$ and $dx/dz = b + 2cz$. Hence

$$\overline{d\sqrt[2]{a + bz + cz^2}} = -\frac{b + 2cz}{2dz\sqrt[2]{a + bz + cz^2}}.$$

The presence of dz in the wrong place under any interpretation of the initial d may be a careless slip.

The final result given in this manuscript in the use of his new calculus is to find the curve such that the sum of the ordinate y and the subnormal BP is xy (Figure 6.3).

[22] In Leibniz (2008), pp. 612–618 the rules are given correctly, but in the Leibniz (1920) translation there are some misprints or transcription errors, e.g., $d1/x^2 = -2/x^2$ and $d1/x^3 = 3/x^2$.

[23] The application of Sluse's rule as presented in Chapter 2 will agree with Leibniz's result.

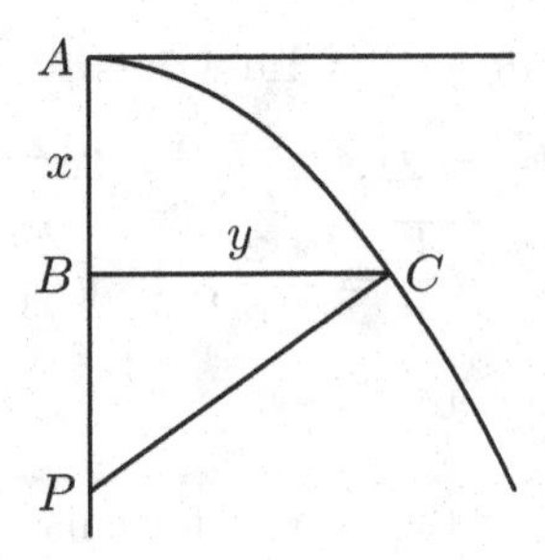

Figure-6.3

Since $BP = \overline{ydy/dx}$, we arrive at the differential equation

$$y + y\frac{\overline{dy}}{dx} = xy,$$

which is one of the very few examples where Leibniz does not require dimensional homogeneity. Canceling y gives that

$$\overline{dx} + \overline{dy} = x\overline{dx},$$

whence by summation of these differences we get that $x + y = x^2/2$ which is the equation of a parabola. Leibniz concludes with the remark "Thus we have the curve in which the sum of $CB + BP$ (multiplied by a constant r) is equal to the rectangle $AB \cdot BC$." (The "r" is probably an afterthought since Leibniz realizes that the same reasoning holds if the left side of the initial equation is multiplied by r.)

Still more progress has been made seven months later in a letter to Oldenburg written on June 21, 1677 (N.S.) as an answer to the *Epistola Posterior*[24] After fulsome praise for Newton's results, Leibniz sketches some of his own recent work. He first states that dx is the difference between successive values of a variable x and dy the corresponding difference in a variable y which depends on x. Thus if $y = x^2$

$$dy = (x + dx)^2 - x^2 = 2x\,dx + (dx)^2 \approx 2x\,dx.$$

The approximation is exact if dx is infinitely small because $(dx)^2$ is infinitely small with respect to dx and may be ignored. Leibniz then explains how the method can be used to find the tangent to $f(x, y) = 0$ where $f(x, y)$ is a quadratic polynomial in two variables and shows that in this case the

[24]We recall that this letter, written October 24, 1676 (O.S.), was not picked up by Leibniz, now in Hanover, until late June 1677 nearly eight months later.

method is equivalent to Sluse's rule. Next Leibniz illustrates what is now called the chain rule by stating that

$$d\sqrt[3]{a+by+cy^2} = \frac{b\,dy + 2cy\,dy}{3(a+by+cy)^{2/3}}.$$

Also a much more complicated example is considered: Leibniz finds the tangent to the curve

$$a + bx\sqrt[3]{y^2 + b\sqrt[3]{1+y}} + hyx^2\sqrt[3]{y^2 + y\sqrt[2]{1-y}} = 0$$

where y is the abscissa and x the ordinate. Leibniz now has also realized that if the ordinate of a curve can be recognized as what we would call a derivative, then the area defined by the curve can be calculated. For example, if

$$\omega = \frac{b + cy + dy^2 + cy^3}{z\left(\sqrt[z]{1 + by + cy^2/2 + dy^3/3 + ey^4/4 + \ldots}\right)}, \tag{6.4}$$

then the area defined by ω between $y = 0$ and some arbitrary y is given by a rectangle with one side equal to $\sqrt[z]{1 + by + cy^2/2 + dy^3/3 + ey^4/4 + \ldots}$ and the other side a line segment of unit length.[25] Ignoring the fact that Leibniz's formula is in error since the derivative of the area function is not given by (6.4) unless $z = 2$ and consequently the radical represents a square root,[26] the above example demonstrates knowledge of the chain rule and the principle behind the FTC.[27] Another interesting area result is given a little later in the letter.[28] Suppose in Figure 6.4 that an unknown curve has a subtangent TB is given by the series $a + bx + cx^2 + \ldots$ where $x = BC$.

[25]Notice in this example that since Leibniz is probably thinking here in terms of the cartesian definition of line segment multiplication, his formulas do not have dimensional homogeneity.

[26]This may be a simple slip. It is unlikely at this time that he thought that $dx^{1/n} = (1/n)x^{-1/n}\,dx$ as it is when $n = 2$, since he had given the correct formula for $z = 3$.

[27]Newton (1959), II, pp. 219–227. Also see Aiton (1985), p. 81 and Hall (1980), p. 70.

[28]Newton, *Ibid.*, p. 223f.

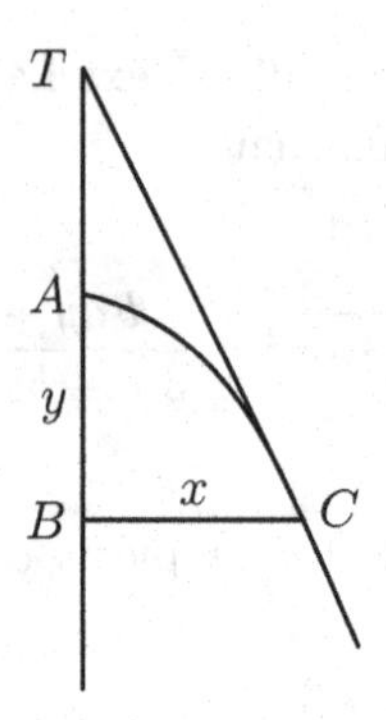

Figure 6.4

Then if $AB = y$, y is given by the series $bx + cx^2/2 + dx^3/3 + \ldots$. No proof is given, but the result obviously follows by writing

$$\frac{dx}{dy} = \frac{x}{a + bx + cx^2 + dx^3 + \ldots},$$

separating variables and integrating. Building on his familiarity with Pascal's characteristic triangle and the knowledge that it permits the calculation of the subtangent by similarity considerations, he can now find complicated differentials and use them to find tangents. He can also find some areas in a general way, very likely, from generalizing a similar principle for series: if $y\,dx$ can be recognized as a difference dz, then $\int y\,dx = z$.

Only in July 1677, seven months after his efforts of November 1676[29] and a month after his letter to Oldenburg did Leibniz succeed, apparently for the first time, in correcting and polishing his previous work on tangents.[30] Here in differential form all the standard rules of calculus are clearly stated. These include rules for the differentials of powers (given by arbitrary exponents not just positive integers), products, quotients, and for taking the differential "implicitly" of an equation in two variables. The chain rule is not explicitly stated but is implicit in the formulas he gives. No derivations, however, are given. They are only found in a manuscript which in the opinion of J.M. Child (the editor and translator of these manuscripts) dates from around 1680 or at least substantially later than the summer of 1677.[31]

[29]Manuscripts in this period, however, may be lost.

[30]Leibniz (1920), pp. 128–134. The manuscript written in French is entitled *Méthod générale pour mener les touchantes des Lignes Courbes sans calcul, et sans réduction des quantités irrationelles et rompues.*

[31]*Ibid.*, pp. 134–144.

The manuscript is obviously intended as a summary of Leibniz's work over the previous years. The title translated is "The elements of the new calculus for differences and sums, tangents and quadratures, maxima and minima, dimensions of lines, surfaces, and solids, and for other things that transcend other means of calculation."[32] Beginning with a discussion of tangents, he has a diagram (Figure 6.5) clearly identifying the characteristic triangle. The hypotenuse of the triangle if the sides are infinitely small is identified with a segment of the tangent, and the fact that the ratio of the ordinate y to the subtangent t is as $dy : dx$ which Leibniz had known since 1674 is derived by a similar triangle argument: "Thus to find differences of series is to find tangents."

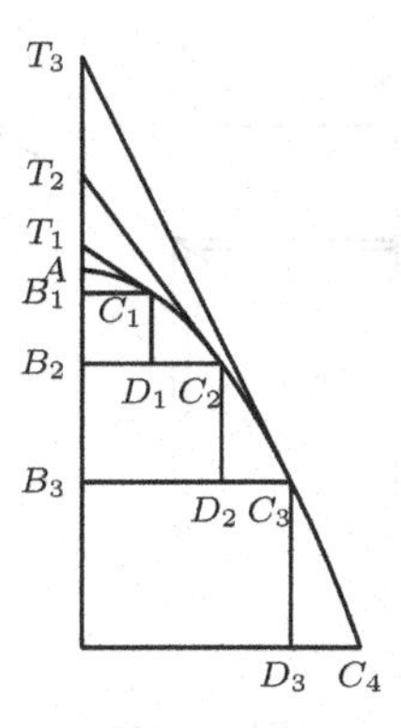

Figure 6.5

This idea is applied to the hyperbola to obtain

$$dy = -\frac{a^2}{x^2}dx,$$

(note the unit segment a again!) from which it follows that the subtangent t is equal to $-x$. Moreover, since an infinitesimal element C_1C_2 of the curve in Figure 6.5 is $\sqrt{dx^2 + dy^2}$, if $y = x^2/2a$ so that $dy = (x/a)dx$ then

$$C_1C_2 = \frac{dx}{a}\sqrt{a^2 + x^2}.$$

It follows that the total length of the curve is given by

$$\frac{1}{a}\int dx\sqrt{a^2 + x^2},$$

[32] *Elementa calculi pro differentiis et summis, tangentibus et quadraturas, maximis et minimis, dimensionibus linearum, superficiierum solidorum, aliisque communem calculem transcendentibus*

"hence $C_1C_2 : dx$ [is] as the ordinate of the hyperbola $\sqrt{a^2+x^2}$[33] to the constant line a; that is, $\frac{1}{a}\int dx\sqrt{a^2+x^2}$, a straight line equal to the arc of a parabola depends on the quadrature of the hyperbola." Moreover, since once again $t : y = dx : dy$, $t\,dy = y\,dx$, we obtain that

$$\int t\,dy = \int y\,dx,$$

which is evidently another transmutation theorem which Leibniz states "enunciated geometrically, gives an elegant theorem due to Gregory."[34]

Of perhaps greater significance are "other things which are immediately evident from a figure." If we consider Figure 6.6,

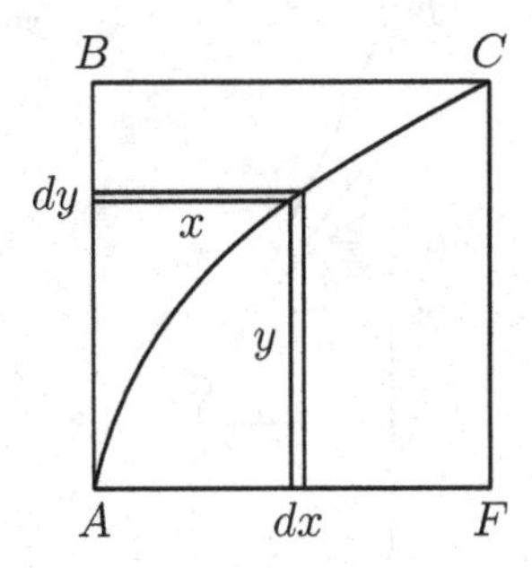

Figure 6.6

then "the figure $ABCA$ together with its complimentary figure $AFCA$ is equal to the rectangle $ABCF$, for the calculus readily shows that $\int y\,dx + \int x\,dy = xy$." This seems to be the earliest statement of the integration by parts formula in modern form, but it is interpreted as a statement about complimentary areas instead of as the inverse of the differentiation formula for products.[35] Also Leibniz notes that to find the volume of the solid of revolution generated by rotating AFC about the x or y axes it is only necessary to find $\int y^2\,dx$ or $\int x^2\,dy$, while the moment about the vertex (does Leibniz mean the y axis?) is given by $\int yx\,dx$. Next he gives a correct

[33] i.e., the hyperbola

$$\frac{y^2}{a^2} - \frac{x^2}{a^2} = 1.$$

[34] Leibniz describes a diagram illustrating this theorem which may be taken from Gregory or as Child in Leibniz (1920) thinks from a diagram in Barrow's *Lectiones Geometricae*, Lecture XI. For the diagrams see *Ibid.*, p. 140 (n.71).

[35] However, the latter interpretation is given a few pages later.

formula for the surface area of a solid of rotation about the x axis which he says is proportional to

$$\int y\sqrt{dx.dx + dy.dy},$$

noting that this integral reduces the calculation of a surface area to the quadrature of a plane figure. For example, if $y = \sqrt{2ax}$ so that $dy = adx/y$, the surface area integral becomes $\int dx\,\sqrt{2ax + a^2}$ which is the area under the parabola $y = \sqrt{2ax + a^2}$.

Leibniz still wants to pursue the analogies between his calculus and his early work in finite differences which motivated it. In a section called "The fundamental principle of the calculus" he defines x as the sequence

$$0 \quad 1 \quad 3 \quad 6 \quad 10 \quad 15 \ldots n(n+1)/2 \ldots$$

so that a dx is one of the sequence of natural the numbers and $\int x$ is one of the partial sums $0, 1, 4, 10, 20, \ldots$. Again, he points out that here and in general $d \int x = x$ and $\int dx = x$. The manuscript ends first with calculations of the differential of a product or a quotient involving two or three variables (where products like $dxdy$, etc., are omitted as being infinitely small in comparison with dx or dy), and then the modern differential and integral formulas for $y = x^r$ for any rational number r.[36]

Leibniz did not publish his calculus until 1684. The differential part is found in Volume 4 of the *Acta Eruditorum* and is entitled *Nova methodus pro maximis et minimis, itemque tangentibus, quae nec fractas nec irrationales quantitates moratur et singulare pro illis calculi genus.*[37] This paper is a highly compressed outline of the calculus together with some applications. No proofs of the basic rules are given What is notable is Leibniz's effort to avoid appeal to infinitesimals. Given a tangent to a curve at (x, y) and subtangent t, Leibniz defines the "difference" dx to be a difference between arbitrary values of the variable x and so is a *finite* line segment. Then the difference dy is a segment such that $dy : dx = y : t$.[38] This is followed by various rules for differentials. First he states that the differential of sums or differences of variables is given by the obvious rule, e.g.,

$$d\overline{z - y + w + z} = dz - dy + dw + dx.$$

[36]The case of $\int x^r\,dx$ where $r = -1$ is not noted, however.

[37]French translations of this and subsequent papers dealing with calculus published in the *Acta* may be found in Leibniz (1989).

[38]If we think of y/t as a derivative (which Leibniz did not), his definition is equivalent to the modern one.

Then he *defines* the differential of product xv or quotient $z = v/y$ as $x\,dv + v\,dx$ or

$$dz = \frac{\pm v\,dy \mp y\,dv}{y^2}.$$

This formula differs from the modern form

$$dz = \frac{y\,dv - v\,dy}{y^2}$$

because to Leibniz dz is positive or negative depending on whether we compute it by setting

$$dz \approx \frac{v(x+dx)}{y(x+dx)} - \frac{v(x)}{y(x)}$$

or

$$dz \approx \frac{v(x)}{y(x)} - \frac{v(x-dx)}{y(x-dx)}.$$

Additionally Leibniz states that it can be determined whether of not the ordinate y is increasing or decreasing by looking the sign of dy, while a local maximum or minimum or minimum occurs when $dy = 0$. If dy is infinite then the tangent is vertical and if $dy = dx$ then it is inclined at 45° with respect to the x axis. Without giving a precise definition, Leibniz then speaks of the "difference of the differences" or second differential ddy whose sign in relation to that of dy governs the convexity or concavity of the curve. (For instance, if dy is positive and $ddy > 0$, then the curve is convex with respect to the x axis.) After giving as an example the calculation of dw when

$$w = \frac{v}{y} + \frac{y}{z} + \frac{x}{v},$$

Leibniz moves on to formulate rules for the differences of powers and roots. For example, dx^a is $ax^{a-1}\,dx$ and

$$d\sqrt[b]{x^a} = \frac{a}{b}\sqrt[b]{x^{a-b}}\,dx.$$

In particular if $w = 1/x^3$, then $dw = -3\,dx/x^4$. He concludes the statement of the rules with the remark that if one knows the

> ... algorithm of this calculus which I call *differential* then one is able to find by ordinary calculation all the other differential equations, those for maxima and minima, and also tangents, with out having to eliminate either fractions, or irrationalities, or other radical signs, which has been inevitable in methods used up to the present.[39]

[39] *Ibid.*, p. 110 (my translation).

After an example of taking differentials "implicitly" for a rather complicated equation,[40]the paper closes with three applications. The first is a derivation of the law of refraction between two media obtained by several contemporary writers on optical theory including Thomas Harriot (1560–1621), Descartes, Willebrord Snell (1580–1626), Fermat, and others. In Figure 6.7 the medium on the side of E and C has respective densities h and r. We seek the path EFC where E and C are fixed points such that $EF \cdot r + CF \cdot h$ is minimized.

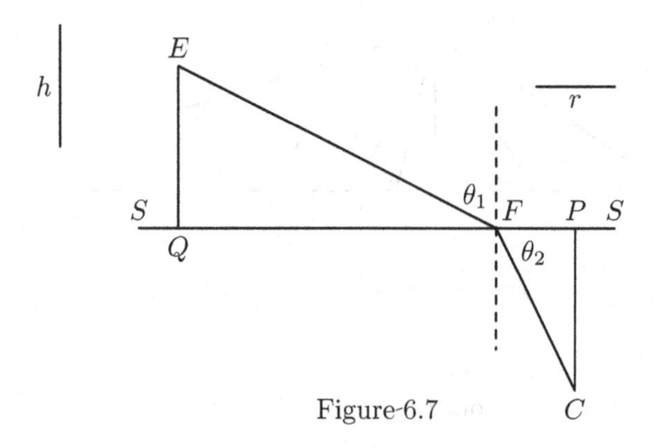

Figure 6.7

To solve this problem, Leibniz lets $PQ = p$, $QF = x$, $CP = c$, $EQ = e$, and $PF = p - x$. Then we want to minimize

$$w = h\sqrt{c^2 + (p-x)^2} + r\sqrt{e^2 + x^2};$$

or, abbreviating $c^2 + (p-x)^2$ by l and $e^2 + r^2$ by m and taking differentials, we have at the minimum of w that

$$h\,dl/2\sqrt{l} + r\,dm\,/2\sqrt{m} = 0.$$

Leibniz now argues that we can take C or E so that $CF = EF$ or equivalently $l = m$, since this change will not affect the angles θ_1, θ_2 of incidence and refraction. Therefore $h(p-x) = rx$ or $h/r = x/(p-x)$ implying since $CF = EF$ that $\sin\theta_1/\sin\theta_2 = h/r$. Leibniz notes that earlier derivations by "very eminent savants" had to pass through "multiple detours" to reach this result [41] while anyone familiar with the calculus can do it in "three lines."

[40]The equation is $x/y+(a+bx)(c-x^2)/(ex+fx^2)+ax\sqrt{g^2+y^2}+y^2/\sqrt{h^2+lx+mx^2} = 0$. *Ibid.*, p. 112.
[41]*Ibid.*, p. 115.

The second example is the derivation of a non-obvious geometric property for a generalization of a class of ovals studied by Descartes via the taking of differentials. Consider the curve (Figure 6.8) such that

$$\sum_{i=1}^{6} l_i = g \tag{6.5}$$

where $l_i = Pf_i$.

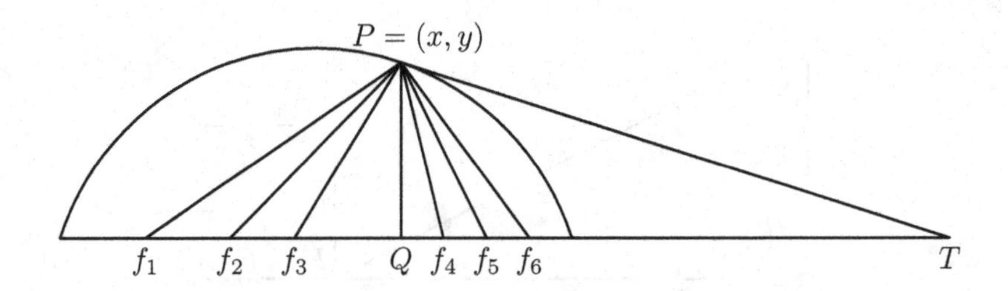

Figure 6.8

Let TP be a tangent at the point $P = (x, y)$ and $t = TQ$ be the subtangent. Then Leibniz states without proof that

$$\sum_{i=1}^{6} \frac{y}{l_i} : \sum_{i=1}^{6} \frac{f_i - x}{l_i} = t : y, \tag{6.6}$$

and remarks that a similar equation holds for any number of segments l_i. (perhaps, ten or even more). Moreover, the proof is easy using calculus, while it may be impossible using existing methods.[42] How did Leibniz derive this peculiar result? We can guess that he did something like the following: Without loss of generality we can assume a rectangular rather than an oblique coordinate system. Then taking differentials of (6.5), after noting that $l_i = \sqrt{(x - f_i)^2 + y^2}$, yields that

$$\sum_{i=1}^{6} \frac{(x - f_i)\,dx}{l_i} + \frac{y\,dy}{l_i} = 0.$$

Dividing by dy and solving for dx/dy gives

$$\frac{dx}{dy} = \frac{\sum_{i=1}^{6} \frac{y}{l_i}}{\sum_{i=1}^{6} \frac{(f_i - x)}{l_i}}.$$

[42] *Ibid.*, p. 116.

Replacing dx/dy by t/y completes the proof of (6.6).

The last application consists is a type of inverse tangent problem. which he had solved as early as July 1676.[43] Suppose we have a curve WW (Figure 6.9) such that the subtangent t is equal to a constant segment a, then what is WW? This problem is easily solved using calculus.

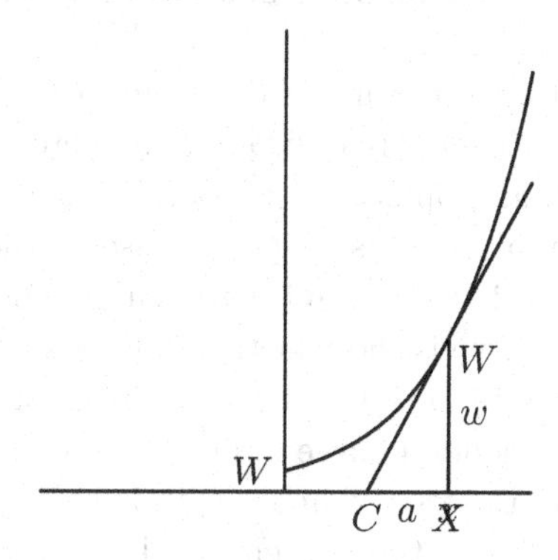

Figure 6.9

First note that the similarity between the triangle CXW and the characteristic triangle gives that

$$\frac{w}{a} = \frac{dw}{dx}.$$

At this point Leibniz, instead of following the modern procedure of integrating the equation to arrive at $w = c\exp(x/a)$, assumes that dx is a small segment b so that

$$w = \frac{a}{b}\,dw.$$

This means that successive values of w dw apart behave like a geometric progression and are ordinates of a "logarithmic curve" or equivalently "if the w are numbers, x will be their logarithm."[44] How does he reach these conclusions? No argument is given but Leibniz's reasoning may have resembled the following. Since

$$dw = w((k+1)b) - w(kb) = \frac{b}{a}w(kb),$$

[43]See *Ibid.*, p. 117, note 70.

[44]*Ibid.* By a "logarithmic curve" Leibniz means what we would call an exponential curve. But exponential functions were only studied in 1697 by Johann Bernoulli (1667–1748).

it follows that $w((k+1)b) = ((b/a)+1)^{k+1}w(0)$. Taking logarithms of both sides gives that

$$\log w((k+1)b/w(0)) = (k+1)b(\log((b/a)+1)/b) \propto x.$$

The above argument is, however, only a conjecture. We present another possibility below in a discussion of Leibniz's concept of the "progression of the variables."

Two years later in 1686 the integral sign makes its first published appearance in the *Acta*.[45] There are only a few concrete results in this article. Much of it is a quasi-philosophical discussion of issues relating to quadrature. Quadrature Leibniz says is a special case of the "inverse tangent" problem where the task is to determine the curve whose tangent has certain properties. If the curve is thought to be algebraic, but it is not known how to express its equation in terms of the usual algebraic operations, Leibniz recommends that a general curve of algebraic form with undetermined coefficients be used, its tangent computed, and the coefficients determined from the given properties of the tangent to the given curve. Leibniz also criticizes Descartes for barring transcendental curves such as the cycloid or quadratrix from geometry. These too are susceptible of quadrature and can be reconstructed from their tangent, by substituting a general algebraic equation with an additional term "v" and proceeding as before to find undetermined coefficients and v. Leibniz gives no concrete examples of this procedure but he illustrates his calculus by proving a theorem due to Isaac Barrow who had shown that the area under a curve defined by the subnormal to any curve is half the square of final value of the ordinate.[46] To state Barrow's theorem exactly and help illustrate Leibniz's procedure, in Figure 6.10 suppose that the curve $HFEV$ crosses the axis VP at V. If $E = (y, x)$ is a point, let EP be the normal to $HFEV$ at E; also let EZ be perpendicular to VP so that $AZ = p$ where p is the subnormal AP of EP. The conclusion is that the area determined by the curve ϕZZZ with ordinate AZ between V and C is half the square on $CH = \bar{y}$.[47]

[45] *De geometria recondita et analysi indivisibilium atque infinitorum*, Vol V, pp. 226–233. See also Leibniz (1989), pp. 126–143.

[46] Barrow's theorem and its complicated proof is Proposition 1 of Lecture XI of Barrow (1916), p. 126.

[47] Following Leibniz we denote values of the abscissa by y and ordinates by x. He gives no diagram. Our diagram is derived from Barrow's. This result is an early one. In the *Historia et origo* Leibniz implies that it dates from around 1674. The argument he gives, although using similar ideas, is more primitive than in the *Acta* paper. See Leibniz (1920), p. 40.

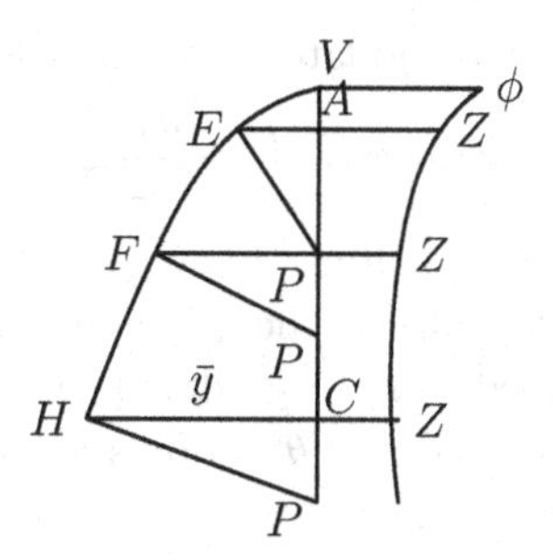

Figure 6.10

A young Scotch mathematician John Craige (1663–1731), who it seems almost alone of the British had been impressed by Leibniz's 1684 article, had attempted to reprove Barrow's result using Leibniz's technique:

> ... but in the course of the calculation he failed in his aim, which is understandable given the novelty of the method; [to give others] access to a method whose resources are manifestly immense,[48]

Leibniz shows how to solve the problem in one line by observing that $p\,dx = y\,dy$, a result obviously following from the similarity of the characteristic triangle and the triangle with sides y and p. Hence the area bounded by $VC = \bar{x}$, CZ, and ϕZZZ is

$$\int p\,dx = \int y\,dy = \bar{y}^2/2.$$

This again is a type of transmutation theorem. For as Leibniz remarks elsewhere, if it is desired to find the area under a curve with ordinate p (in this case ϕZZZ), all we have to do is find another curve $NFEV$ so that p is its subnormal. This curve he calls the "quadratrix" of the original one.[49]

Here in its first published form we see a casual use of the inverse relation between the operations of summing and taking differences.[50] There is almost nothing in this paper similar to the table of difficult "fluents" assembled by Newton as early as 1666, but Leibniz does claim that

$$a = \int \frac{dx}{\sqrt{2x - x^2}} \qquad (6.7)$$

[48] Leibniz (1989), p. 137.
[49] *Historia et origo*; Leibniz (1920), p. 41. Leibniz claims that he wrote down a "large collection of theorems using this result of Barrow's.
[50] But for Leibniz the "limits of summation" are given by context; he has no notion of what we call a "definite integral."

where x is the versedsine of a, and later in the article he exhibits the formula

$$\arcsin x = \int \frac{dx}{\sqrt{1-x^2}}. \tag{6.8}$$

This indicates that by 1686 Leibniz knew how to take differentials and "sum" some (and probably all) of the trigonometric and inverse trigonometric functions.[51] To verify (6.7) by what probably was his method, we use the abbreviations $x = \text{vs}\,\theta$ and $\theta = \text{arcvs}\,x$ for the versed sine of θ or the inverse versed sine of x. Then if $r = 1$, it follows that

$$d(\text{vs}(\text{arcvs}\ x) = \sin\theta\, d(\text{arcvs}\ x) = dx.$$

Since $\sin\theta = \sqrt{1-(1-x)^2} = \sqrt{2x-x^2}$,

$$d(\text{arcvs}\ x) = \frac{dx}{\sqrt{2x-x^2}}.$$

The same technique works for (6.8). Notice that Leibniz's approach is probably different from that of Newton who would have expressed the integrands of (6.7) and (6.8) as series, integrated them term by term and then identified the result with arcvs x and $\arcsin x$.

From the 1680s until his death in 1716 Leibniz made many applications of his calculus to physical problems including dynamics, planetary motion, optics, the trajectory of a projectile in a resisting medium, the derivation of the catenary, and the determination of centers of gravity. Many, but not all of these efforts, were solutions using calculus of problems which had been solved or posed earlier by others using traditional methods. At least two were stimulated by Newton's *Principia.* Leibniz claimed to have become aware of Newton's *magnum opus* after reading a review of it in the *Acta Eruditorum* in the summer of 1688, but to have only read the book in 1689. Based on this awareness, he submitted two papers to the *Acta* which were published in January and February of 1689.[52] The first was on the motion of a projectile in a medium whose resistance is proportional to either to its velocity or to the square of its velocity. In the first case the motion is rectilinear and Leibniz arrives at the equivalent of a first order linear differential equation (but given in terms of differentials not derivatives). Leibniz's conclusion agrees with Newton, but his treatment of resistance proportional

[51] The versed sine of an angle θ is $r(1-\cos\theta)$.

[52] These papers were original, but the influence of the *Principia* on them may have been greater than Leibniz admitted. Bertoloni Meli (1993) has argued that Leibniz's testimony is incorrect and that he had actually studied the *Principia* in 1688. The papers are analyzed in Guicciardini (1999), pp. 147–156 and Aiton (1972). For Meli's argument see pp. 7–9, 96ff, and Appendix I. This work also contains a translation of the *Tentamen*, pp. 126–142.

to the square of the velocity is mistaken. The second paper *Tentamen de Physicis Motuum Coelestium Rationibus* attempts to derive Kepler's laws of planetary motion from a Cartesian-like vortex point of view. Like many others Leibniz opposed Newton's conception of gravity as a mysterious force acting instantaneously across space as an "occult" quantity, similar to the meaningless scholastic forms, introduced into physics; all motion Leibniz believed was the result of contact. In the *Tentamen* Leibniz supposes, like Descartes, that the planets are carried around the Sun in a fluid vortex, but his vortices are much more complicated than the Cartesian ones which Newton demolished in the *Principia.* Leibniz supposes that the vortex is constructed in such a way that the primary motion of a planet is what he calls "harmonic," that is to say, its tangential velocity is proportional to its distance from the Sun. From this assumption alone it is possible to prove Kepler's second law that equal areas are swept out in equal times, even without any assumptions concerning the force of attraction. To get an elliptical orbit (Kepler's first law) Leibniz supposes that the center of the vortex oscillates to and from the Sun caused by the changing balance of two forces: the "solicitation" of gravity and a centrifugal "conatus" (force). The first results from certain properties of the fluid and the second from the constrained harmonic motion. Leibniz calls this motion the "paracentric velocity" of the planet and he shows that if the planet moves in an elliptical orbit with the Sun as one focus the gravitations solicitation varies as the inverse square of the distance of the planet to the Sun. The paper ends at this point. Leibniz has given a quasi-explanation of Kepler's first two laws in terms of vortices, but not explanations of the properties of the fluid causing the motion or of Kepler's third law that the square of the planetary orbital period if proportional to the cube of the major axis of the ellipse. These Leibniz admits:

> ... remain to be accounted for ... But since these matters have to be re-examined more deeply, they cannot be included within the brevity of this essay. What seems fitting to us will be explained separately in a more appropriate fashion.[53]

Leibniz begins his essay with a brief history of astronomical ideas from the Pythagoreans to the present without mentioning Newton who is referred to only later in the paper:

> I see that this proposition [the inverse square law of gravitation] was also known to the renowned Isaac Newton as it appears

[53] *Tetamen* in Meli (1993), p. 142.

from the review in the *Acta*, although from the review I cannot determine how he attained it.[54]

This possibly dishonest "putdown" of Newton—given that Leibniz may have already thoroughly examined the *Principia*—would become one of the damming factors against him during the later priority controversy. It is certain that Newton was not pleased by Leibniz's article.[55]

Our emphasis here is on Leibniz's mathematics, not his physics. Detailed discussions of the latter can be found in the books and articles of Meli, Aiton, and Guicciardini cited above, and the references contained therein.[56] But to give some idea of the mathematical flavor of Leibniz's approach to mechanics we cannot resist looking at a short paper "Concerning the isochronic curve on which a heavy body descends without being accelerated and a controversy with the abbot D.C."[57] published in the *Acta Eruditorum* in April 1689.[58] The problem is to find a curve C such that a body sliding along it without friction under the influence of gravity falls vertically without acceleration.

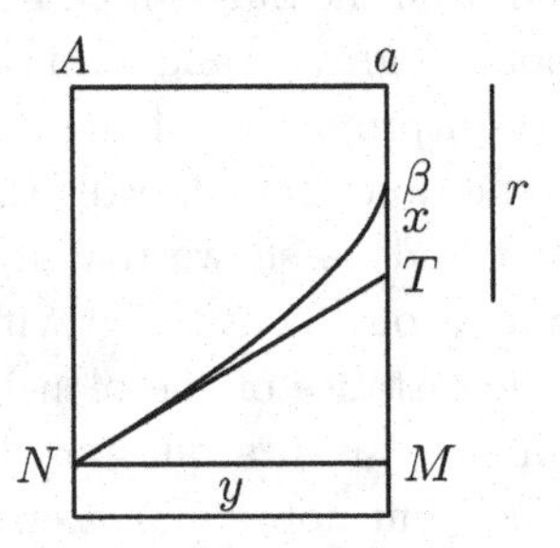

Figure 6.11

Leibniz gives no analysis by from which such a curve can be discovered. Instead he rather mysteriously presents the curve and verifies that it works. Leibniz claims that C is a "square-cubical" parabola such that the square of NM multiplied by a parameter $r = aP$ where P is a point on the extension

[54] *Ibid.*, p. 138.

[55] See the discussion of the controversy in Chapter 10.

[56] With the publication of two articles in 1694 and 1695 on the nature of force, Leibniz launched a quarrel with the Newtonians about how force should be measured that lasted almost to the nineteenth century. This is known as the *vis viva* controversy. We sketch a brief history of the dispute and explain the issues involved in Appendix D.

[57] *De linea isochrona in qua grave sine acceleratione descendit et de controversia cum Dn Abbe D.C.*

[58] A translation in to French may be found in Leibniz (1989), pp. 160–165.

of aM below M is always equal to the cube of the height βM. If a body initially falls from a a distance $a\beta = 4r/9$ and then slides without friction on $\mathcal{C}$ (see Figure 6.11), the vertical component of its velocity according to Leibniz will be constant. Leibniz gives no equation, but if we take the origin at β and (following Leibniz's usual custom) call the abscissa y and the ordinate x, then in modern notation the curve will have the equation $ry^2 = x^3$. Leibniz's proof that $\mathcal{C}$ is the correct curve is rather sparse. He gives few details. Let N be a point on $\mathcal{C}$ and call aM x and NM y. Assuming that a body first falls from a to β and then slides on $\mathcal{C}$ without friction, He then asserts that the vertical component of its velocity at any point of $\mathcal{C}$ will be the same as its velocity at β. To prove this, let NT be the tangent to $\mathcal{C}$ at N, cutting AP at T. Then it is well-known for square cubical parabolas such as $\mathcal{C}$ that

$$\frac{TM}{NM} = \sqrt{\frac{a\beta}{\beta M}},$$

and consequently[59] setting $h = a\beta$,

$$\frac{TM}{TN} = \sqrt{\frac{h}{h + \beta M}} \equiv \sqrt{\frac{h}{aM}}.$$

Now

$$\frac{TM}{TN} = \frac{v_x}{v}$$

where v_x is the vertical component of the velocity v of the body after it starts to slide on $\mathcal{C}$. Hence

$$\frac{v_x}{v} = \sqrt{\frac{h}{aM}}.$$

But by Galileo's laws of falling bodies

$$\frac{v}{v_1} = \sqrt{aMh}$$

where v_1 is the velocity of the body after it has fallen vertically the distance h. Therefore, putting the last two equations together we have

$$v_x = \sqrt{\frac{h}{aM}} v_1 \sqrt{aMh} = v_1,$$

[59]The reader may enjoy deriving the relations between TM, NM, and TN by differentiating the equation and using the fact that $a\beta = 4r/9$. The details can be found in *Ibid.*, p. 162f.

which was to be proved. Leibniz also remarks that for a fixed point N there is an infinite family of square-cubical parabolas with different parameters r such that a body sliding on the curve will have the isochrone property and pass through N.[60]

How Leibniz actually found C is unclear, but using his calculus we can derive it and clarify the seemingly arbitrary relation between the parameter r and the initial height $4r/9$. Let ds/dt be the velocity of the body sliding on C. From Galileo's results concerning falling bodies we can see that ds/dt is independent of C and depends only on the height x fallen. Specifically,

$$\frac{ds}{dt} = \sqrt{k^2(x+h)}$$

where k is a proportionality constant. Since there should be no vertical acceleration, the vertical velocity dx/dt is constant and equal to v_1. Hence

$$\frac{ds}{dx}\frac{dx}{dt} = \sqrt{k^2(x+h)},$$

or

$$\frac{ds}{dx} = \sqrt{1+\left(\frac{dy}{dx}\right)^2} = \frac{\sqrt{k^2(x+h)}}{v_0}.$$

Therefore

$$\frac{dy}{dx} = \sqrt{(k^2/v_0^2)(x+h)-1}.$$

Setting $(k^2/v_0^2)(x+h)-1 = z$, $dz = (k^2/v_0^2)dx$ and

$$dy = (v_0^2/k^2)\sqrt{z}\,dz.$$

This implies that

$$y = (2/3)(v_0^2/k^2)z^{3/2} = (2/3)(v_0^2/k^2)((k^2/v_0^2)(x+h)-1)^{3/2} + C.$$

If the summit of this curve is at the origin,

$$C = -(2/3)(v_0^2/k^2)(k^2/v_0^2h-1)^{3/2}.$$

Since $h \equiv v_0^2/k^2$, $C = 0$. So

$$y = (2/3)(v_0^2/k^2)((k^2/v_0^2)x)^{3/2}.$$

Squaring we get that

$$\begin{aligned} y^2 &= (4/9)(v_0/k)^4(k/v_0)^6x^3 \\ &= (4/9)(k^2/v_0^2)x^3 \\ &= (4/9h)x^3. \end{aligned}$$

[60] *Ibid.*, p. 163.

If we set the parameter $r = 9h/4$, the initial height fallen will indeed be $4r/9$.

How should we characterize the fundamental aspects of Leibniz's calculus. The notation is certainly familiar, but at the same time his calculus is rather different from the modern version. As Henk Bos and others have noted there are no derivatives in Leibniz's differential calculus. The expression dy/dx certainly occurs but it itself is not a function; it is rather a ratio of two finite (or infinitely short) segments such that $dy : dx = y : t$ where t is the subtangent associated with the tangent drawn to the curve at the point (x, y). Furthermore, although they may be characterized by some relation between variable line segments denoted by x and y, curves are purely geometrical entities and not graphs of functions. The algebraic calculations in Leibniz's work may seem familiar and modern, but this is misleading. Like Viète his algebra is largely an algebra of geometric objects—line segments, rectangles, cubes, etc. Only objects of like dimension can be added or subtracted and as in the classical Euclidean theory ratios can only hold between magnitudes of the same kind, line segments to line segments, for example, never a line segment to a square. It is of course true, as we have already seen, that Descartes had shown that the product, quotient, or root of line segments can be expressed by another line segment. This meant that it was not necessary to preserve homogeneity in expressions involving geometric objects; one ought to able to write, for instance, $\sqrt{a - x^2}$ instead of $\sqrt{ab - x^2}$. The price of such a procedure this, however, was often felt to be too high, since one had to choose a unit line segment in advance. But such a choice introduced an arbitrary feature which was felt to have no place in a general geometric treatment. Relations between lines, areas, or other objects should not depend on how they are scaled. Descartes' innovation applied conveniently to finding roots of polynomial equations for they are numbers and can be represented by line segments, but if possible it was to be avoided in pure geometry.

The foundations of Leibniz's calculus, as we have repeatedly shown go back to his earliest work on series in 1673: the sum of a sequence of differences $a_{i+1} - a_i$ is $a_{n+1} - a_1$. In his calculus these differences are "differentials"; dx is the difference between two infinitely close values of x. For more complicated expressions involving x, for example x^3

$$d(x^3) = d(x + dx)^3 - d(x^3) = 3x^2\,dx + 3x\,(dx)^2 + (dx)^3 \approx 3x^2\,dx,$$

the approximation becoming exact if dx is infinitely small. And if we write $\int 3x^2\,dx$ for the sum of all (infinitely many) terms $3x^2\,dx$ between, say, 0 and $\bar{x}$, just as in the case of summing finitely many terms of a sequence of differences, we have

$$\bar{x}^3 = \int d(x^3) = \int 3x^2\,dx,$$

so that $\bar{x}^3/3 = \int x^2\,dx$ and in genera if $y\,dx$ can be expressed as a difference dz then $\int y\,dx = \int dz = z$. In this way the complicated summing techniques common to Leibniz's contemporaries (and still inflicted on second semester calculus students in the evaluation of Riemann integrals) to find areas or volumes can be avoided. This simple idea is the foundation of Leibniz's calculus and leads almost immediately to many other results. For instance, since $d(xy) = x\,dy + dx\,y$ we have that

$$xy = \int d(xy) = \int x\,dy + \int y\,dx,$$

which is the integration by parts formula. And as Leibniz announced in his 1684 article, the differential alone tells us many things about the curve with ordinate y: if $dy > 0$ or $dy < 0$, evidently y is increasing or decreasing and if $dy = 0$ for some x y is neither increasing or decreasing at x and therefore may attain a local maximum or minimum there.

Leibniz also speaks of higher differentials. If dx is an infinitely small part of x then ddx is a difference of differences and a part of dx which becomes infinitely small with respect to dx. For example,

$$\begin{aligned} dd(x^3) &= ((x+2dx)^3 - (x+dx)^3) - ((x+dx)^3 - x^3) \\ &= 6x^2\,(dx)^2 + 6(dx)^3 \approx 6x^2\,(dx)^2, \end{aligned}$$

In modern notation $ddy/(dx)^2$ is the second derivative, and so if $ddy > 0$ or $ddy < 0$ the curve is concave up or concave down.

Another key notion of Leibniz's calculus, but one which has not survived in its original form, is "progression of the variables." In the simplest case if dx is constant, he says that x "progresses arithmetically," which implies that $ddx = 0$. But this need not always be so; one can always find some other difference scheme to define a differential. To take a geometric example, if to find the arc length of a curve we approximate it by a polygon with side ds, we can require that x, y, or s progresses arithmetically so that in the characteristic triangle dx, dy or ds is constant. This in general will impose different progression on the other two variables. The same will be true with any specified relation $f(x, y)$. Specifying which variable progresses arithmetically will determine the "progression" of the other variable. For example, if $y = x^2/a$ and dx is constant, dy will not be constant; instead

y will progress so that $dy = 2xdx/a$, and if dy is constant dx will be determined so that $2xdx/a$ is constant. Often Leibniz will determine the progression by supposing that $dy/f(y)$ or $dy/f(x)$ is constant where f is some algebraic expression in x or y. In our language this corresponds to choosing one of the variables to be "independent" or to defining one or both variables in terms of a parameter progressing arithmetically. Thus to say ady/x is constant is equivalent to saying that $ady/x = dt$ so that x and y are functions of t such that

$$\int a\,dy = \int x\,dt$$

or equivalently $ay = \int x\,dt + c$. In other cases given a differential dx corresponding to an arithmetically progressing x, Leibniz will define new differentials, for example, $dw = \sqrt{adx}$ or $dw = (dx)^2/a$. The first of these is infinitely small when dx is infinitely small, but at the same time infinitely large compared to dx, while the second is infinitely small with respect to dx. So even for first order differentials (generated from first differences) there is no privileged class of differentials. But in spite of this indeterminacy of differentials it will always the case that $\int dw = x$ and that $\int y\,dw = \int y\,dx$. Sometimes Leibniz speaks of two linked variables x and y where y progresses "geometrically" while x progresses arithmetically. In this case we would think of y being an exponential function of x. But such a definition violates homogeneity.[61] Instead Leibniz would define y by the equation $a^2dy = ydxa$ (where again a is a unit line segment). This implies that ax is the area under the hyperbola having ordinate a^2/y between $y = a$ and $y = x$. By the work of Gregory St. Vincent and Mercator, Leibniz knows that this entails that y progresses geometrically, i.e., if we take equal increments Δx of x the ratio $y(x + \Delta x) : y(x)$ is constant.[62]

There is an ironic contrast between the mathematical styles of Newton and Leibniz. Since for Newton, following Barrow, a curve is generated by a moving point his "fluxion" usually is a rate of change or velocity; hence a function of time. Although line segments can represent it in a geometrically constructed argument, fundamentally it is not a geometric object; it can be measured by numbers according the units in use, "feet per second," for

[61]That is, how can y, a variable line segment, be of the form b^x which has no geometric definition?

[62]Recall from Chapter 3 that Gregory St. Vincent had shown the logarithmic properties of the hyperbola. Needless to say, this complicated and somewhat unclear notion of progression of a variable introduces ambiguity into Leibniz's conception of a differential which is particularly the case for higher order differentials. For various examples of progression and a more detailed analysis, see Bos (1974), pp. 24–31 and pp. 42–52.

instance. This means that the homogeneity requirements of the traditional geometric treatment are absent from Newton's work. Leibniz would hardly ever have written that if

$$\frac{105a^3x^{2n} - 192b^3x^{5n}}{x\sqrt{ax^n + bx^{2n}}} = \frac{q}{p},$$

then

$$\frac{70aax^n - 56abx^{2n} + 48bbx^{3n}}{n}\sqrt{ax^n + bx^{2n}} = y.^{63}$$

From Leibniz's geometric point of view such an equation made no sense, and it is moreover unclear whether he ever integrated an expression as complicated even when suitably modified. However, as Newton grew older, while still using algebra to calculate fluxions or "integrating" or finding the "fluent" from a given fluxion in a near modern manner, he viewed the procedure with increasingly disfavor because it departed from the beautiful geometrical style of the ancients, which he felt was the only form suitable for published mathematics. Consequently he recast everything in geometrical form. We have noted at the end of Chapter 2 that Newton had such a mastery of geometrical methods that he thought as naturally using them as he did using algebraic techniques. The result is that the *Principia* is not merely a body of results obtained from "fluxions" and then hidden by a geometric veil. Some of it, to be sure, is of this form. But while several of Newton's theorems may have been established in this way, others were not. Newton derived them by pure geometry and they are can be very difficult to obtain by calculus.[64]

On the other hand, Leibniz's methods are symbolic. They have fatally compromised the geometric paradigm by making it almost unnecessary to refer to any but the simplest diagrams. Reasoning almost becomes mechanical with the new notation doing the work and eliminating the need for the awesome ingenuity shown by a Barrow or Huygens. This is all very modern. But at the same time, as we have said, Leibniz's calculus is not like its functionally oriented modern version. Instead, it is a calculus of geometric objects firmly grounded in the traditional paradigm so that the quotient dy/dx only represents the ratio of two line segments, namely the ordinate y to the subtangent t. One consequence of this approach is that to analyze convexity Leibniz must take a second differential which as early critics noted rests on the obscure notion of something infinitely small in

[63] "To resolve problems by motion" in Newton (1962), p. 25. Recall that q and p are the vertical and horizontal components of velocity, so that q/p is equivalent to dy/dx.

[64] See Guicciardini (1999) for a thorough discussion of these issues.

comparison to something already infinitely small. Newton by contrast can take a fluxion of a fluxion which is a perfectly good *function* which in the Newtonian context represents an acceleration. It was not until Euler's work in the 1740s that Leibniz's differentials were finally eliminated and calculus began to assume its modern form.[65]

[65]See Bos (1974), Chapter 5. Euler's two volume textbook of 1748 *Introductio in analysin infinitorum* might even be used in a modern calculus class, save for the fact that it is written in Latin.

Chapter 7

Logic

Besides his work in infinitesimal analysis, Leibniz made repeated investigations in logic, beginning in the Mainz period and continuing throughout his life. Leibniz's approaches to logic and mathematics were quite similar. In both fields he wanted to create a "calculus," centering on the construction of a new notation which by its manipulation would make reasoning and the discovery of new truths nearly automatic. And as we will see in Chapter 8, he considered most of his results in these areas to be mere examples of a vast scheme to find a way of mechanizing all thought, which would both serve as an aid to discovery and be capable of settling religious, political, and scientific controversies. Both Leibniz's mathematics and his logic also share another characteristic. They are simultaneously revolutionary and are still in many respects embedded in a traditional paradigm. His calculus of tangents and quadratures, as we have shown, was not our familiar functional and algebraic calculus which uses his notation. On the contrary, like Descartes he was still heavily influenced by the dominant geometrical orientation of his time; his calculus consequently was one of geometric objects: subtangents, subnormals, and the like, not one of functions. In a similar fashion, his many attempts to construct a "calculus" of logic are mixtures of both revolutionary insight and of Aristotelian traditionalism. There are aspects to Leibniz's logic which were not surpassed until the work of George Boole, William S. Jevons, Gottlob Frege, Schröder, (1841–1902) and Bertrand Russell (1872–1970) in the mid-nineteenth and early twentieth centuries. Indeed, he was so far ahead of his time that some of his work could be mistaken for arguments in a contemporary axiomatic modern algebra text, while others (such as the "alphabet of human thoughts") have not yet been implemented. At the same time he was even more firmly rooted in Aristotelian traditionalism than he was in what we have called the "geometric paradigm." In some ways, in fact, he was less advanced logically than the Stoics or later scholastics such as William of Ockham. This conservatism was probably a consequence of both his educational experience and the philosophy he developed. Leibniz was a superbly trained seventeenth century Central European university graduate. He had thoroughly mastered the university curriculum and its Aristotelian values were deeply imprinted in him. He very much respected Aristotle's logic, and some of his own core logical ideas were shaped by his own metaphysics which had a significant inheritance from Aristotle. Unfortunately, unlike his contributions to mathematics, Leibniz's work in logic had no influence either during his life time or later. He published nothing in the field, and almost all of his writings (as is the case as well with many of his other interests) are rough drafts or fragments and contain various imperfections and inconsistencies. After his death in 1716 his manuscripts were more or less forgotten. Nineteenth century historians such as Karl von Prantl (1820–1888) misinterpreted and denigrated his logical ideas which they did not understand.[1] And by the time his worth was discovered in the late nineteenth century, the bulk of his work been surpassed by contemporary logicians such as Frege and Schröder.[2]

We have seen that his Master's Thesis *Dissertatio de Arte Combinatoria* written in 1666 contained some of his earliest work in mathematics, a theory of permutations and combinations. From hindsight, this was a harbinger of things to come. He had demonstrated a degree of mathematical talent, especially since he worked out many of the results by himself even though almost none of them were new and could be found in the existing literature, with which he was mostly unfamiliar. Realizing this in later years, he considered it an immature work by "a young man just out of school." But it also marks the beginning of his logical investigations and interest in the "Universal Characteristic." These efforts were far more original than the mathematics presented in the thesis. In particular, we find here the earliest statements of three fundamental logical theses which he would pursue and elaborate for the rest of his life. They were not original with him, but are central to the original work in logic (some of which is already embryonically present in the *Dissertatio de Arte Combinatoria*) which he will develop in the following decades.

[1]See e.g., Prantl's *Geschichte der Logik im Abendland* published from 1855 to 1870.

[2]The scholars primarily responsible for this belated recognition were Louis Couturat (1867–1914) who edited and published many of his logical manuscripts and Bertrand Russell (see especially Russell (1964).

To understand these theses and the consequences Leibniz derived from them, we need some background. In almost all of his work Leibniz focuses on categorical propositions. These are propositions having a subject and predicate connected by a copula. They have the basic form "A is B" or "A is not B" possibly modified by words such as "all," "every," "no," "not" or, "some." The subject A is a noun or noun substitute denoting an object or class. One can think of B either as another class or as a property predicated of A. In the first or "extensional" case the proposition asserts that some or all of the class A is (or is not) contained in the class B. But we can also think intensionally: here A as defined by its properties, and if the proposition is true, then the concept or definition of B is part of the concept or definition of A. Thus the proposition "Man is an animal" viewed extensionally says that the set of men is a subset of the class of animals, while from the intensional interpretation it says that the concept of "animal" is part of the concept of "man." We shall see that for metaphysical reasons Leibniz almost always favored the intensional point of view. Categorical propositions also have two additional properties technically called "quality" and "quantity." The quality refers to whether of not the proposition affirms or denies, and the quantity refers to whether the proposition is universal, particular, singular, or indefinite. For example, to quote from a later work of 1679,[3] "The pious man is happy" (meaning "every pious man" or "all pious men") is a universal affirmative proposition while "The wicked man is not happy" or equivalently "No wicked man is happy" is a universal negative proposition. On the other hand, "Some wicked man is wealthy" and "Some pious man is not wealthy" are particular affirmative or negative propositions. A singular proposition is different from a particular proposition because it has a specific individual as a subject; for example, "Caesar crossed the Rubicon." Finally, indefinite propositions such as "Cretans lie" are ambiguous as to whether they are universal or particular. It is also important to note that categorical propositions seem not to express a relation such as "Theatetus is taller than Socrates;" nor do they appear to involve hypothetical or disjunctive forms like "If A then B" or "Either A or B" where both A and B are propositions. (In fact, Leibniz hardly ever considers in their own right propositions whose components are other propositions.) This description of various types of categorical propositions was basic to the Aristotelian tradition and was presented in the university courses in logic which all students including Leibniz studied.

[3]Leibniz (1966), p. 25ff.

Leibniz's first thesis is a belief expressed in *Dissertatio de Arte Combinatoria* that *all* propositions (including those of relational, hypothetical, or disjunctive type) either can be transformed into or are logically equivalent to one or more propositions of categorical form.[4] We will see below that this was a far reaching and contested claim that powerfully interacted with Leibniz's philosophy.

The second thesis also found in the *Dissertatio de Arte Combinatoria*, is that there exists a list of absolutely primary indefinable concepts which functions as an "alphabet of human thoughts." As in an alphabet these basic concepts would be clearly signified and function as "letters" in the in the sense that all complex concepts can be reduced to some subset of them just as words can be viewed as a collection of letters. Unlike the spelling of words, however, the order of the "spelling" of complex concepts would not be important. Hence if "Rational" and "Animal" were two primitive concepts, "Rational Animal" or "Animal Rational" would be the same complex term, namely "Man." As we will show in Chapter 8 this conception of an alphabet of human thought was in itself, like Leibniz's conception of a proposition, not especially original. It was very much part of the contemporary intellectual climate, and similar ideas had occurred to a number of Renaissance and seventeenth century thinkers. But what Leibniz did with the idea *was* original. In *De Arte Combinatoria* to represent concepts by n-tuples of numbers he first thinks of numbering the list of primary concepts and representing complicated concepts by (unordered) $n-$tuples, so that if 3, 6, 7, and 9 are numbers assigned to four primary concepts, we would have (3, 6, 7, 9), (7, 9, 6, 3), and all other permutations representing the same complex idea. This conception of "spelling out" complex terms by numbers allows Leibniz to apply his theory of combinations to solve various problems. He asks, for instance, how many complex ideas can be generated from n primary terms. As we have seen, his combinatorial theory implies that the number is $2^n - 1$. He also considers the following problem: given a complex term S which is the subject of a categorical proposition, we want to find all the possible complex terms P which when predicated of S make the proposition "S is P" true. To give an example, suppose S is represented

[4]For a discussion of this feature of Leibniz's thinking see Russell (1964), Chapter I, Mates (1986), Chapter III, or Parkinson (1965), Chapter I (especially pp. 33–35). For example, Leibniz reduced hypothetical propositions of the form "If A is B then C is D to the form "the C-ness of D is a predicate of the B-ness of A," or in keeping with the definition of the truth of a proposition given below "The [concept of] B-ness of A *contains* the [concept of] C-ness of D." In a similar way Leibniz reduces the disjunctive proposition "Either A or B" first to hypothetical and then to categorical form.

by (3, 6, 7, 9). The simplest predicates P for which the proposition S is P obviously correspond to the primary components of S taken singly 3, or 6, or 7, or 9. Thus we can generate the propositions "S is (what is represented by 3, 6, 7, or 9. But we can also generate predicates by taking the components of S two, three, or four at a time (the latter generating the trivial proposition "S is S"). In Leibniz's terminology this requires finding the "com2nations" and "com3nations" of (3,6,7,9). There are six or the former and four of the latter. So we can have "S is (6,7)," "S is (3,7,9)," etc., through all possibilities. Leibniz also discusses the converse problem of finding all subjects S that make the proposition S is P true for a given predicate P. Of course to make a meaningful problem we need some restrictions. The primary terms of P must be a subset of the primary terms of any S and we might think of a possible S for a given P as a subset containing the primary terms of P with the additional terms chosen from a larger fixed set of primary terms L, the determination of the number of such S being a combinatorial problem which Leibniz solves.[5] A fourth logical problem Leibniz considers in the *Dissertatioo de Arte Combinatoria* is the determination of all the figures and moods of a syllogism.[6] By a combinatorial analysis using the fact that there are three propositions in any syllogism, each having two possible qualities and four quantities, he finds that there are five hundred twelve possible simple moods and two thousand forty eight figures for a syllogism of which only eighty eight are useful.[7]

The project of finding all predicates for a given subject and all subjects for a given predicate which determine true propositions already suggests Leibniz's third thesis. This is an intensional concept of truth which is peculiarly Leibniz's own and which he had fully developed by 1679.[8] It was a basic and unchanging element in most of his work on logic and metaphysics for the rest of his life. He expressed it as "*Praedicatum inest subjecto.*" In other words, the belief that every true categorical proposition has its predicate somehow contained in the subject; "or else I do not know what truth is."[9] Let us call this the "Predicate-in-Subject Principle."[10] At the

---

[5]Couturat (1961), p 43.

[6]See Appendix C for a discussion of the basic concepts of classical logic.

[7]Knecht (1981), p. 38. What Leibniz means by "figure" here, however, is unclear. In classical logic there are only four figures. See Appendix C.

[8]*Elementa calculi.* See Leibniz (1966), pp. 17–24.

[9]Letter to Arnauld, July 1686. [Leibniz (1969), p. 337.] An earlier statement can be found in a letter to Conring, March 1678. (*Ibid.*, p. 188.)

[10]This terminology is due to Mates (1986), p. 84. Broad (1975), p. 6, calls it the "Predicate-in-Notion Principle."

simplest level there are propositions which are necessarily true from the definitions of their terms such as "Bachelors are unmarried" or "All men are animals." The truth of these propositions follow at once from the fact that by definition a "bachelor" is an "unmarried male," and a "man" is a "rational animal," hence in particular an "animal." In either case, we can say that the predicate is contained in the subject (although strictly speaking the inclusion holds between the *concepts* of the predicate and subject). There can be, of course, more complicated examples such as the theorems of geometry where the inclusion is much more involved and non-obvious. But all such cases can in principle be analyzed and reduced to identities such as A is A or AB is A, etc., in a finite number of steps.[11] More problematically, Leibniz asserts that this concept of truth also holds (in a certain sense) for what are commonly called "contingent" (or after Kant "synthetic") propositions whose denial is not an apparent self-contradiction. As an example, Leibniz considers "Caesar crossed the Rubicon." The denial of this does not seem to be a contradiction. One can imagine a possible world where Caesar did not cross the Rubicon, became a life long friend of his son-in-law Pompey, and abided by the decree of the Senate terminating his proconsular authority in Gaul and ordering him to return to Rome. But to Leibniz the real difference between "Caesar crossed the Rubicon" and "Bachelors are unmarried" is that the concept of Caesar, known only to God, is what Leibniz calls "complete" and includes *everything* ever predicated of him—even minor facts such as his having diarrhea at 5:01 in the afternoon September 16, 54 B.C. due to spoiled liver pâté at a banquet for allied tribal chieftains near Alesia. But to unravel this concept of Caesar would require an infinite analysis beyond the powers of any human mind.[12] To say this differently

[11]It might be thought that this definition of truth is only workable for universal affirmative propositions. We will see shortly how Leibniz applies it to the particular affirmative case. In the case of the universal negative proposition "No A is B" which *excludes* the concept of B from the of A, he will write "Every A is not-B" so that the concept of "not-B" is included in that of A.

[12]Whether or not this idea of a complete concept for an individual in reality abolishes the distinction between necessary and contingent propositions and is incompatible with Leibniz's belief in human freedom has caused great controversy among commentators. Mates (1986), Chapter VI, argues that it does. C.D. Broad (1975), pp. 25–36, and (1949) is more sympathetic with Leibniz's position. Both would agree, however, that God's creation of Caesar having the predicates he does is not necessary; instead it represents God's freely chosen plan to create the best of all possible worlds in that actualizing a Caesar who did not cross the Rubicon or who did not have this particular episode of diarrhea would have resulted in a world which *on the whole* would have been worse than the one which actually exists, even if we cannot see how this can be so. On the other hand, propositions like "All men are animals" or the truths of geometry hold in every

and to use post-Kantian language, *all* true propositions except those asserting existence of substances other than God are analytic in this extended sense.[13]

By the Spring of 1679 the early ideas of *De Arte Combinatoria* are much more developed.[14] Still fundamental is the idea of an alphabet of human thought, but now the primary terms are represented by prime numbers and complex ideas are no longer just tuples of integers. Both a subject S and a predicate P of a proposition will be represented by products of the prime numbers associated with the simple ideas composing them.[15] With this representation how do we know that the proposition "S is P" is true? Because of the Predicate-in-Subject Principle it is obvious that this is the case if, and only if, P is represented by a sub-product of the factorization associated with S; or equivalently if, and only if, the product representing P should divide that representing S, a technique which gives a simple, mechanical way to judge whether or not a universal affirmative proposition is true. The next step is to generalize this method so that it applies to particular or negative propositions such as:

$$\text{Some } S \text{ is } P \tag{7.1}$$

$$\text{No } S \text{ is } P \tag{7.2}$$

$$\text{Some } S \text{ is not } P. \tag{7.3}$$

To determine when these propositions are true, we need to find a more refined criterion than just the divisibility of the characteristic number of S by that of P. For (7.1) Leibniz gives the example "Some experienced man is prudent." Here the concept of "prudence" is not necessarily contained in the concept of "experienced man"; equivalently, the universal statement "All experienced men are prudent" may not be true since if the numbers associated with "man," "experienced," and "prudence" are respectively 6, 7, and 30, 30 does not divide 42. Leibniz gets around the difficulty by claiming that the "some" represents a subspecies of the subject obtained by modifying the concept of the subject by the addition of a new concept. Thus the above statement will be true if there is a minimal sub-species S_1 of "experienced men," perhaps those defined as possessing "natural judge-

possible world and are independent of God's decrees.

[13]Leibniz accepts (with some modification) Anselm of Canterbury's (c. 1033–1109) and Descartes' ontological proofs of God's existence. The property of existence is necessarily part of the definition of God. Therefore He exists.

[14]*Elementi calculi*, Leibniz (1966), pp. 17–24.

[15]Because of the logical identity $S \cdot S = S$ none of these prime factors need to be raised to a higher power.

ment," such that "S_1 is P" is true. In general (7.1) will hold when there is a product of primes x so that Sx is divisible by P so that $Sx = Py$ where x and y can be chosen so that they are relatively prime.[16] This idea also solves another difficulty. In normal usage the term "some" in this example indicates the *existence* of at least one prudent man. Therefore even if the appropriate universal affirmative proposition is true, we may not be able to infer the particular case if the concept of the subject applies to no existing individuals. In a much later work (after 1690) Leibniz gives an example of what the scholastics called "conversion *per accidens*": "Every laugher is a man, therefore some man is a laugher."[17] This seems an invalid deduction since the premise may be true in Leibniz's sense, but the conclusion may be false unless some man actually laughs. Classical logic avoids this difficulty by assuming that the class of laughers is non-empty, or to say this more formally, universal affirmative propositions have existential import. This is certainly a natural assumption given an extensional point of view. Leibniz, rejects this strategy because he wants a logic that applies to possible worlds as well as to the actual one. By treating "laughers" as a subspecies of man (whether or not it actually exists) or in general speaking only of concepts and their modifications, such inferences are valid in "the region of ideas." Next negative universal and particular propositions like (7.2) or (7.3) can be reduced to a consideration of affirmative ones. Thus the proposition "Some silver is not soluble in common *acqua fortis*" is equivalent to denying the universal affirmative proposition by "All silver is soluble in common *acqua fortis*." In the same way, a universal negative proposition is the denial of a particular affirmative one. Unfortunately, Leibniz's treatment of particular affirmative proposition has a flaw which he soon noticed. It turns out that *every* particular affirmative proposition is true at least in the "region of ideas." Since there is no rule against combining actually incompatible concepts to define imaginary subspecies of S, a characteristic number x may always be found so that Sx is divisible by P. For instance, the statement "Some man is a stone" would be "true" in Leibniz's sense. Another consequence would be that no universally negative proposition of the form (7.2) could be true.[18]

Leibniz attempts to handle such problems in another manuscript of

---

[16]If x and y had a common prime factor p, p would represent a common concept in both the subject and predicate in the proposition and therefore could be discarded.
[17]*Some logical difficulties*, Leibniz (1966), p. 115. In general "conversion *per accidens* is the deduction "Some P is S" from "All S is P."
[18]Couturat (1961), p. 330.

April 1679.[19] He introduces the minus sign as a form of negation in the characteristic numbers of S and P. Each term will now be represented by an ordered pair of characteristic numbers $[S_1, -S_2]$ and $[P_1, -P_2]$ where the first number is positive and the second negative.[20] The second element of the pair evidently is intended to solve the difficulty we have just pointed out in the previous system, by ruling out impossible properties. For example, if S_1 is the characteristic number of a concept C, S_2 is the characteristic number of properties incompatible with C. Now suppose R is another term represented by $[R_1, R_2]$. In analogy with the previous system a restriction of S which we will write as SR is represented by $[S_1 \cdot R_1, -S_2 \cdot R_2]$. Leibniz then postulates that, given this representation, "S is P" is true if, and only if, $P_i|S_i$, $i = 1, 2$. This new system also entails that the pairs representing terms be relatively prime. For if $T|S_i$ then the proposition "S is P^*" where P^* is represented by the contradictory predicate $(T, -T)$ would be true. Leibniz now extends his scheme to handle other types of propositions. First a particular negative proposition like (7.3) will be true if, and only if, its universal affirmative version if false. Next he states that a universal negative proposition like (7.2) will considered to be true if, and only if, "two numbers of different signs and different terms have a common divisor."[21] In other words, if, and only if, either S_1 and P_2 or S_2 and P_1 are not relatively prime. For example, if $[10, -3]$ represents "pious man" and $[5, -14]$ "unhappy," then the proposition "No pious man is unhappy" is true since 2 divides both 10 and -14. To see why this definition makes sense, suppose that (7.2) is false or equivalently that its particular affirmative version (7.1) is true. This would mean that the "subspecies" represented by $S^* \equiv [S_1 \cdot P_1, -S_2 \cdot P_2]$ is possible. But if $(S_1, P_2) = R$. The proposition S^* is R^* would be true where R^* is represented by $[R, -R]$. A similar argument may be applied if $(S_2, P_1) = R$. This argument also implies that the criterion of truth for the particular affirmative proposition (7.1) (which is the negation of (7.2)) is that both $(S_1, P_2) = 1$ and $(S_2, P_1) = 1$. Thus if $[11, -9]$ and $[5, -14]$ respectively represent "wealthy men" and "unhappy," "Some wealthy man is unhappy" will be true exactly when "No wealthy man is unhappy" is false, which is the case since $(11, -14) = (5, -9) = 1$. Finally, the third

[19] *Rules from which a decision can be made, by means of numbers, about the validity of inferences and about the forms and moods of categorical syllogisms*, Leibniz (1966), pp. 25–32.

[20] Leibniz, however, uses no brackets. He writes S_1, $-S_2$, etc. We add the brackets for clarity.

[21] *Ibid.*, p. 27.

proposition (7.3) will be true if its denial (7.2) is false, in other words, if P_1 does not divide S_1 or P_2 does not divide S_2.

This dual characteristic number system also justifies some classical logical relationships. For instance, the universal affirmative proposition "S is P" implies its particular affirmative form (7.1).[22] Also the standard conversions in classical logic continue hold: the proposition "No S is P" is equivalent to "No P is S," and "Some S is P is equivalent to "Some P is S." But perhaps of greatest significance is that the scheme can be used to verify certain classical syllogisms. As an example consider *Darapii.*[23] It has the premises:

$$S \text{ is } P$$
$$\text{Some } S \text{ is } R.$$

We can conclude that "Some R is P" Making the usual conventions, this follows because $P_i|S_i$ and the fact that $(S_1, R_2) = (S_2, R_1) = 1$ implies that $(R_1, P_2) = (R_2, P_1) = 1$.

As impressive as Leibniz's system is, there are still serious problems. Although the logical analysis is correct, to use it satisfactorily one would have to assign characteristic numbers to all concepts in such a way that *all* true propositions satisfy the divisibility criterion and that there is no inconsistency in the numbering system. This would be an enormous job and if accomplished would seem to render Leibniz's method for discovering true arguments pointless, since one would have to know in advance all true propositions and their mutual relationships in order to assign the numbers correctly. What can happen if the assignment is not done properly is illustrated by Leibniz himself.[24] The syllogism:

every pious man is happy
some pious man is not wealthy
therefore some wealthy man is not happy

is invalid. Yet if "pious man" is represented by $[10, -3]$, "happy" by $[5, -1]$, and "wealthy" by $[8, -11]$, the conclusion is true since 5 does not divide 8.

Leibniz never entirely abandons characteristic numbers and the corresponding notion of an alphabet of human thought; but, evidently sensitive

[22]Since $(S_1, S_2) = (P_1, P_2) = 1$ and $P_1|S_1$, $P_2|S_2$ we have that $(S_1, P_2) = (S_2, P_1) = 1$.
[23]A brief survey of the types of syllogisms in classical logic is given in Appendix C.
[24]Couturat (1961), p. 246 and Leibniz (1966), p. xxii.

to the resulting problems, he searches for alternative formalisms, not depending on the divisibility of numbers. His goal, as we have already pointed out, is to find a way to reduce logic to a computational, mechanical system. He never abandoned this project and constantly experimented with schemes to achieve it. In lieu of characteristic numbers and the use of divisibility to judge the truth of a proposition, his general strategy will be to symbolize certain logical relations and treat them as axioms. Then he will manipulate the notation to deduce consequences. We can only describe a few of his most important efforts here. Many are quite similar to each other and differ only in notation.[25] One of the more interesting schemes is contained in the essay *Inquisitiones de analysi notionum et veritatum.*[26] As before, the proposition "A is B" where A is the subject and B the predicate will be true if the concept of B is contained in that of A.[27] If both the propositions "A is B" and "B is A" are true Leibniz writes $A = B$. Next logical multiplication is defined by saying that the propositions:

$$\begin{aligned} &A \,\text{is}\, B \\ &A \,\text{is}\, C \end{aligned} \tag{7.4}$$

are equivalent to the single proposition "A is BC." In other words, the union of the concepts of B and C is contained in the concept of A. Next the negation of A which Leibniz writes as "Non A" is defined formally as satisfying the identity:

$$\text{Non-non}\, A = A.$$

Using this symbolism Leibniz states three logical principles which he regards as axioms:

$$\begin{aligned} &A \,\text{is}\, A \\ &A = AA \\ &AB \,\text{is}\, A. \end{aligned} \tag{7.5}$$

He points out that (7.4) can always be transformed into an identity since "A is B" is equivalent to $A = BY$ where Y is a property modifying B so that the result is identical to A. To give a simple example, if $A =$ "man" and $B =$ "animal," then "A is B" is true; and if $Y =$ "rational," the identity $A = BY$ follows. In many cases Y will be undetermined, but Leibniz claims

[25] See Couturat (1961) and Leibniz (1961), (1966), and (1969) for additional details and examples.

[26] Leibniz (1961), pp. 356–399 and (1966), pp. 47–87. The paper is analyzed in Couturat (1961), pp. 354–62.

[27] Sometimes he substitutes "contains" for "is", writing "A contains B."

that it always exists, since we can make the trivial choice $Q = A$ and write the identity $A = AB$. Although this identity is fairly obvious if A contains B he gives a formal proof using the axioms. First $B = BB$ by Axiom (7.5) which Leibniz calls "The principle of tautology." Now also $A = YB$ for some undetermined Y. Then by the tautology principle and two substitutions we have $A = YBB = AB$.

Leibniz now uses this symbolism to deduce some of the syllogisms of Aristotelian logic. For example, suppose the premises:

$$\text{all } A \text{ are } B$$
$$\text{all } B \text{ are } C$$

are symbolized as

$$A \text{ is } B$$
$$B \text{ is } C.$$

Then they are by what has just been shown equivalent to the identities

$$A = AB \tag{7.6}$$
$$B = BC.$$

Now substitute BC for B in (7.6), obtaining

$$A = ABC.$$

By (7.6) it follows that $A = AC$ which is equivalent to A is C or "All A are C." This syllogism in the scholastic terminology of classical logic is called "*Barbara*," and is the first mode of the first figure. Now consider the second mode of the first figure or "*Celarent*":

$$\text{no } B \text{ is } C \tag{7.7}$$
$$\text{all } A \text{ are } B \tag{7.8}$$
$$\text{therefore no } A \text{ are } C.$$

The method to prove this is exactly the same as that for *Barbara* if we write (7.7) and (7.8) as "All B are not C" and "All A are not C." Similar reasoning allows Leibniz to derive the syllogism "*Darii*":

$$\text{all } B \text{ are } C$$
$$\text{some } A \text{ are } B$$
$$\text{therefore some } A \text{ are } C.$$

Using "Q" to particularize A, the second premise and conclusion can be written as "QA is B" and "QA is C." The proof of *Barbara* may then be applied. The same method works for the syllogism *Ferio*:

no B is C

some A are B

therefore some A are not C

since it can be transformed into the mood *Celarent*.

Leibniz goes on to discuss several methods to symbolize particular affirmative propositions such as "Some A are B." First, he claims that they can always be reduced to logical identities since we can choose concepts X and Y to particularize A and B so that $AX = BY$. Another way to represent particular affirmations is to assert that something to which both A and B applies exists, or in Leibniz's symbolism "AB *est*." A third method is to regard them as denials of universal negative proposition. If, for example, we have the proposition, "Some men are wise," we can write it as the denial of "No men are wise." To do this, he introduces the sign $\neq$. Thus if $A =$ "men" and $B =$ "wise," the denial can be written "$A \neq Y \text{non} - B$." That is to say, no modification of "ignorance" is contained in the concept of "Man." The proposition "Some men are ignorant," on the other hand, would be the negation of the universal affirmative proposition "$A = BY$" which is "$A \neq BY$," so "wisdom" is also not a property included in the concept of "Man."

Still other ways to write propositions as identities or nonidentities are shown in two fragments of 1690 *Primaria Calculi Logici fundamenta* and *Fundamenta Calculi Logici*.[28] The universal affirmative "A *est* B" becomes "A non-$B = 0$" where 0 denotes the absence of predicate. Similarly, the universal negative "No A is B" is written "$AB = 0$." Thus if A is the term "Man" and B "Rational Animal," the universal affirmative and negative are equivalent to saying respectively: (1) no being which is both a man and something other than a rational animal or (2) no being which is both a man and rational animal can exist as a subject of a true proposition. With these conventions the particular affirmative or negative propositions "Some men are rational" or "Some men are irrational" can be represented as "$AB \neq 0$" or "$A \text{non} - B \neq 0$."

In a fragment of 1690[29] Leibniz introduces a new logical operation, which he denotes by "+". Here an extensional interpretation of propositions

[28] Leibniz (1966), pp.90–94, and see the discussion in Couturat (1961), pp. 358–361.
[29] Couturat (1961), p. 362f.

is given. Propositions are now statements about classes. To say that "All men are animals" is equivalent to asserting that the class of men is included in the class of animals. If H represents "Men" and A "Animals," this in turn can be symbolized by "$A = H + X$" where X is some other undetermined class and "+" now represents the union of classes. And just as in the intensional case where we can write "All A is B first as "$A = BY$" and then as "$A = AB$," "$A = H + X$" becomes "$A = A + H$." As before we can express particular affirmative or negative propositions by the use of $\neq$ and "non," so that, for example, "Some A is not B" can be expressed as "$B \neq A + B$" or "No A is B" as "non-B = A + non B." However, Leibniz soon reverts to treating propositions intensionally and in various later manuscripts and fragments just uses $A + B$ or $A \oplus B$ as a substitute for AB. In several fragments written after 1690[30] he introduces the notion of logical inequality and subtraction. He writes $B < A$ if the concept of A is contained in the concept of B or alternatively $B + X = A$ where X may be an undetermined concept. He goes on to prove such theorems as $B < A$ holds if, and only if, $A + B = A$; and if $A < C$, $B < C$, then $A + B < C$. In this context two terms A and B are said to be "communicating" if they share some common property; otherwise they are said to be "diverse" or "non-communicating." With these definitions he is able to show that if A, B and C, D are diverse, then the premises $A + B = C + D$ and $A = C$ imply that $B = D$. Next the equation $A - B = C$ is said to hold if $B < A$ and C is that part of the concept of A which is not in B (that is, B and C are non-communicating); this definition is equivalent to saying $A = B + C$ where C and B share no concept in common. "0" is also introduced as a term having no concept so that $A + 0 = A$ and $A - A = 0$.[31]

One undated manuscript entitled *De formae logicae comprobatione per linearum ductus* which is rather different from the rest and especially deserves mention.[32] This again is one of a small number of manuscripts where Leibniz adopts an extensional interpretation of propositions. However, in this manuscript he abandons a notational approach to test the validity of syllogistic reasoning and substitutes a diagrammatic method. Decades before Leonhard Euler's invention of "Euler diagrams" to illustrate class relations and syllogistic logic, Leibniz does exactly the same thing. As we have seen, Leibniz is interested in easy and systematic demonstrations of

[30] *Ibid.*, pp. 367–385. Also see Leibniz (1966), pp. 122–130.

[31] Note that the requirement that B and C have nothing in common is necessary; for otherwise we could write $A = A + A$ which implies $A = A - A = 0$.

[32] Leibniz (1961), pp. 292–320.

all possible syllogisms and introduces two kinds of diagrams to do this. The first uses circles to illustrate the various possible relations between classes and the second line segments. In Figure 7.1, for example, Leibniz uses both schemes to justify the syllogism in the Third Figure called "*Disamis.*" This is the argument:

some C is B

all C is D

therefore Some D is B.

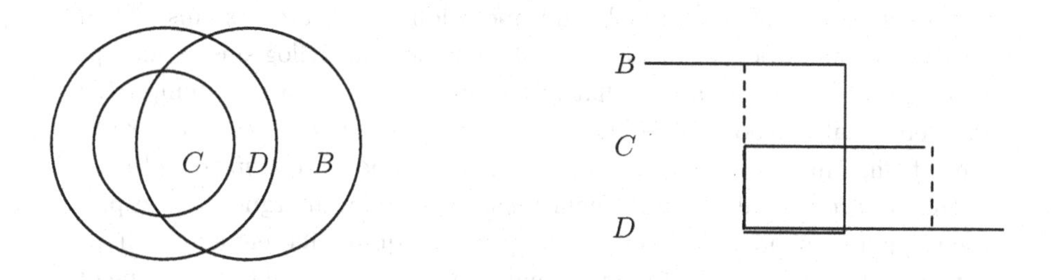

Figure-7.1

To give another example, Figure 7.2 shows Leibniz's two diagrams for the syllogism *Fresisom* in the Fourth Figure which is

no B is C

some C is D

therefore some D is not B.

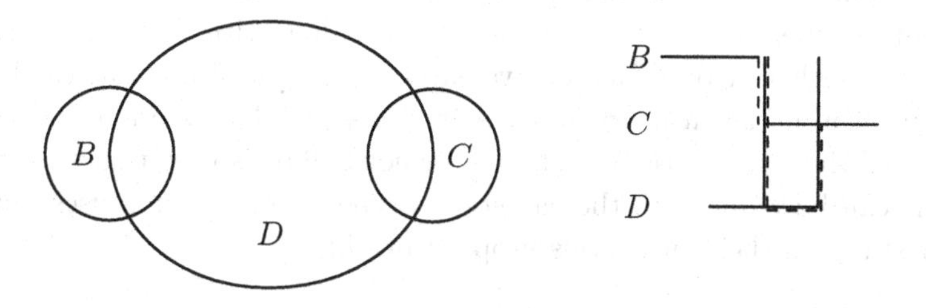

Figure-7.2

Leibniz draws similar diagrams for twenty four syllogisms in the four Figures.

There are many other logical fragments and notes by Leibniz which we have not discussed. Almost all of them give various notations to represent universal or particular affirmative or negative categorical propositions and still more ways to symbolize and test the validity of syllogisms.

Clearly, Leibniz is the most original logician between Aristotle and the nineteenth century. In fact, in some ways he at least as advanced as much later logicians such as George Boole. Boole created a new logical algebra, in many ways similar to the ordinary kind, which allows him to derive conclusions from premises using traditional algebraic techniques. Leibniz, on the other hand, defines symbols and operations and states axioms. Then he gives formal proofs of various logical identities and syllogisms by manipulating the notation in a way that might almost belong in a contemporary mathematical textbook. We know of no other seventeenth century examples of this kind of abstract reasoning where symbols are *defined* and used in this way to formally prove conclusions. One can even argue that despite their apparent similarities, Leibniz's approach is quite different and perhaps closer to the abstraction of modern mathematics than Boole's technique of manipulating algebra to compute results. But Leibniz was also severely limited by a certain traditionalism, to the point of even ignoring advances made by the scholastics. In the first place, as we have seen, for both Aristotle and Leibniz a term such as A is characterized by its properties rather than the class to which it belongs, so that to say that "Man is an animal" is for Leibniz equivalent to stating that the concept of "animal" is part of the definition of "man" and not that man is a subclass of the *class* of animals. This habit is actually contrary to a main current of scholasticism which took the extensional or class point of view. As a consequence, his formalism for the most part applies only to "logical multiplication." AB, for example, denotes a term made up of concepts represented by A and B. In contrast there are only a few fragments dealing with "logical addition" meaning "either A or B" or equivalently the *union* of the classes A and B, which on occasion he symbolizes by "$A + B$." Furthermore, he never combined these two notions. Whether or not Leibniz's view that all propositions can be reduced to the categorical type is correct, it causes him to never study on their own terms propositions like:

> Wealth consists of things transferable, limited in supply, and either productive of pleasure or preventive of pain,

or

> Either Fabius was born at the rising of the dogstar, and will not perish in the sea; or he was not born at the rising of the dogstar and will perish in the sea; or he was not born at the rising of the dogstar, and will not perish in the sea,

both of which may be found in Boole's *Laws of Thought.*[33] Leibniz's logic thus is not propositional in the modern sense. He is mostly only interested in traditional syllogistic reasoning. As we have pointed out, he wants only to easily derive and show the validity of the traditional figures and moods. In order to do this he concentrates on various notations to symbolically represent universal, particular affirmative, or negative propositions.[34] His propositions are basically about collections of concepts. Of course, he knows of situations when the subject and predicate are combinations of other propositions, but he has no real interest in them except occasionally to transform them into categorical form. In summary, Leibniz deals only with what Boole called "primary propositions" which are about things rather than "secondary propositions" which are about other propositions. In this setting valid reasoning is necessarily of syllogistic form. Since Leibniz had probably read nearly all the existing logical literature, he must have been familiar with the embryonic propositional logic due to the Stoics as well as the more developed form created by the scholastics; but except for a few remarks about hypothetical syllogisms, he never applies his techniques in this direction. In this respect, he is less advanced than certain medieval and later logicians. In William of Ockham's *Summa Logicae* (c. 1323) and other writings, for example, one can find verbal equivalents to and analysis of many standard propositional tautologies and inferences including statements equivalent to De Morgan's laws which assert that

$$\neg(A \vee B) \Leftrightarrow \neg A \wedge \neg B$$
$$\neg A \vee \neg B \Leftrightarrow \neg(A \wedge B),$$

as well as many other standard propositional inferences.[35] Propositional arguments involving relations and conditional or hypothetical arguments

[33]Boole (1951), pp. 59, 138–140, and 180.

[34]In fairness, he is rather successful in accomplishing this and can be favorably compared with Boole, whose analysis of the syllogism is much more complicated and opaque than Leibniz's. See, e.g., Boole (1951), Chapter XV, especially pp. 233–235.

[35]"Opposita contradictoria copulativae est una disiunctiva composita ex contradictoris partium copulativae," quoted in Boehner (1952), p .67. See also pp .54–75. Kneale (1964), p. 291 and p. 294f gives an extensive list of such tautologies and inferences found in the medieval literature.

not reducible to standard syllogistic form can also be found in the philosopher and mathematician Joachim Jungius's (1587–1657) textbook *Logicae Hamburgenis* and other standard texts. Leibniz was certainly aware these aspects of logic for he had studied the scholastics at an early age and had read and admired Jungius.[36]

What were the reasons for these self-imposed limitation? Couturat suggests that they resulted from Leibniz' excessive respect for the Aristotelian tradition.[37] But since there is no evidence that Leibniz ever blindly accepted the authority of any philosopher, his metaphysical requirements were probably the most significant factor.[38] In the first place, recall that universal affirmative propositions interpreted extensionally have existential implications. In traditional logic "some" implies existence; so the only way to get a particular affirmative proposition from a universal one is to assume the existence of the classes implied by the universal proposition. This means that we cannot deduce propositions like "Some horse-like animals with one horn in their forehead are unicorns" from "All unicorns are horse-like animals with one horn in their forehead." Since for metaphysical reasons Leibniz wanted to *avoid* existential import,[39] he deliberately chose the intensional interpretation, which in turn naturally led to a view making truth a matter of the relationship between concepts, not between propositions or between a proposition and the actual world.[40] Secondly, as Russell has maintained, Leibniz's belief that all propositions are or can be transformed into categorical form also is closely related to his philosophy. In several places Leibniz himself admits that key features such as the nature of substance follow from his logic,[41] but the implication goes in the reverse direction as well. Some of this key metaphysical ideas seem to imply his conception of propositions and what it means for them to be true. To take one example, as a Nominalist, Leibniz rejected the intrinsic reality

[36]De Risi (2007), p. 35; see also the documents in Jungius (1971). Russell (1964), p. vi states that Leibniz explicitly refers to De Morgan's Laws.

[37]Couturat (1961), p. 386.

[38]We agree here with Parkinson (1965). See especially pp. 17–20.

[39]Sometimes, however, he writes "A is B" as "AB is," or AB is an entity." But it is not clear if the last two forms assert actual existence, or an abstract possibility of existence in that AB is not a contradiction. See Mates (1986), pp. 54–57 for a detailed discussion of these issues. Kneale (1964), p. 323 claims that Leibniz accepted the existential import of universal statements. The balance of evidence, however, seems otherwise.

[40]To use modern terminology Leibniz had a coherence rather than a correspondence theory of truth.

[41]Parkinson (1965), p. 123. See also several of the passages quoted in Russell (1964), pp. 205–221, as well as his letters to Arnauld and his *Discourse on Metaphysics*. Russell was the first to argue that Leibniz's logic determined his metaphysics.

of relations and their role in argument; relations, he says, are "a merely mental thing."[42] The only existing things are individual substances and their properties.[43] This means that Leibniz never incorporates relational propositions into his various attempts at logical symbolism. He only wants to somehow force them into his simple categorical form. Thus, propositions such as "Peter is similar to Paul..." or "Titus is bigger than Caius" are reduced to "Peter is A now and Paul is A now" for some property A or "Titus is big in as much as Caius is small."[44] Unfortunately, Leibniz's interpretation of relations also causes him—with the exception of occasional spasms of interest–to omit a class of arguments like "Because a horse is an animal, the head of a horse is the head of an animal,"[45] even though similar arguments are analyzed in the *Logica Hamburgenis*. More fundamentally, Leibniz view of the Universe as an aggregate of unextended substances or monads which are characterized by their internal states and completely isolated from each other[46] suggests that the only propositions worth analyzing are categorical ones, i.e., statements about the "states" or "predicates" of a monad. Clearly, Leibniz's view of truth as an internal relation between subject and predicate is a natural consequence of this position. It is possible, therefore, that Russell's thesis that Leibniz's metaphysics is a consequence of his logic needs to be reversed. Leibniz' metaphysics may be motivated by factors which are prior to his logic. We agree with Garber that it may result from a desire to somehow preserve the Aristotelian notion of substantial form as a causal agent or principle of force and activity which to Leibniz was present in all substances.[47] He then worked out the logical consequences of such a view and only afterwards argued in the reverse direction as a way of grounding his metaphysics on what he conceived to be the more solid ground of logic. One can speculate that had Leibniz developed a different metaphysics, his logic would have also been different and,

[42]Quoted in Mates (1986), p. 210.
[43]*Ibid.*, p. 209. This is true at all stages of Leibniz's philosophical development; by the 1690s the substances become monads.
[44]*Ibid.*, p. 180 and Kneale (1964), p. 324. Leibniz's sometimes murky views on relations has been the subject of a rather substantial literature (and much dispute) among commentators. The dominant opinion is that his analysis of relations was not successful.
[45]The example is due to de Morgan and is quoted in Parkinson (1965), p. 19.
[46]That this appears to be otherwise is due to God's pre-established harmony. The sequence states $S_1, S_2 \ldots$ of monad M_1 and $T_1, T_2 \ldots$ of monad M_2 are designed to be consistent with each other, in such a way that in the case of humans at least the illusions of the usual phenomenal world, causal effects, etc., are preserved. But in reality there is no causal contact between M_1 and M_2.
[47]Garber (2009), Chapters 3 and 4.

given his penchant for symbolism, would have allowed him to anticipate or perhaps even surpass the logicians of the nineteenth century. But the investigation of such questions is really the subject of a different work.

Chapter 8

The Universal Characteristic

Even in his mature period one has the impression that Leibniz's knowledge of Scholastic and Renaissance philosophy was deeper than his detailed knowledge of either ancient or contemporary mathematics. This was undoubtedly true up to his second visit to London in 1676 and perhaps well beyond, and is certainly understandable. He had read almost every major work in philosophy beginning well before his teens and had also a rigorous academic training in it; by age thirteen or so he could read, he says, the scholastic philosophers with "no ordinary delight."[1] On the other hand, he was twenty six before he began in 1672 to study mathematics seriously under the tutelage of Christiaan Huygens. During the Paris years he rapidly lost his earlier near total ignorance of mathematics. By 1674 or 1675 he certainly had, as he testified, mastered the geometry of Descartes and had additionally read under Huygens direction Pascal's *Lettre de A. Dettonville* (1659), Gregory St. Vincent's *Opus geometricum* (1647), James Gregory's *Geometriae pars universalis* (1668), and Wallis' *Arithmetica Infinitorum* (1656). Leibniz also speaks of an acquaintance with Cavalieri work and a "close study of Slusius."[2] All this was enough to allow him to reach the research frontier in infinitesimal analysis. But to repeat some earlier judgements, it is doubtful that he ever had the kind of deep mastery of the classical geometric tradition characteristic of Huygens, Barrow or Newton, all of whom could clinch an argument by a citation from one of the ancient masters. Leibniz was certainly very gifted mathematically and his early discoveries are impressive. But considered only in isolation they are no more impressive than results found by others. We have already pointed out, for

[1]Mackie (1845), p. 24.

[2]See Leibniz's letter to James Bernoulli of 1703 in Leibniz (1920), pp. 11–20 and Antognazza (2009), p. 143.

example, that his arithmetical quadrature of a circle or conic sections was just one of a family of infinite series/infinite product representations found earlier by mathematicians of the calibre of Gregory, Wallis, or Newton. His quadrature of the cycloid was similar to what had been done by Roberval or Huygens. And we certainly can find in Barrow or Huygens results at least as ingenious as his various transmutation theorems, some of which he may, in fact, have "borrowed" from Barrow.[3] And in respect to problem solving or the ability to handle or synthesize involved mathematical arguments, he was probably equal to the more gifted of his contemporaries, but ranked no higher. Mathematically, he is perhaps broader than Newton since he made discoveries in several unrelated areas (some of which we discuss below), but in pure geometric power and mathematical depth he is distinctly below Newton who in the seventeenth century was in a class by himself.

We can also argue, as most historian have, that much of the specific content of Leibniz's calculus, specifically the recognition that:

- tangent and quadrature problems are (in some sense) inverses of each other
- the characteristic triangle can be used to find tangents
- a curve might be considered a polygon with infinitely many infinitesimally short linear sides
- area is the sum of infinitely many infinitesimally thin rectangles

had been anticipated by several mid-seventeenth century mathematicians, or even in some cases by the Greeks. Yet no other mathematician, not even Newton, had constructed an algorithmic approach to infinitesimal analysis quite like Leibniz's. It is not simply the fact that Leibniz's notation dx, dy, $\int y\,dx$ is convenient. Its real advantage is that by manipulating the notation by itself according to simple rules, without much attention to the underlying geometric structure, one is led almost without conscious thought to new approaches and solutions to what had been considered difficult problems. In several cases in the previous two chapters we have seen how Leibniz could give in a few lines solutions to problems that, say, Barrow or Huygens could certainly have also solved, but in several pages of complex argument.

There is a paradox at the heart of the mathematical revolution which

[3]In the priority dispute with Newton he was accused of using the same transmutation techniques as Barrow. While there is some evidence for this accusation, we find it ambiguous and capable of less damaging interpretations. The priority dispute and the originality of Leibniz's transmutation theorems will be discussed in Chapter 10.

Leibniz began. To use an Aristotelian metaphor, while the "matter" of his achievement derives from contemporary work on tangent and quadrature problems, its motivation, spirit, and uniquely symbolic "form" own much to nonmathematical projects and ideas which he had absorbed prior 1672. Their pursuit became a life-long obsession and were just as (or possibly even more) important to him than any concrete metaphysical or mathematical accomplishment for which he is remembered today. The most significant were a cluster of ideas operating on several levels which Leibniz variously called the "Universal Characteristic," or "*Ars Inveniendi.*" Leibniz believed it possible to invent a special language or *Lingua Philosophica*which would serve both as an international language and as a tool for the discovery and verification of knowledge. Additionally the structure of this language would make reasoning a mechanical process. Indeed its very form would either render logical mistakes impossible, or at worst they would be like errors in arithmetic. In this way it would increase our mental powers just as a telescope increases our visual power.[4] For example, he writes:

> Those who write in this language will not make mistakes provided they avoid errors of calculation, barbarisms, solecisms, and other errors in grammar. In addition this language will possess the wonderful property of silencing ignoramuses. For people will be unable to speak or write about anything except what they understand, or if they try to do so... either the emptiness of what they put forward will be apparent to everybody or they will learn by speaking and writing.[5]

In another famous passage Leibniz claims that it would be an infallible tool to settle sectarian controversies:

> ... there will be no more need of a disputation between two philosophers than between two accountants. It will in fact suffice to take pen in hand, to sit at the abacus and—having summoned, if one wishes, a friend—to say to one another '*calculemus.*'[6]

There very many similar passages in the multitude of Leibnitzian manuscripts from the 1660s until the end of his life.

How should this marvelous *Lingua Philosophica* be constructed? An early effort was associated with Leibniz's logical scheme of "factoring"

[4]Eco (1995), p. 281.
[5]Mates, (1986) p. 186.
[6]*De arte characterista ad perfiendas scientias ratione nitentes* (1688) quoted in Antognazza (2009), p. 244.

complex concepts into simple ones represented by prime numbers which we have analyzed in the previous chapter. Recall that the central attribute of this system was that the truth of the proposition "A is B" then could be checked by seeing whether or not the number representing B divided that representing A. In a fragment *Lingua generalis* written in early 1678 Leibniz imagined that it could form the core of a universal language basically consisting of numbers. To speak it, Leibniz devised a method of representing the numbers by letters in a euphonious way. The numbers 1 through 9 would be represented by the first nine consonants *b, c, d, f, g, h, l, m, n*. To make this method produce pronounceable words he used the vowels *a, e, i, o, u* to indicate powers of 10. For example, 81374 could be written as *Mubodilefa*. Even better, the same number could be written in many ways: 81374 would also have the representations *Bodifalemu*, i.e., $1000+300+4+70+80,000$. This phonetic flexibility Leibniz believed would aid in the composition of poetry and beautiful songs in music.[7]

As we have seen, Leibniz soon realized that this idea of using products of primes to represent concepts and judging the truth or falsehood of a proposition according to whether or not the predicate is a "factor" of the subject suffered from fatal defects, and so he replaced it by symbolic approaches. Leibniz was always fascinated by ideographic languages such as Chinese and Egyptian hieroglyphics which represented objects and words by symbols or pictures instead of letters. But he felt that the symbols in these languages were inadequate because they were not "Real Characters;" that is, they did not directly represent concepts and were not combined in a rational way.[8] It would be better, he felt, to first isolate a set of primitive concepts—or what Leibniz called an "alphabet of human thought" (an idea which, as we have seen, first appears in the *De Arte Combinatoria*) whose combinations would generate all possible concepts. Symbols might then be assigned to the primitive concepts and combined in such a way that the structure of all complex concepts would be clear. Then valid reasoning would be transparent and mistakes could be easily found. Or in Leibniz's words:

> Profoundly meditating on this state of affairs, it immediately appeared as clear to me that all human thoughts might be entirely resolvable into a small number of thoughts considered as primitive. If then we assign to each primitive a character, it is possible to form other characters for deriving notions, and we

[7] Couturat (1961), p. 62f.
[8] Rossi (1989), p. 274.

> would be able to extract infallibly from their prerequisites and the primitive notions composing them; to put it in a word we could always infer their definitions and values, and thereby the modifications to be derived from their definitions. Once this and been done, whoever uses such characters in their reasoning and in their writing would either never make error, or, at least would have the possibility of recognizing his own (or other people's) mistakes using the simplest of tests.[9]

Leibniz frequently hints rather expansively, but perhaps dishonestly, that he is either in possession of or on the verge of completing this wonderful project. We have already seen a sample of this rhetoric in an August 17, 1676 letter to Oldenburgh, quoted in Chapter 5. And in another passage he writes:

> My invention includes the whole use of reason, a judge for controversies, an interpreter of notions, a balance of probabilities, a compass which will pilot us through the ocean of experience, an inventory of things, a general calculus, a guiltless kind of magic, a kind of writing that everyone will read in his own language, finally a language which will circulate worldwide, everywhere supporting true religion.[10]

We could exhibit similar examples almost *ad infinitum*. But although he wrote what must be hundreds of thousands of words in his unpublished manuscripts describing its uses and benefits, the actual construction of such a Characteristic remained an unfulfilled dream throughout his life. He realized that the task would be vast and could only be completed after the completion of what he called the *Encyclopedia*. The Encyclopedia would exhibit all human knowledge: scientific, historical, and philosophic in a logically organized manner. Examination of any subject included in it would show its connections to any other subjects. Its logical structure, implications, and the assumptions which entailed it would be made clear. All the basic and primitive concepts which it contained would be defined. The aggregate of these concepts would serve as the generators of all human thought. This structure of the Encyclopedia would make it more than a classification of existing knowledge; it would have demonstrative value and be an aid to new discoveries. And once the Encyclopedia was finished it would determine the structure of the Characteristic which would be its symbolic isomorphic image. But to do all this was beyond the powers of any

[9] *De scientia universalis seu calculo philosophico*, quoted in Eco (1995), p. 280f.
[10] Rossi (1989), p. 288f.

one individual. Leibniz imagined that large teams of scholars, working cooperatively under the patronage of some German ruler, would be necessary for it to be constructed.

But however unattainable the dream of the *Lingua Philosophica* was in practice because of the absence of the underlying Encyclopedia, Leibniz believed that special cases of it were already in existence or could be found prior to the complete Characteristic. Algebra, the science of undetermined numbers, was already one such example. Like the Characteristic it was composed of simple terms denoted by letters. The letters were the analogues of the alphabet of human thoughts. Complex terms were the formulas composed of the letters according to fixed rules. These formulas then could be manipulated using signs for operations such as addition, multiplication, etc., and their possible transformations would lead to new truths in an automatic fashion. Algebra, therefore, can be regarded as a logic of number, just as logic is an algebra of concepts. But existing algebra was too restricted being limited to numbers, and Leibniz thought that even within mathematics there was a yet undiscovered generalization of it which he called *Logistic* or a Universal Mathematics. Algebra, mechanics, geometry, and other mixed or applied sciences were branches of this Universal Mathematics which would be a general science of magnitude or quantity. It would resemble algebra in being able to determine unknown from known quantities. Sometimes also, Leibniz spoke of it as the *Ars Combinatoria* or "Combinatorial Art" which would be a general science of mathematical forms or structures. Like its generalization, the full Characteristic, it would be an aid to discovery and invention. New theorems could be discovered and proven by incorporating known theorems into equations which could be added, multiplied, or otherwise manipulated. Of course, this Universal Mathematics, like the general *Lingua Philosophica* or Characteristic, was never (and probably could not be) completed by Leibniz. But besides writing enormous amounts of propaganda for it, he was motivated to make concrete progress on partial realizations of his dream. It is this motivation, more than his technical talent, that lies behind his mathematical originality. Calculus was certainly the most complete and successful example of this effort. Leibniz's approach to logic, although his achievements were unrecognized for more than two centuries, was a second. We have already commented at length on his many attempts starting with the *Dissertatio de Arte Combinatoria* both to symbolize syllogistic logic and to turn its employment into a quasi-mechanical process. However, he also constructed at least two other symbolic schemes or "calculi" in areas other than calculus and logic which are now mostly forgotten. In both, as in the calculus, the reasoning is done by manipulating a suggestive notation in a formal manner. The first is his *Analysis Situs* communicated in a letter to Huygens in 1679 which attempts a symbolic study of incidence, congruence, and similarity properties. He writes:

> I have discovered certain elements of a new characteristic which is entirely different from algebra and which will have great advantage in representing to the mind, exactly... even without figures, everything that depends on sense perception. Algebra is the characteristic for undetermined numbers... but does not express situation, angles, and motion directly. Hence it is often difficult to analyze the properties of a figure by calculation... But this new characteristic cannot fail to give the solution...[11]

In other words, Leibniz wants to develop a system that analyzes geometric figures and loci problems directly without the need to reduce geometric problems to algebraic ones using the methods of Viète or Descartes. He begins by centering on the notion of congruence. Two triangles ABC and DEF he says are congruent if they can be superimposed on each other. He denotes such a congruence by $ABC \breve{\circ} DEF$. Two arbitrary points are also congruent to each other and this fact can be used to describe loci. Let the initial letters of the alphabet such as A, B, C, etc. denote given points and letters at the end such as X, Y, Z denote unknown points. Then $A \breve{\circ} Y$ signifies all of space, i.e., the set of all points Y congruent to A. In a similar fashion, $AC \breve{\circ} AY$ describes the surface of a sphere, $AY \breve{\circ} BY$ a plane perpendicular and passing through the midpoint of the segment AB, $AY \breve{\circ} BY \breve{\circ} CY$ a straight line perpendicular to the intersections of the perpendicular bisectors of the sides AB, BC and CA of the triangle ABC, and $ABC \breve{\circ} ABY$ a circle perpendicular to the line segment AB with radius CD and center D where D is the midpoint of AB, and CD is the perpendicular from C to AB. From these definitions Leibniz is able to prove some simple geometric facts in a symbolic mechanical way. As one example he shows that the intersection of a plane and a sphere is a circle. To see this, recall that the "equation" of a sphere is $AC \breve{\circ} AY$ and that of a plane is $AY \breve{\circ} BY$ which implies setting $Y = C$ that $AC \breve{\circ} BC$. Since the congruence relation is transitive and we have that $AC \breve{\circ} AY$, it follows that $BC \breve{\circ} AY$. Next, because $AY \breve{\circ} BY$, we can conclude that $BC \breve{\circ} BY$. Joining this congruence with the "identity congruence" $AB \breve{\circ} AB$ gives $ABC \breve{\circ} ABY$ which, as we have

[11] Leibniz (1969), p. 384. Also see Couturat (1961), p. 394 for related quotations.

seen, is the equation of a circle.[12] The reader will note that this reasoning via symbolic manipulation strongly resembles that of some of the logical schemes described in the last chapter. Leibniz goes on to consider similarity. Similar triangles, he says, are triangles that cannot be distinguished when looked at singly. He then proves that triangles are similar if, and only if, they are equiangular, which in turn is equivalent to the corresponding sides of similar triangles having equal ratios. His arguments to establish these equivalences are designed to bypass the complicated reasoning in Book VI of the *Elements*, and they have a more "metaphysical" flavor than in his consideration of congruence. For instance, if $ABC \sim LMN$ and:

> ... if some two sides were given, such as AB and BC which have a ratio to each other which no two sides of LMN have, then one triangle could be distinguished from the other, even viewed singly.[13]

Unfortunately this particular "characteristic" did not progress much further than a treatment of congruence and similarity and did not impress Huygens. For as he points out in a reply of November 22, 1679, Leibniz's examples were well known and his methods could only solve trivial problems.[14] Huygens was correct. The examples Leibniz used to illustrate his characteristic consisted of only the most elementary facts of geometry. However, Huygens missed the point and did not appreciate the "modern" formalistic quality of Leibniz's reasoning. Some historians[15] have viewed the *Analysis Situs* as a forerunner of the quasi-vectorial systems of Hermann Grassmann (1809–1877) (who, in fact, gave some acknowledgement to Leibniz) and August Möbius (1790–1868), but aside from introducing the idea of symbolic representation of geometric entities Leibniz's work was quite different from theirs and contained no true vectorial concepts.[16]

A second "calculus" is embodied in his efforts to "mechanize" the finding of solutions to n linear equations in n unknowns. We have already seen in his thesis of 1666 *Dissertatio de Arte Combinatoria* simple applications of combinatorial reasoning. Leibniz was fascinated by combinatorics which he felt would be a critical ingredient of his grand project of a Characteristic. This feeling was so profound that he often spoke of the "Combinatorial Art" as a synonym for it. As with other aspects of the Characteristic the

[12] *Ibid.*, pp. 385–390.
[13] *Ibid.*, p. 394f.
[14] Huygens (1899), p. 243.
[15] Couturat (1961), in particular see p. 430. Also Echeverria (1979).
[16] Crowe (1994), p. 4f.

application of combinatorics remained a dream rather than an actuality, and just what he meant by or intended to do with this Art remains unclear. But this general vagueness did not prevent him from making specific applications. In one instance he used combinatorial reasoning to arrive at the concept of the determinant of a linear system of equations. If, for example, we have three homogenous linear equations in two unknowns x, y (and using ij, $1 \leq i \leq 3$, $0 \leq j \leq 2$ to signify the coefficient in the i-th row and j-th column):

$$10 + 11x + 12y = 0$$
$$20 + 21x + 22y = 0$$
$$30 + 31x + 32y = 0,$$

Leibniz states that this system will have nontrivial solutions if, and only if,

$$10 \cdot 21 \cdot 32 - 10 \cdot 22 \cdot 31 + 11 \cdot 22 \cdot 30 - 11 \cdot 20 \cdot 32 + 12 \cdot 20 \cdot 31 - 12 \cdot 21 \cdot 30 = 0,$$

which is exactly the condition that the row vectors of the system be linearly dependent. Leibniz goes on to define the general "determinant" and in a manuscript of 1678 to state "Cramer's Rule" 72 years before Gabriel Cramer (1704–1754) published it.[17] In this way the computation of solutions, etc., for systems of linear equations would become an automatic though lengthly computation.[18]

We should also mention two other mathematical discoveries of Leibniz. One was his discovery of binary arithmetic. Using only 0 and 1 he showed how to express any number using base 2 and to do arithmetical calculations using this scheme in an unusually simple way.[19] In Leibniz's eyes the binary system probably was that part of the Characteristic which yielded a simplified "calculus" for arithmetic. He also drew religious implications out of this invention. The possibility of doing calculations using only 0 and 1 was a symbol of the Christian dogma that God created the world out of nothing.[20] In another application of combinatorial reasoning Leibniz generalized the notion of the binomial coefficients $\binom{n}{j}$ which arise from

[17]*Specimen Analyseos novae*, summarized (together with other notes) in Couturat (1961), pp. 482–494. Cramer published his rule in *Introduction à l'analyse des lignes courbes algébriques* (1750). Colin Maclaurin (1698–1746), also seems to have discovered it independently and published it in 1748 in his *Treatise on Algebra.*

[18]Couturat (1961, pp. 483–485.

[19]*Ibid.*, p. 473.

[20]Mackie (1845), p. 187. In correspondence with a Jesuit missionary to China, Leibniz sent a description of his binary arithmetic in the hope that it might aid in the conversion of the Chinese to Christianity.

computing $(x+y)^n$ to the multinomial case. He showed that

$$(a_1 + a_2 + \dots a_k)^n = \sum_{k_1 \dots k_n} C(n, k_1 \dots k_n) a_1^{k_1} \dots a_n^{k_n}$$

where the integers k_i are non-negative, $\sum_{i=1}^{n} k_i = n$, and

$$C(n, k_1 \dots k_n) = \frac{n!}{k_1! \dots k_n!}.$$

Using this identity Leibniz is able to prove a famous "little theorem" of Fermat (which Fermat had stated without proof in a letter to a friend):

$$a^{p-1} \equiv a (\operatorname{mod} p) \tag{8.1}$$

where p is a prime number and a is any integer. Moreover, if we assume that $(a,p) = 1$, then also the congruence $a^{p-1} \equiv 1 (\operatorname{mod} p)$ is true. To show these results Leibniz first notes that

$$a = 1 + \cdots + 1 = 1^p + \cdots + 1^p.$$

Hence

$$a^p = a + \sum_{k_1 \dots k_p} C(p, k_1 \dots k_p)$$

where $\sum k_i = p$, $0 \leq k_i < p$. He was aware that $C(p, k_1 \dots k_n)$ is an integer. Also trivially

$$k_1! \dots k_p! C(p, k_1 \dots k_p) = p!.$$

Because p is prime and $k_i < p$ for $i = 1, \dots p$, it follows that $p \mid C(p, k_1 \dots k_p)$, and so $p \mid a^p - a$. Also if $(a,p) = 1$, p must divide $a^{p-1} - 1$ which completes the proof of (8.1).[21]

Leibniz seems to have thought of all of his specific "calculi" as just partial implementations of the Universal Characteristic. While these implementations were original, it is important to realize that his general conceptions were much less so. Similar ideas were very much in vogue in the seventeenth century. For instance, the idea of "Real Characters" or conventionally defined symbols representing concepts rather than just objects or words is due to Francis Bacon (1561–1626) and the various artificial philosophical language projects of George Dalgarno (1626?–1687), Comenius (1592–1671), Seth Ward (1617–1689), John Wilkins (1614–1672), Athanasius Kircher (1601–1680), and Cave Beck (1623–c. 1706) show many features in common with Leibniz's Characteristic. And they are only the

[21] *Ibid.*, pp. 495–500.

most prominent among many lesser known examples. In the writings of the individuals just listed there are many passages not dissimilar from what one can find in Leibniz. To get a feeling for what these projects entailed, we look in some detail at one of the most prominent: John Wilkins, Bishop of Chester, was one of the founders of the Royal Society and became its Secretary after the death of Oldenburg in 1677. In 1668 he published the *Essay towards a Real Character and a Philosophic Language.* This was the most complete project for a philosophical language produced in the seventeenth century.[22] and Leibniz is known to have read it in 1670.[23] Wilkins in his own words attempted to construct a system of "marks or notes" having:

> such a dependence upon and relation to, one another, as might be suitable to the nature of the things which they signified and so likewise if the names of things could be so ordered as to contain such a kind of affinity or opposition in their letters and sounds ... we should, by learning the character and the names of things, be instructed into their natures.[24]

Paralleling Leibniz's idea of a universal classification of knowledge via the construction of an Encyclopedia, Wilkins began his scheme by following an Aristotelian/scholastic tradition of definition. First he assigns every concept to one of 40 genera. Then particular members of the genus would be specified by assigning them to a sequence of subclasses. Each subclass would be contained in a prior class and determined by some difference. For example a dog/wolf belongs to the "Animate" subclass of "Substance" and is contained three other classes. It is "Sensitive" rather than "Vegetative." It is further in the subclass "sanguinous." In this class it is not a fish or bird but a "beast" which is divided into two species "viviparous" and "oviparous." A dog is viviparous. But viviparous beasts may be "whole footed," "cloven footed" or "clawed." A dog is clawed. The clawed subspecies may be "rapacious" or "not rapacious." A dog is rapacious. It does not belong to the "cat kind," but the "dog-kind." It is "European," not "exotic." It is "Terrestrial," not "Amphibious" (which includes seals). Finally, a dog or wolf is "Bigger" not "Smaller" which includes Badgers and Foxes. In this manner, using 270 pages, Wilkins believed that he had managed to classify every known concept in terms of the 40 genera, 251 differences and 2,030 primitive species. Next, having designed a simplified grammar, Wilkins invents Real Characters or symbols to write his language. These

[22] Eco (1995), p. 238.
[23] Antognazza (2009), p. 94.
[24] Quoted in Rossi (1979), p. 124.

consist of a system of dashes and hooks to stand directly for concepts corresponding to his system of classification. Grammatical ideas are expressed by other symbols which according to Umberto Eco are "extremely difficult to read." In this form the language is designed to be read not spoken, and by its use Wilkins hoped to be able to express every possible human thought in a way independent of all existing natural languages. In order to speak the language each genera is assigned two letters with differences expressed by consonants and species by vowels or two diphthongs. To give a few examples of Wilkin's procedure: if *Di* signifies "stone" and *b* "vulgar." *Dib* is a "vulgar stone" and *Diba*" is the subspecies of "vulgar stone" which is "rock," and if *Ska* is "religion" *Skam* is the restriction which is "grace" while *Skaν* is the restriction "happiness." Following this method the first two words of the Lord's prayer become in Wilkin's language *Hai Coba.*

Other writers expounded different systems of Real Characters and phonetic assignments but they all basically followed an Aristotelian classification system similar to Wilkins in terms of nested genera and species, each distinguished from the class that contained it by some difference. Besides differing from Wilkins in the way the world was divided up into genera, some systems were even more elaborate than his. George Dalgarno, a Scottish school master who taught grammar at an Oxford private school, in his *Ars signorum* (1661), for instance, believed that all substances could be defined as a collection of properties which were probably infinite in number. Even restricted they would yield between 4,000 and 10,000 species while Wilkins had only 2,030. He devised an elaborate phonetic system to describe objects in terms of the properties characterizing them. For example, "mule" was written "*Nνksofpad*" where *Nνk* means animal with uncleft hoof, "*sof*" 'deprived of, and "*pad*" sex.[25]

Leibniz treated many of these projects respectfully. He copied Dalgarno's, classifications in a rough draft concerning his encyclopedia.[26] Also the method, which we have already briefly described, of assigning euphonious sounds to products of primes which represented concepts resembled the methods of Wilkins and Dalgarno to vocalize their systems.[27] In the last six months of 1671 and early 1672 he constructed an elaborate table of definitions modeled on Wilkins, whose book he discussed with Oldenburgh. In the Mainz period he had corresponded with Athanasius Kircher,

---

[25]For detailed descriptions of Dalgarno's and Wilkins' and systems, see Rossi (2006), pp. 160–167, Eco (1995), Chapters 11 and 12, and Maat (2004).
[26]Rossi (2006), p. 166.
[27]*Ibid.*, p. 183.

a German Jesuit living in Rome who thought he had decoded Egyptian hieroglyphics and had published in 1663 *Polygraphia nova et universalis ex combinatoria arte detecta.*[28] However, Leibniz also had serious reservations the rival projects. He felt that they amounted to the construction of artificial international languages and fell short of the true Characteristic which he wanted to construct. Nevertheless most of these projects including Leibniz's own shared a common set of fundamental ideas. Let us recapitulate the most important:

(i) *The Alphabet of human thoughts* We have already seen this in Leibniz's writings in logic and the Characteristic as early as 1666. It is also found in most of the other writers mentioned here. For every branch of knowledge or even to understand the entire universe it is possible to isolate a set of primitive or fundamental concepts, relationships, and principles from which all other concepts can be derived in a systematic rule bound way. The aggregate of all concepts so generated would comprise a new and very precise language without any of the vagueness and confusion of normal human languages. If this could be done Descartes wrote in a letter to Martin Mersenne in November 1629:

> I believe that such a language is possible and that it is possible to discover the science on which it must depend, a science through which peasants might judge the truth better than philosophers do today.[29]

(ii) *Real Characters* As we have mentioned, this idea was originally due to Francis Bacon who was inspired by Egyptian hieroglyphics. For him hieroglyphics did not stand for words but were emblems representing and having "something in common with the thing signified." This suggested to him the possibility of devising artificially constructed symbols or "Real Characters" which were not limited to being emblems but which would "represent not just letters and words, but things and notions." Their use would be established by convention. Once the primitive concepts or "alpha-

[28]This was a set of two dictionaries involving Latin, Italian, Spanish, French, and German by which anyone fluent in one of these languages could encode a message using Roman and Arabic numbers from the first dictionary and which could be decoded using the second dictionary. The system was very complicated and probably unusable. See Antognazza (2009), p. 94 and Eco (1995), pp. 196–200 for details about Kircher. Ultimately Kircher wanted his system to handle Hebrew, Geek, Bohemian, Polish, Lithuanian, Hungarian, Dutch, English, Irish, Nubian, Ethiopic, Egyptian, Congolese, Angolan, Chaldean, Arabic, Armenian, Persian, and several other languages including "Canadian."

[29]Eco (1995), p. 218.

bet of ideas" had been isolated, the Real Characters assigned to them could be combined to denote complex concepts according to definite rules. Since the Real Characters would reflect reality and would communicate their meaning and relation to the other Real Characters directly by their structure without recourse to words or any natural language, a scholar would be able to understand, discover, and demonstrate all possible knowledge in any area covered by the language. As a secondary benefit the resulting *Lingua Philosophica* would replace Latin as a means of scholarly communication.[30]

(iii) *The Encyclopedia.* Before constructing the new language one should completely classify everything in the world and list all knowledge in an orderly way–a guiding metaphor here was that more specialized subjects were related to the more general like the branch structure of a tree. The Encyclopedia would be a mirror of this *arbor sciential.*

(iv) *The Art of Memory.* The *Lingua Philosophia* would also be a new "Art of Memory" substituting for the original version that went back to classical times which in general depended on the association of concepts to be remembered with "images" or statues in imaginary "rooms". But knowledge of the proper Real Characters would make this unnecessary, for they would automatically call to mind what was to be remembered. John Wilkin's views are typical. Since the Real Characters resemble the things or ideas they signify:

> ... if the *Names* of things could be so ordered as to ... be in some way answerable to the things they signified; This would yet be a farther advantage superadded by which, besides the best way of helping the *Memory* by natural Method, the *Understanding* likewise would be highly improved.[31]

(v) *Combinatorics* From the late Renaissance onward many writers interested in constructing an artificial language besides Leibniz were interested in the theory of combinations, by which the number of complex ideas formed from simple ones could be computed and their structure exhibited.

All these ideas are found in several writers of the sixteenth and early seventeenth centuries in addition to Leibniz and in fact have a long history which we shall explore in the next chapter.

[30]The quotations are from Bacon's *Advancement of Learning* in Rossi (2006), pp. 109, 145.
[31]Quoted in Rossi (2006), p. 167.

What then was original in Leibniz's conception of the Characteristic or *Lingua Philosophica*? Although John Wilkins and others imagined that their languages, being free of the imprecision of existing languages and consisting of Real Characters directly signifying concepts rather than words, would increase the clarity of communication, Leibniz more than anyone else stressed that his *Lingua Philosophica* should both assist in discovery, i.e., be what he called an *Ars Inveniendi* and render reasoning a *mechanical* nearly automatic process similar to the way algebra handles numbers. Concerning Wilkins' and Dalgarno's projects he says:

> ... it seems that these excellent men have not completely understood the magnitude or the true use of the project. For their language, or writing, only accomplishes that people who speak different languages can easily communicate; but the true Real Characteristic, as I conceive it, must be reckoned among the most effective instruments of the human mind, as it will have the greatest power for invention, retention and judgement. For it will achieve in every subject matter, what arithmetical and algebraic characters do in mathematics.[32]

Most of his various "calculi," we have described, or specific results have in common what Umberto Eco has called "blind thought," or what Leibniz himself called "*cogitatio caeca;*" that is to say, one can obtain results by following the rules or by manipulating the notation without having to keep in mind (or even understand) the underlying meaning.[33] In this feature Leibniz's mathematics resembles, say, "long division" where in elementary school the student is taught the algorithm and can obtain the correct answer without understanding what he is doing.[34] This is in radical contrast to the classical geometry of Leibniz's contemporaries where one must understand in depth every step in the argument or construction. This ability to mindlessly obtain results in Leibniz's various "calculi" is not found in the artificial language projects of his contemporaries and is the main (and perhaps the only) reason that makes it possible to teach calculus to the average college student in the average US public university. For instance, one can compute $\int \log x \, dx$, by letting $u = \log x$ and $dv = dx$. Then using

[32]Quoted in Maat (2004), p. 300. This book gives an extraordinarily detained account of the language projects of Wilkin's and Dalgarno.
[33]Eco (1995), p. 279–281.
[34]The division algorithm is usually not really explained until a college course in number theory or modern algebra.

the formula

$$\int u\,dv = uv - \int v\,du, \tag{8.2}$$

we get that

$$\int \log x\,dx = x \log x - \int x \cdot \frac{dx}{x}$$

so that

$$\int \log x\,dx = x \log x - x.$$

Provided we blindly memorize the "integration by parts" formula (8.2) and a few indefinite integrals, we don't have to understand anything else. Similarly, the chain rule when written as $(f \circ g)'(x) = f'(g(x))g'x)$ is not at all obvious, but if we write

$$\frac{df}{dx} = \frac{df}{du}\frac{du}{dx}$$

where $u = g(x)$ it seems trivial! We have something behaving like the product of fractions, the denominator of one being the numerator of the other. Again, we can perform this calculation automatically without having a clue to the meaning of what we are doing. Certainly, his symbolism behaves the way the *Ars inveniendi* should (except that students still find creative ways to make mistakes in it); it is equally tempting to conclude that calculus was the only part of the Characteristic he perfected—at least to the satisfaction of others—and that such symbols as dy/dx or $\int_a^b f(x)\,dx$ are examples of "Real Characters." Leibniz indicates as much in a manuscript entitled *De la beauté* des théorèmes where he says:

> Theorems are only intelligible by means of their symbols or characters. Images are a type of character. When the characters can be similar to things, so much the better. The beauty of theorems consists in the arrangement of their characters[35]

.

[35]Leibniz (2008), p. 439. (My translation.)

Chapter 9

The Baroque Cultural Context

Where did the ideas discussed in the previous chapter come from and what motivated them? The search for perfect philosophical languages, an *Ars combinatoria*, etc. which would serve as a universal and infallible means of communication and obtaining knowledge did not spring up *ex nihilo* in the seventeenth century. The political and religious situation of the period was almost certainly a potent psychological factor in stimulating it. Thinkers like Wilkins, Dalgarno, or Leibniz still embodied the medieval heritage of rationalism. They were completely untouched both by the modern awareness of the historical mutability of human thought or by the postmodern sense of relativism which it eventually entailed. They felt that human reason properly directed ought to be able to settle any disputed question whether in it was in religion, philosophy, or politics, and to which there was only one correct answer. Everywhere, however, in the mid-seventeenth century there was intellectual strife, especially in religion. Men and women could die horribly (by fire) over esoteric points of religious dogma such as predestination or the nature of the Eucharist. Europe had just ended the Thirty Years War, the most destructive conflict (even in comparison with the Napoleonic wars) until the World Wars of the twentieth century, and Britain its bloody Civil War. Yet there were sincere intellectuals arguing every side of these conflicts.

But since these individuals did not make mistakes in mathematics (or if they did knew how to correct them), perhaps the problem lay in the imprecision of the natural languages in which this warfare was expressed. Therefore if a language could be designed on rational principles, so that the conclusions it was possible to draw using it were obtained by procedures more like those of mathematics, such conflicts either might be avoided or easily settled. Leibniz sincerely believed in this possibility, and his view was shared by others: John Wilkins argued that his philosophical language, being perfectly unambiguous, would clarify all religious differences. Cave Beck thought his language would help the progress of religion. William Petty (1623–1687), a partisan of Cromwell and charter member of the Royal Society, thought that the exact definitions of theological terms would reveal that the terrible conflicts between sects were caused by terminological differences. John Amos Comenius, the Moravian educational reformer and refugee from Hapsburg persecution, thought that if a perfectly clear and rational universal language could be constructed then "all men would belong to one race and one people."[1]

However, political or social motivations need intellectual substance to work on. Few major ideas are completely new inventions; they may be mutations, but there has to be something already available to mutate. We find that the intellectual environment that stimulated Leibniz, Wilkins, and other proponents of universal/philosophic languages developed out of a whole complex of Renaissance ideas. This debt is somewhat paradoxical, since the effort to construct an infallible language seems hyper-rational; yet, as is well known, the period from about 1500 to the early seventeenth century which preceded it was significantly less "rational" (at least according to modern standards) than either the earlier scholasticism or the science that followed. The core beliefs of this period have become nearly extinct since the eighteenth century Enlightenment. They are now so alien that their nature and importance have been difficult to interpret even for specialists in the period. Until relatively recently they have been ignored or dismissed by scholars who concentrated on elements of Renaissance culture such as art and Civic Humanism which resonate more strongly with modernity. For we are dealing with an age heavily invested in magic, talismans, astrology, alchemy, witchcraft, exotic forms of Neoplatonism, and whose central intellectual commitment shared by most thinkers was belief in the primordial wisdom of the mythical Hermes Trismegistus or "thrice great Hermes" who was considered to have lived before Moses, to whom he may have imparted his wisdom.[2]

In this atmosphere individuals such as Marsilio Ficino (1433–1499), Cornelius Agrippa, Pico della Mirandola (1463–1494), or even (in some aspects)

[1] Rossi (2006), p. 155f.

[2] Hermes Trismegistus was the purported source of occult wisdom contained in the so-called Hermetic Corpus, a collection of Greek texts reputed to be of vast antiquity. Actually it reflected late Neoplatonic and gnostic currents and was probably compiled between the first and third centuries A.D., although parts may go back to late Pharaonic Egypt (c. 500 B.C.).

Johannes Kepler and Isaac Newton seem in significant ways more intellectually distant from us than rationalists like Galileo, Descartes, Leibniz, or for that matter their medieval precursors such as Thomas Acquinas (1225–1274) or William of Ockham. We can understand and appreciate the ideas of the latter even when they are mistaken much more easily than we can astral magic, Newton's alchemy, Kepler's attempt to embed the planetary orbits in the five regular solids and uncover celestial musical harmonies, Jerome Cardan's (1501–1576) casting the horoscope of Christ, Pope Urban VIII's (1568–1644) consumption of astrologically distilled liquors as a defense against horoscopes cast by Spanish cardinals predicting his own death, or Benvenuto Cellini's (1500–1571) assistance in calling forth demons in the Roman Colosseum. Much valuable work on the period has been done by such scholars as Frances Yates, Lynn Thorndike, D. P. Walker, Paolo Rossi, and others.[3] However, the period is so complex that a completely satisfying intellectual history has yet to be written and certainly cannot be attempted here. But the Renaissance elaborated on certain fundamental ideas inherited from the Middle Ages, and the results form the soil in which the seventeenth century linguistic project took root. Among the most significant was a cabalistic/Neoplatonic vision of the world as a Book which God has written in a divinely constructed language which was probably spoken by Adam. Consequently we can "read" this Book and comprehend the world if we know this language. Many thinkers, especially in the early sixteenth century, inspired by the Cabala identified this language with Hebrew. Later candidates, as we shall see, were ancient Egyptian or even Chinese. But whatever it was, in its structure the language directly reflected the structure of reality.

This view was reinforced by the characteristic Renaissance preoccupations with resemblance, similarity, emblems, and analogy. A typical example was the doctrine of "signatures," where it was believed that every earthly object mirrors in some way some other object or state. For example, in the case of medicine a herbal remedy will necessarily bear a sign indicating the disease for which it is a cure. Also since there is a bond between heaven and earth, between the macrocosm and microcosm, in many cases the object's signature indicates one or more celestial bodies which cause its properties. The nature of this astral relationship can be understood by the proper interpretation of marks or signs instilled by God on the object. This being so, it was natural to believe that the divine language, mirrored

[3]See, for example, Yates (1964), (1966), and (1982); Thorndike (1923–1958), Walker (1958), Rossi (2006), and Bouwsma (1957).

the universe. Certainly, the names of objects in it were the correct ones, given by God, and reflected their inner essence; and if we knew these names we could perfectly understand and control the objects. Perhaps with the proper vocabulary, we could even command the stars and manipulate the forces they exercise on man. Cabalistic reasoning exploited these ideas to the fullest extent possible. Since Hebrew probably was the divine language, esoteric truths could be obtained by permuting the letters of words in the Torah or assigning numerical values to their letters and performing various arithmetical operations on them.

But by the late sixteenth and early seventeenth centuries, this exalted view of Hebrew began to suffer competition. An increasing fraction of late Renaissance thinkers, influenced by the Hermetic tradition, substituted Egyptian hieroglyphics in its place as a repository of secret, even magical, knowledge. For example, the *Hieroglyphia* by Horapollo which was a late and rather fanciful Egyptian treatise on hieroglyphs, probably translated into Greek around the fifth century A.D. and discovered in 1419, became extremely popular in the sixteenth century and was published in at least nine editions beginning in 1505. Its very obscurity suggested a profound and hidden wisdom. The German Jesuit Athanasius Kircher thought of Egyptian as a repository of ancient wisdom codified by Hermes Trismegistus. In the service of this idea he published elaborate and completely incorrect "translations" of hieroglyphics. (So many similar works on hieroglyphics were produced that one can speak of an "Egyptomania" in the sixteenth century.)[4] The fact that ancient Egyptian appeared to be an ideographic and not alphabetic language intrigued contemporary thinkers. As we have seen in the previous chapter, they inspired Francis Bacon's concept of Real Characters signifying not words but concepts. By the mid-seventeenth century another candidate for a language containing wisdom unknown to Europe was Chinese. This resulted from reports by Jesuit missionaries who began to visit China in relatively large numbers and send back reports from the 1570s onward. Kircher, for instance, thought that hieroglyphic methods had been given to the Chinese by Ham, the son of Noah, whom he also identified with Zoroaster.[5] Leibniz was also fascinated by Chinese ideographs and wondered that if suitably modified they might form the foundation of his Characteristic.[6]

[4]Rossi (2006), p. 78.
[5]Eco (1995), p. 158f.
[6]Antognazza (2009), p. 433f. He corresponded from 1697 onward with a French Jesuit Joachim Bouvet (1656–1730) who had been sent to the Imperial Court in Peking by Louis XIV.

All these ideas, however, were soon to undergo a transformation. The connection between some sacred language which had actually existed and arcane Hermetic wisdom gradually became less prominent (except among Rosicrucians). There may never have been a language with the wonderful properties assigned to it by Renaissance thinkers, yet it should be possible to construct a substitute. But in significant ways the earlier conceptual structure was preserved—the *Lingua Philosophica*, besides being a vehicle of universal communication, would give insights, just as the Divine language or hieroglyphics did, into truth and reality which were not possible using the natural languages. Thus it would serve essentially the same role as an actual sacred language. One can think of Leibniz's *Lingua Philosophica* and the related projects of Wilkins and Dalgarno as "modernized" seventeenth century versions of the Renaissance tradition we have sketched.

A third extremely prominent intellectual force in the Renaissance besides a belief in a primordial language or the Hermetic corpus as a source of hidden wisdom, which was both closely connected to the cabbalistic, quasi-magical reasoning characteristic of the period and helped to shape the cultural climate that influenced Leibniz, originated in the work of a medieval Catalan scholar Raymond Llull (c. 1232-1315) four hundred years earlier. Llull is worth a digression since the mind-set he and his followers created helped to motivate Leibniz's Characteristic and even some specific areas of his mathematical research. Llull is considered the founder of Catalan prose literature and perhaps the greatest medieval author from Catalonia. His works are vast in number and bulk. 263 titles have been catalogued and they total well over 3,000,000 words. Many were first written in Arabic or Catalan and then translated by him or others into Latin.

Born in the capital Ciutat de Mallorsques[7] of Majorca, Llull was a poet, mystic, novelist, missionary, student of Arabic, author of novels and philosophical treatises, he had a colorful and complicated life. His father, who came with James I of Aragon when he conquered Majorca in 1229, was either a merchant or more likely a member of the minor nobility. In either case he was wealthy and had considerable property in Majorca. Early in life Llull appears to have written troubadour poetry and become a seneschal at the court of the King of Majorca, James I. He married in 1257 and had two children. In a later autobiography (which is the main source of information concerning his life) he relates that at this time he was "reasonably

[7]Known as Palma after the sixteenth century. I am indebted to Bonner's (1985) edition of Llull's writings for much information about Llull's life and for some aspects of his philosophy.

well-off, licentious and worldly." In his early thirties in 1266 he relates that he had a transformative spiritual experience: the figure of Christ crucified and suspended in the air appeared to him on five successive evenings. This led him to renounce his previous life and to dedicate himself to serve Christ, especially by converting Muslims and Jews by demonstrating the errors of their faiths. At this point he realized that he was too ill-educated for such a mission. He had not attended a university and testified that he hardly had knowledge of grammar, and he was already over thirty which was considered rather elderly in this period. These realizations and the encouragement of a Dominican priest whom he had met in Barcelona[8] led him to an intensive course of self study. In the next few years he taught himself Arabic, mastered Aristotle and Plato, as well as medicine and theology. For most of his long life he engaged in constant travel to Rome, Sicily, Cyprus, North Africa, Paris, Avignon, Majorca, Vienne, Montpelier, and many other places. He sought out audiences with Kings and Popes (both in Rome and Avignon) in order to persuade them to back his mission and more generally to found schools to teach Arabic and other languages to missionaries, and to instigate another Crusade to recover Jerusalem. Although he received friendship and some recognition from James I of Majorca, his son James II, and Philip the Fair of France, for the most part his requests were rejected. Only near the end of his life in 1311 did the Council of Vienne honor one of them by decreeing the creation of schools of oriental languages at some major European universities. Although Llull seemed quite otherworldly and dressed in a plain hairshirt while fulfilling a vow of poverty, he was quite familiar with contemporary politics and, for instance, supported Philip the Fair's ruthless destruction of the Templars and the burning of its Grand Master at the stake. Some of Llull's travels were dangerous. In 1291 he sailed to Tunis where he was arrested and then expelled for preaching against Islam. In 1307 he was beaten and imprisoned for six months in Bougie before once again being expelled. On both occasions he came close to execution. There is a tradition which is probably unfounded, but which is still sometimes repeated that he returned to Bougie in 1315, was stoned and died of his wounds at the age of age 83 or 84. He seems, however, to have died either in Tunis or on the voyage home in late 1315 or early 1316. He is buried in the church of San Francisco in Palma.

Intellectually Llull is so remote from our secular age that he is an almost unintelligible figure today, and therefore he and the movement he originated

[8] St. Ramon de Penyafort, retired Master-General of the Dominican Order who was then about ninety.

have not have not received the kind of study merited by Kepler, Galileo, or the "Mechanization of the World Picture." He belongs to a Platonic–Augustinian tradition of medieval thought. Since he trained in Majorca far from any major university he had little contact with the latest developments in scholastic philosophy based on Aristotle. His point of view was "old fashioned" and was more appropriate to the twelfth century than to his own time.[9] He was an extreme Realist at the opposite pole from the subtle and technically intricate Nominalism of William of Ockham and other scholastic philosophers. For him the Dignities of God: Goodness, Greatness, Eternity, Power, Wisdom, Will, Glory, Truth, and Glory are real in the sense that Platonic Forms are real and could be used to demonstrate that other universals exist independently of the individual soul. God's created universe mirrored these Dignities and consequently could be viewed as a book or mirror through which one can learn about God. Llull shared with others of his time and most educated men until the seventeenth century a belief derived from Neoplatonism that the universe was a hierarchy of creation becoming less perfect the more distant its elements were from God. The most perfect divine emanation was the Empyrean beyond the sphere of fixed stars. Below this sphere was a hierarchy of lesser spheres: first those of the seven planets and the Sun (between Venus and Mars), then that of the moon, and finally, below the moon, the four spheres associated with the elements fire, air, water, and earth. The Divine Dignities descended through these spheres from the Empyrean to the earth, and through the mixture of the four elements and "under the influence of the celestial bodies, the bodies of men and all inanimate things were constituted and governed, though the souls of men could free themselves from their control."[10]

Aside from his novels and poetry written in Catalan which are still valued today, Llull is remembered for his "Art" which was a grandiose system by which symbols denoting universals could be mechanically manipulated in such a way that all possible conclusions could be drawn. He imagined that it would be an infallible tool for the discovery and demonstration of truth in Astrology, Medicine, Law, Theology, Philosophy, and for the conversion of infidels. It would be easy to use. In a month one could achieve more with it than after an entire year in the study of logic.[11] It would be "based on the actual structure of reality, it was a logic that followed

[9] Hilgarth (1971), p. 10.
[10] *Ibid.*, p. 14.
[11] Rossi (1961), p. 186.

the true patterns of the universe."[12] This was an idea of stupendous originality although it almost certainly had connections with Spanish-Judaic Cabalism.[13] The system he constructed, described and improved in dozens of manuscripts with titles like *Ars compendiosa inveniendi veritatem* (c. 1274), *Ars universalis* (c. 1274–8), *Lectura super figuras Artis demonstrativae* (c. 1285–7), and *Ars generalis et ultima* (1308) to realize these goals depended on the formal manipulation of rotating concentric rings adorned with symbols.

This work does not seem to have been much appreciated in his own time. Llull, who had no academic degree, was very much an outsider among conventionally Aristotelian Arts faculties. Around 1287 he lectured in Paris on the Art, but he hints that the lectures were not a success. Towards the end of his life, however, he made one enthusiastic disciple. This was Thomas le Myésier (d.1336), a medical doctor and academic at the Sorbonne. His main contribution to the survival of Llullism was to bequeath Llull's manuscripts to the Sorbonne. The fate of the Llullian Art in the fourteenth century is obscure. By the end of his life there were two Llullist centers in Paris: one at the University, the other at the Chartreuse of Vauvert (now in the Luxembourg Gardens).[14] There were also groups studying the Art in Catalonia, Majorca, and Italy. It soon, however, survived partially underground for it was condemned for various heretical errors by the Inquisition of Aragon in 1376[15] and later at the instance of Jean Gerson (1363–1429) by the Faculty of Theology of the University of Paris in 1390.[16] By the late fifteenth and early sixteenth centuries, however, Llullism began to explode in influence. According to Dame Francis Yates Llullism became one of the "major forces"[17] of the Renaissance. There were nearly a hundred editions of Lull's works which had appeared in Europe prior to 1582.[18] Often a prime ingredient in a mix of ideas that included the Cabala, alchemy, astral magic, artificial memory, occultism, the revelations of Hermes Trismegistus, fascination with hieroglyphics, it and its intellectual descendants emerge as central obsessions in sixteenth and early seventeenth century thought. Pico della Mirandola (1463–1494) noticed its similarities to the Cabala and dreamt of harmonizing the two; he also confessed that

[12] Yates (1954), p. 117 as quoted in Hilgarth (1971), p. 14.
[13] Yates (1966), p. 188.
[14] Hilgarth (1971), p. 156.
[15] Llull (1985), Vol. I, p. 76.
[16] *Ibid.*, p. 72.
[17] Yates (1954), p. 166.
[18] Bouwsma (1957), p. 79.

his own system was inspired by the Art. It was accepted by Nicholas of Cusa (1401–1464) who collected and copied Llull's manuscripts and was greatly elaborated by Giordano Bruno who believed the Art to be divinely inspired. The French humanist Lefèvre d'Étaples (1455–1536) directed a resurgence of interest in Llull in France.[19] The magus Henry Cornelius Agrippa von Nettesheim wrote an influential exposition and commentary on the Art which was cited by Leibniz.[20] Other commentators included the Byzantine theologian Cardinal Bessarion (1403–1472) and John Dee (1527–1608), who was one of the most influential occult philosophers of the Elizabethan Renaissance and a Llullist. As late as the mid-seventeenth century Athanasius Kircher produced a huge nearly unreadable book, the *Ars Magna Sciendi sive combinatoria* (1669) which he viewed as a perfection of the Art.[21] By this time the Art's original purpose of converting unbelievers was forgotten. It and its later developments was now seen as a *clavis universalis*, a key that would enable the adept to master all knowledge and unlock the hidden structure of reality. To Agrippa the Llullian Art was queen of all the arts. Using it men without any previous knowledge would be able to eliminate all error and find "the knowledge and truth of all knowable things."[22] And for Kircher his modified Llullist system which was deeply intertwined with Hermeticism was "the great art of knowledge or combinatorial art, through which the broadest door is opened for quickly acquiring knowledge in all the arts and sciences"[23] There is a vast literature from the sixteenth and seventeenth centuries with similar adulation of Llullism in language that could word for word been written by Leibniz referring to his Universal Characteristic.

What exactly was the Art and how did it work? There are no unanimous answers; much of the task of interpretation remains to be done. But we can distinguish at least three of its general characteristics. The first, as we have already pointed out, is Llull's extreme Neoplatonic metaphysical realism and "exemplarism". We have seen that Llull thought of the world as a divinely constructed hierarchy progressing from the world of generation and corruption of the sublunar sphere, through the planetary orbs and *primum mobile*, to the divine archetypes beyond the sensible universe. This was a fairly commonplace idea in the Middle Ages and Renaissance

---

[19] *Ibid.*

[20] Agrippa, however, later in life turned against Llull and attacked the Art in his *De vanitate scientiarum* published in 1600.

[21] Couturat (1961), pp. 541–543.

[22] Rossi (2006), p. 30.

[23] Quoted in Fauvel and Wilson (1994), p. 52. See also Rossi (2006), p. 141f.

which he shared with conventional Aristotelians. However, he added the Neoplatonic claim that each inferior layer was a collection of metaphors and symbols encoding the structure of the world and imprinted as a seal imprints wax by the superior members of the hierarchy above it. These symbols could then be put into correspondence with the symbols of the Art. By manipulating the latter one could understand the former and thereby grasp the essential structure of reality. As developed further by Renaissance occult thinkers such as Agrippa or Dee, one could even work backward and influence the higher supersensible levels of the hierarchy by the correct manipulation of symbols and metaphors in our earthly world—in this way Egyptian priests were supposed to have called down influences of the stars to animate the statues in their temples. As we have already hinted there are analogies between Llullism and Cabalism, not only in the metaphysical preconceptions of both, but also in their common belief in the efficacy in the circular manipulation of letters or symbols. For example, the early (200–600 A.D.) Hebrew text *Sefer Yezirah* (The Book of Creation) was widely known and often reprinted by the Renaissance taught that God had constructed the universe by manipulating the Hebrew alphabet which represented its fundamental principles. Further "the letters were placed on a revolving sphere, wheel, or circle and that creation occurred through the process of revolution"[24] These similarities between Llullism and Cabala were noted as indicated above by Pico della Mirandola as well as many other commentators. It is hard therefore to avoid the suspicion that the examples of *gematria* such as the *Sefer Yezirah* may have in part inspired Llull to create his system. A second feature of Llullism is its formalism. Llull attempted to isolate in each scholastic subject a small number of fundamental ideas, anticipating the belief of Leibniz and others in the existence of an alphabet of human thought. For instance, in moral theology he lists 9 attributes, 9 relations, 9 questions, 9 subjects, 9 virtues, and 9 vices. These are designated by letters and arranged in compartments on the concentric wheels. Sometimes colors are also used to indicate compatibility or incompatibility of the fundamental concepts and rules are given for understanding various combinations. By mechanically exploiting the rules of the Art, in other words by lining up the compartments of the concentric wheels in various ways one can generate a vast array of combinations of concepts and easily "solve" the philosophical or theological questions of the day, and in fact indicate the solution by a symbolic formula. For example, in one of

[24]Coudert (1999), p. 123.

his accounts of the Art, the *Ars Demonstrativa, Quaestio (39)* concerning the soul Llull poses the question "Whether in each power of the soul there is action and passion" The "solution" is the alphabetical string

[DL — EI — MR — SS — VV — YZ — AS — SX]

which represent combinations of compartments on various wheels.[25] By "decoding" these symbols according to his rules one has the solution. This is one of hundreds of "questions" posed in his works not only in theology, but in medicine, law, as well as astrology, and solved in a similar way.[26] Leibniz, although he criticized the specifics of the Llullian art and was contemptuous of contemporary Llullists,[27] was deeply impressed with the algorithmic aspect of it we have described. The automatic way in which an adept of Llullism was able to obtain conclusions seems to have been the inspiration for Leibniz's desire to import the same features into his Characteristic. The emphasis that Llull put on exploring all combinations of conceptual possibilities by rotating and lining up the compartments of his concentric wheels also was the reason why a synonym for his and related systems was the *Ars combinatoria.* Llull's methods certainly stimulated new interest in combinatorial analysis in the seventeenth century especially on the part of Martin Mersenne (1588-1648), Bruno, Kircher, Christopher Clavius, and others. This, in turn, was certainly the motivating factor in Leibniz's *De Arte Combinatoria*, some of which resembles rather closely the Llullist writings in the same period by Mersenne and Kircher. Leibniz indicated this influence in a 1714 letter to Nicolas Remond, who was the chief counselor of the Regent (the Duke of Orleans):

> When I was young, I found some pleasure in the Llullian art, and I said something about this in a little schoolboyish essay called On the Art of Combinations, published in 1666 and later reprinted without my knowledge.[28]

It is responsible as well for his enduring belief in the importance of the "Combinatorial Art" to his project for a Characteristic, and therefore indirectly for the combinatorial aspects of his mathematical research. It was probably also behind his idea that all complex concepts were combinations

[25] Llull (1985), Vol. I, p. 474.
[26] It is this aspect which has caused some critics to believe that Llull, however crudely, has anticipated some aspects of modern logic or even computer science. See e.g., Gardner, (1982).
[27] Maat (2008), p. 273.
[28] Leibniz (1969), p. 637.

of elements of an "alphabet of human thought." And just as Llull and his followers had manipulated the compartments on concentric wheels to find fruitful combinations of concepts, so Leibniz imagined that some sort of combinatorial process would serve as an *ars inveniendi* or logic of discovery and invention.[29] Again, this reduction of knowledge to a small number of principles parallels the Cabalistic reduction of reality to the Hebrew alphabet; however, the algorithmic and especially logical aspect of Llullism seems an original contribution and which transcends the Art's possible roots in Cabala.

Another aspect of the Art which seems to anticipate the seventeenth century mathematization of Nature has been summarized by Dame Francis Yates. Although Llull's work, aside from his interest in very elementary combinatorial results, was not deeply mathematical in a technical sense, he is full of Neoplatonic and Pythagorean conceptions, holding that the circle, triangle, and square were fundamental to the understanding of reality:

> On its deepest metaphorical level, in its most secret application, the art works out the structure of the universe in terms of the circle, the triangle, and the square...the practice of Llullism through the centuries must have helped to form a habit of mind which sought for mathematical explanations or demonstrations of reality. As is well known, Descartes told his friend Isaac Beekman (1588–1637) that his new universal system of knowledge, based on analytical geometry, was to take the place of the Art of Ramon Llull.[30]

One is reminded here of Galileo's famous claim (and challenge to the Aristotelians) that:

> Philosophy is written in that great book which ever lies before our eyes–I mean the universe–but we cannot understand it if we do not first learn the language and grasp the symbols in which it is written. This book is written in the mathematical language, and the symbols are triangles, circles, and other geometrical figures, without whose help it is impossible to comprehend a single word of it . . . [31]

Yates may have exaggerated the contributions of Llullism to the view that Nature has a mathematical structure. But given that it was already a conventional idea found in many strands of Renaissance thought that Nature

[29] See the discussion in Eco (1995), pp. 273–279 or in Fauvel and Wilson (1994).
[30] Yates (1954), p. 155.
[31] Quoted from the *Two New Sciences* in Burtt (1932), p. 75.

is a Book which can be understood if we can decode the Natural Language in which it is written, the achievement of Kepler, Galileo, Descartes, and Newton was to replace one kind of Natural Language (the Llullian Art, Hebrew or Hieroglyphics) by another—mathematics. Seen in this light the Mathematical Platonism considered by Alexander Koyré to be a fundamental aspect of post-Galilean science is a development and not a rejection of Renaissance ideals. For Galileo geometrical diagrams and for Leibniz differentials and integrals and differential equations are just replacements for the occult symbols, "signatures" of a Pico della Mirandola or Paracelsus (1494–1541), or the combinations of Llull in expressing the inner essences of things. But further investigations along these lines, however fascinating, would take us too far afield.

Of course there always had been detractors of Llull; his followers had been satirized by Rabelais (1494–1553) (and later by Jonathan Swift (1667–1745) in his account of Laputa).[32] Descartes who had little regard for the Art speaks with amused contempt of meeting an old man at an inn who boasted that by means of Llullism he could speak for one hour on any topic whatever, on a second topic for an hour and so on for twenty hours.[33] By the middle of the seventeenth century the great days of the Art were over; few took it literally and by the nineteenth century it was dismissed by scholars such as Prantl as "pure nonsense".[34] Only in this century has it been seen (perhaps too optimistically) as a distant precursor of modern logical ideas. However, one can view Llullism as part of a Renaissance inoculation which together with Hermeticism and the search for a divine language deposited the germs of a great many commonplace ideas in the seventeenth-century mind, creating the general intellectual climate that stimulated a quest for a philosophic language.

From hindsight Leibniz's work seems uncannily modern. There are the beginnings as we have seen in Chapter 7 of symbolic logic, unappreciated until the time of Frege and Russell. There is his unique approach to what we now call calculus that made it possible to do difficult problems in a unified mechanical manner. The notation he created almost makes computation automatic; the way it fits together and its very appearance help to give results that are opaque in a geometrical setting. We notice in all his work—whether in the calculus, the *Analysis Situs*, or logic—the faith that it is possible by invention of the proper symbols and rules for their

[32]Llull, 1985, Vol. I, p. 87.
[33]Gaukroger (1995), p. 102.
[34]Yates, *Ibid.*, p. 117.

manipulation to both discover new results and to organize old ones. In this way, we can view Leibniz and his followers as revolutionaries who invented a radically new kind of mathematics, overturning the contemporary geometric paradigm, and preparing the way for the Bernoullis, Euler, Cauchy (1789–1857), and Gauss (1777–1855). This, perhaps even more than the physical discoveries of Newton, made the modern world possible. But if the consequences were "modern" ironically inspiration behind them was curiously archaic, deriving from a set of long forgotten ideas. For as we have tried to show, the familiar notation we have described at the beginning of this paper was for Leibniz just a subset of the Real Characters from the Universal Characteristic which were appropriate to mathematics.[35] It is perhaps an overstatement to claim as Yates (1966) did[36] that:

> The mature Leibniz, the supreme mathematician and logician, is still emerging straight out of Renaissance efforts for conflating the classical art of memory by using the images of the classical art on the Lullian combinatory wheels.

For this is only a small part of a tradition from which Leibniz drew inspiration. Yet Yates' fundamental claim of connections between Renaissance thought and Leibniz is almost certainly correct.

We have now reached the most problematic point in this long discussion of external cultural influences on Leibniz's achievement. The relationship between the intellectual environment and the creative act of an individual is an insoluble metaphysical puzzle similar to the mind-body problem that obsessed seventeenth-century Occasionalism. John Wilkins and many others were exposed to the same kind of ideas as was Leibniz. Wilkins for instance, created an overcomplicated artificial language and wrote a justly forgotten book *Mathematical Magick* (1648), but he did not create calculus. Leibniz may have failed to construct a workable *Lingua Philosophica*, but he produced examples of a "calculus," one of which survives today. Just why there is a creative response by one individual and not another to cultural or environmental stimuli has long fascinated psychologists, but

[35] Rossi (2006), pp. 208–211) has also conjectured that the influence of the symbolic climate we have been describing was not confined to Leibniz. He feels that it has connections to the general improvement of mathematical notation and algebraic methods in the sixteenth and seventeenth centuries. It is interesting to note that Newton owned eight volumes of Llull [Llull, 1985, Vol. I, p. 85, n. 75] and that he also thought about constructing an artificial language [Shapiro (1969), p. 221]. In the writer's opinion this whole question needs further investigation for it is probably no coincidence that a usable algebraic notation was invented in the emblem soaked culture of the late Renaissance.

[36] p. 383.

will probably never have a compelling answer. We can only explain it by tautologies involving the individual's "genius" just as the medical student in *Le Malade Imaginaire* speaks of the "dormative virtue" of opium. But the cultural environment can at least make that act intelligible. In Leibniz's case we cannot talk about direct mathematical connections between the tangled nest of Renaissance ideas we have sketched and actual results. Leibnizian notation may seem a system of "signatures" incorporating mathematical analogies just has Dee's mysterious symbol, explained in his *Monas hieroglyphica*, incorporates analogies to the structure of the cosmos, but Leibniz was a creative *mathematician* and Dee, a fairly conventional Renaissance magus who knew his Euclid, was not. Leibniz is also novel where Dee, Kircher, and others are almost interchangeable. His calculus is a mutation which is appropriate to his cultural milieu, but cannot be reduced to it; at most that milieu provided Leibniz with metaphors and analogies to the grandiose but failed projects of others which helped to shape his mathematical motivation and mind-set: in fact, his very desire to invent a "calculus". The relationship is similar to an analogy (which is today almost a cliche in cultural studies) pointed out by Gertrude Himmelfarb over forty years ago between Malthusianism and the ideology of nineteenth-century capitalism and Darwinian evolutionary ideas.[37] Darwin and Wallace's (1823–1913) theories of natural selection are creative achievements of the highest order and are not reducible to the Gadgrindian ideology of Malthusian economists; but this fashionable set of ideas made it reasonable to ask certain questions and think in certain directions. Likewise the prior speculations of Lamark (1744–1829), Erasmus Darwin (1731–1802), Buffon (1707–1788), Cuvier (1769–1832), Geoffroy Saint-Hilaire (1772–1844) may have formed a necessary but by no means sufficient substrate for Darwin, just as work on tangent and quadrature problems by Descartes, Fermat, Barrow, Sluse, Pascal Cavalieri, and others did for Leibniz. But in neither case was the creative achievement of Darwin or Leibniz a "synthesis" or any kind of direct extension of previous work. In both cases previous intellectual foundations were radically changed. Darwin destroyed the teleological assumptions shared by both evolutionary and non-evolutionary thinkers since classical times and substituted a new kind of essentially nihilistic biological thinking. Leibniz, although still strongly influenced by the traditional geometric paradigm, fancied that he was inventing a calculus for the geometric problems of quadratures and tangent construction; but

[37]Himmelfarb, (1967).

in so doing he unwittingly began the destruction of a two thousand year old primacy of geometry and the construction of a radically new kind of mathematics almost "orthogonal" to the previous tradition; but in doing this, we can argue that however indirectly and at the level of metaphor some fairly curious– indeed "irrational"–ideas helped shape the future development of Western mathematics.

We also suspect that there are a multitude of other links between Leibniz's philosophy and a whole gamut of Renaissance ideas: Hermeticism, Cabalism, occultism, astral magic, fascination with hieroglyphics, and symbolic schemes of every kind. The monadology itself (the word "monad" may be borrowed from Bruno) seems an echo of Renaissance pan-psychism; moreover the idea that each monad mirrors however imperfectly the state of the universe without being acted upon by that universe is an abstraction of the common Renaissance idea of the relation between microcosm and macrocosm or more generally that every existing thing reflects, resembles, or symbolizes in innumerable ways the whole hierarchy of being. This system may reflect the technical imperatives of Leibniz's logic (or vice-versa) and be designed to ward off the materialism of Spinoza or Hobbes; but there are shadows here of Agrippa, Pico, and the Cabala.[38]

Historians like Paolo Rossi and Francis Yates have begun to unravel the tangled web of late Renaissance thought, but the work has barely been begun. Aside from the difficulty that most of the material is still in untranslated, unread Latin tomes, the Zeitgeist is so different from post-Enlightenment patterns of thought which are still second nature to us that it is inaccessible to us in ways that later and more familiar intellectual currents are not. For example, in her analysis of the Llullian system Yates (1954) like other commentators such as Bonner,[39] Rossi (2006) and Eco (1995) can only give us a vague idea of how it actually worked in detail. Llull's "thought style" from our vantage point is so peculiar that no one—even a Renaissance expert—can use it to "solve" Llull's "problems." This is to be contrasted with the work of for instance a thinker such as Archimedes who lived nearly 1500 years before Llull. The former's arguments are difficult even for a professional mathematician, but at least we can *formally*

[38]In her interesting book *Leibniz and the Kabbalah* Coudert (1999) attempts "to show that van Helmont's kabbalistic philosophy had a decisive influence in shaping Leibniz's *monadology*". Leibniz had a life long friendship with Francis Mercury van Helmont (1614–1698) and had a surprisingly favorable opinion of the type of Christian Cabalism of which van Helmont was one of the chief exponents. For connections of the Cabala to Renaissance thought generally, see Coudert (1998).

[39]The editor of Llull (1985).

understand them even if as we have argued we cannot *use* them in a creative manner. However, even this level of understanding seems difficult for Llull even though he was quite accessible to and taken seriously by many Renaissance thinkers. We can appreciate the tight almost mathematical precision of Scholastic thought—Duns Scotus (1265–1308), for instance, reminds the present writer of a treatise by the Bourbaki group (although the subject matter is very different, similar habits of mind and intellectual styles are present in both)—or the Oration on the Dignity of Man, and Italian Civic Humanism. But what can we possibly make of Hermes Trismegistus, astral magic, or Llullism? It is debatable whether or not the Kuhnian phenomenon of "incommensurability" really exists during periods of revolutionary change in the hard sciences, but a strong case can be made that it exists in major portions of Renaissance thought in comparison with our own. It is just this feature that has led many scholars beginning with Pierre Duhem, continuing with R. Lenoble, Marshal Clagett, A. C. Crombie, John Herman Randall, Jr., and David Lindberg to locate the origins of modern science in the Middle Ages rather than the Renaissance. Francis Yates in all of her books takes the contrary position that the various occult or hermetic influences of the Renaissance are important factors in the development of science. Although the magus starts by operating on demons; his intellectual descendants eventually operate on Nature:

> Quite apart from the question of whether Renaissance magic could, or could not, lead on to genuinely scientific procedures, the real function of the Renaissance magus ... is that he changed the will. It was now dignified and important for man to operate ... It was this basic psychological reorientation towards a direction of the will which was neither Greek or medieval in spirit, which made all the difference.[40]

Whatever the general truth of this assertions, we hope that we have attempted to show that some fairly esoteric Renaissance ideas nourished the seventeenth century climate that made Leibniz's calculus possible.

[40] Yates (1964), p. 156.

Chapter 10

Epilogue

We have seen that Leibniz was unable to obtain a permanent position in Paris after the death of his employer the Elector of Mainz Johann Philipp von Schönborn, and so for want of other options had to enter the service of the Duke of Hanover Johann Friedrich of Brunswick-Lünberg and to accept the Duke's request that he reside in Hanover. Until his death in 1716 he was the ducal librarian employed in an extraordinary number scientific, genealogical, political, and historical researches, most of which have long been forgotten. At the same time he was able to pursue a vast correspondence with practically every major (and minor) intellectual figure in Europe, refine his calculus, formulate dynamical theories from a point of view quite different from Newton's, and develop the intricate philosophy of monads for which he is primarily remembered today. Although he was soon recognized as a European philosopher and mathematician of the first rank, we shall see that this was accompanied by a gradual loss of status and increasing neglect at home.

Leibniz's situation when he arrived in Hanover in December 1676 was inauspicious. He had been offered a position as court councillor which he expected would soon lead to membership in the privy council. He found, however, that, that the Duke wanted him to function merely as the curator of the ducal library, a position inferior to what he had held in Mainz and Leibniz felt not commensurate with his achievements. Only in 1678 after repeated requests was he formally recognized as a councillor and his pay raised to 600 talers per year. This was a substantial salary, but still only about a third of what a member of the privy council earned. Moreover, Leibniz felt intellectually isolated. He had left the most intellectually exciting city in Europe for a provincial backwater of only around 14,000 residents compared with a half million in Paris, and he complained that there was hardly anyone one worthy of serious conversation. Fortunately, Johann Friedrich himself was a cultured man who gave a sympathetic hearing to Leibniz's many projects which included schemes for the federation of the German princes, the improvement of the ducal government and the social welfare of the population, the creation of learned societies, and the reconciliation between the Catholic and Protestant churches—a goal he had pursued since the Mainz period. Leibniz also proved to be an excellent librarian. He doubled the size of the collection by acquiring 3600 volumes from an estate sale for only about half of what the volumes were worth.

Had Johann Friedrich lived, Leibniz probably would have been promoted to greater things; but in December 1679 the Duke died and was succeeded by his brother Ernst Augustus (1629–1698) who had little interest in things of the mind. The only proposal of Leibniz the new Duke supported was a practical one which Leibniz had submitted in December 1678 to Johann Friedrich. The silver mines in the Harz mountains were an important source of ducal revenue. Yet they often flooded, and the hydraulic pumps used to drain them failed in dry years. Leibniz argued that a plan to remedy the situation using windmills by the Court Mining Councilor Peter Harzingk, a Dutch engineer, was worthless and proposed a new plan using more efficient pumps and windmills of his own design. The Duke thought of Leibniz as the true inventor of the windmill scheme and decided to support him against the advice of his own mining engineers;[1] a contract with the Mining Office was worked out in September 1679 whereby Leibniz would receive a life time stipend of 1200 talers per year if a test, made at his own expense, succeeded in keeping one of the mines dry for a year. This agreement was reconfirmed in October 1680 but Leibniz's share of the expense was reduced to a third, the rest being shared equally between the Duke and the Mining Office. Given full control of the project, Leibniz visited the mines more than 30 times between 1680 and 1686, spending a total 165 weeks on site. There was constant conflict between him and the local engineers and mining officials who argued that his windmill was impractical. By 1683 2,270 talers had been spent on the project which was seven times the original estimate, but the project was still unsuccessful. According to a 1684 report by the Mining Office Leibniz's windmill was unreliable working less often than a local windmill used for grinding grain, and when it did function it pumped much less water than the standard hydraulic pump; in fact, the wind was a very unreliable source of power in the area. But Leibniz

[1] Wakefield (2010), p. 180f. Harzingk also waged a bitter priority dispute claiming that Leibniz had simply appropriated his own ideas.

continued to struggle with the project and to make expansive promises. He gave elaborate abstract arguments based on his theories of force and motion. Instead of the vertical windmill he now wanted to build a horizontal one. In 1684 Ernst Augustus had cut off funding but allowed Leibniz to continue at his own expense. After some final tests which were only partially successful, the Duke ordered that Leibniz discontinue the project, which according to the Mining Office had cost 128,100 talers in lost potential income. Leibniz, however, was tenacious; he made another attempt between 1693 and 1696. This time the windmill scheme was abandoned, and he attempted to improve the existing pumps. Unfortunately this effort also failed. The whole sorry affair could not have enhanced Leibniz's stature in the mind of the Duke.[2]

As early as 1677 Leibniz had made some investigations of the Guelfs, an ancient and historically important family from which the Braunschweig-Lüneburg dynasty of the Dukes of Hanover descended and in early 1680 proposed to Johann Friedrich that he write a brief but thorough history of the Guelf House. After the failure of the Harz project Ernst Augustus, trying to find a useful and practical assignment suited to Leibniz's talents, resurrected the proposal. This was an important project. If suitable links were found, the genealogy would increase the prestige of the Ducal family. This was a desirable goal since Ernst Augustus was trying to become an Elector of the Empire, an office which would allow him to join a select group of eight other Electors who had the right to determine the succession to the imperial throne.[3] Leibniz was formally asked to undertake the task in August 1685. He was given financial support including travel expenses, a secretary, and was excused from routine work of the Court. A particularly beneficial aspect of the job was the opportunity to travel. Since arriving in Hanover, Leibniz had little chance to travel. But since documents relating to the Guelfs were in libraries scattered throughout Germany, Austria, and Italy, Leibniz could now make lengthly research trips to interesting places, a side benefit being the opportunity to meet other scholars and intellectuals. At last Leibniz would have someone with whom to talk! Between 1687 and 1714 he visited and collected material from archives in Berlin,

---

[2]The details of the Harz project are given in Aiton (1985), pp. 87–90 and 107–114, Antognazza (2009), pp. 211–213 and 227–230, and Wakefield (2010), pp. 171–188. It is a matter of some dispute whether or not Leibniz was obstructed by ignorant officials at the Mining Office or if he simply lacked the practical engineering experience to be successful and promised more than he could deliver.

[3]The promotion of Hanover to Electoral status came in March of 1692. Leibniz participated in the final negotiations leading up to it [Aiton (1985), p. 177].

Vienna, Augsburg, Marburg, Munich, Wolfenbüttel, Frankfort-am-Main, Venice, Rome, Naples, and Modena. Some of these visits were for extended periods; he remained in Vienna nine months (1708) and in a second visit (1712–1714) for nearly two years.

Leibniz was a very hard worker. He left thousands of pages of manuscripts and notes, but at the same time he had difficulty completing his more ambitious projects. They tended to become unmanageable. This had happened already, possibly due to technical factors, with the Harz project. A similar situation developed with the projected history of the Guelfs. Leibniz certainly did a lot of research, but he wished to be thorough and had set excessively high standards for himself. The project assumed a vast scope: he felt that before even getting to the Guelfs there should be preliminary work on the history of the earth from the Creation based on geological research. (Here Leibniz had some modern ideas; one of these was his belief that the earth had originally been a molten ball whose surface had cooled.) This should be followed by a history of Lower Saxony, exhibiting the connection between its ancient inhabitants and the migrations that gave rise to the present ethnic composition of Europe. Only then, Leibniz felt, could one deal satisfactory with the Guelfs.[4] Ernst Augustus had imagined a project taking a few months or at most a few years, but Leibniz worked on it for more than three decades and only *began* the actual writing a few years before his death in 1716. In the intervening years there were endless requests to the Duke for leave and financial support to visit archives. Although he made progress on the Guelf history—establishing an interesting connection between the Braunschweig-Lüneburgs and the Este family of Italy, there was never any sign of the project's completion.

Ernst Augustus died in January 1698 and was succeeded by his eldest son George Ludwig (1660–1727). This Duke was far less indulgent to Leibniz than his father and uncle had been. He viewed Leibniz's travel as a waste of time and money. In 1703 in a letter to his mother George complained that he never knew where Leibniz was and when asked what he was doing, Leibniz would reply that he was working on "his invisible book" whose existence would take as much trouble as to prove the existence of the "lost book of Moses." To show his displeasure George soon canceled the raise of 400 talers which Leibniz had received after promotion to a higher bureaucratic rank in 1696 and turned down many of Leibniz's requests for foreign travel. Nevertheless Leibniz was somehow able to find excuses to

[4] Antognazza (2009), p. 325.

travel anyway and to prolong his unauthorized absences from Hanover.

George Ludwig's temperament was probably not improved by his unhappy arranged marriage in 1682 to his cousin Sophia Dorothea of Celle (1666–1726). This Sophia was the illegitimate daughter of George Ludwig's older brother George Wilhelm (1624–1705) by his mistress. Sophia resisted the marriage, exclaiming that she would not marry a "pig's snout." The two soon proved to be extremely incompatible. In 1694 she was accused of adultery with Count Philip Cristoph von Königsmark (1665–1694) whom she had known since the age of sixteen, and after George nearly strangled her in a marital quarrel, the two were divorced. Sophia was then confined for the last thirty years of her life in the castle of Ahlden in Lower Saxony, isolated from her two children: George, later George II of England, and Sophia (1687–1757), mother of Frederick the Great and Queen of Prussia from 1713 to 1740.[5] On her deathbed she wrote a letter to her ex-husband placing a curse beyond the grave upon him.

In August 1714 after the death of Queen Anne (1665–1714) George Ludwig inherited the British throne as George I, the first of a dynasty of British rulers from Hanover. As a great grandson of James I he happened to be the closest living *Protestant* relative of Queen Anne. Although there were several closer relatives, they happened to be Catholics, and the Act of Settlement of 1701 barred them from the throne. In George's absence his younger brother, also named Ernst Augustus (1674–1728), was appointed to oversee the Duchy.

As for most people Leibniz's fortunes in life were a complicated function of talent, luck, and personality. Concerning his talent there can be no doubt. There is a strong case that he was the most widely gifted European intellectual in history. He knew and could be productive in almost every field and seemed to only lack interest or ability in artistic or musical creation. He had enjoyed both good and mediocre fortune. He was fortunate in his friendships with Oldenburg, Huygens, and Boineburg. He was unlucky in encountering Hooke and Pell, in having to exchange Paris for Hanover, and in taking on the Harz project. Doubtlessly also, his position in Hanover would have been happier without George Ludwig.

If it is hard to fully understand the personality of our living friends and associates, it is naturally much more so for an individual who has been dead for three centuries. In Leibniz's case a psychological analysis is even more difficult because there is evidence pointing to contradictory features

[5]The Count disappeared in the same year; he was probably murdered as a result of a failed attempt to help Sophia escape. They may have planned to elope together.

of his personality. Some were attractive and some unattractive, but they certainly must have affected his interactions with others. On the positive side, he seems to have been friendly and engaging. He could talk easily with people from all ranks in society including the uneducated:

> He did not mind simple women and lost track of time when he had the opportunity to converse with them. Indeed, he could adapt his speech in such a manner that one would not have taken him to be a philosopher at all. ... He talked to soldiers, courtiers, and statesmen, artists and similar people as if he was one of their profession, so everyone liked him, except people who did not understand such behavior. He spoke well of everyone, saw the bright side of everything, and even indulged his enemies, whom he often could have harmed at court.[6]

He was also a consummate networker and insider (which he had to be in order to get anything done). Leibniz was no shy shrinking violet and impressed people who mattered in the intellectual and political world. Despite the Harz disaster and his failure to complete the history of the Guelfs, he had demonstrated his value in Hanover in many other ways, one of the most significant being his successful negotiation to obtain Ernst Augustus' elevation to the Electorate. During his final twenty one month residence in Vienna he dealt confidently with the Emperor and his ministers, was promoted to near ministerial rank, and was involved in a multitude of diplomatic projects including a scheme for organizing piracy against France and Spain, enemies of the Empire, in the waters off America.

In his correspondence with some of the most prominent men of the age he shows a supreme self-confidence which is combined, it must be admitted, with a certain deviousness. He had a tendency to write what his correspondents wanted to hear, shaping his opinions to suit the recipient. To give one example, Spinoza was a bête noire to the religious and philosophical establishment because of his perceived atheism and materialism. Publicly, Leibniz enthusiastically agreed with this judgement. In a review Leibniz's old teacher Jacob Thomasius had called Spinoza's *Tractatus Theologico-Politicus* "godless," and a certain Johann Gravius of the University of Utrecht had branded it a "most pestilential book." To Thomasius in a letter of 1670 Leibniz replies that "you have treated this intolerably impudent work on the liberty of philosophers as it deserves." In his reply to Gravius he says that "I deplore that a man of such evident erudition

[6]Testimony of Leibniz's secretary Johann Georg Eckhart (1664–1730), as quoted in Antognazza (2009), p. 558f.

should have fallen so low," and in a letter to the theologian Antoine Arnauld (1616–1698) Leibniz calls it "one of the most evil books in the world." In another letter to Thomasius in 1672 he refers contemptuously to Spinoza as "a Jew thrown out of the synagogue on account of his monstrous opinions."[7] But then in spite of these damming opinions, we find Leibniz in 1671 initiating a friendly correspondence with Spinoza himself, addressing him as "Illustrious and most honored sir," and very likely in letters (now lost) praising the *Tractatus*.[8] This correspondence eventually resulted in a visit by Leibniz to Spinoza at The Hague in November 1676 on his trip from London to Hanover where the two philosophers had friendly conversations, discussing among other things Leibniz's version of the ontological proof of the existence of God.

There is also often a tone of extravagant self-praise, boastfulness, and hunger for recognition (and especially the financial rewards recognition brings) in Leibniz's letters: as impressive as his achievements actually were, they are made to seem even more so, and in a letter written in the autumn of 1671 to Duke Johann Friedrich of Hanover, his future employer, he lists at least ten projects ranging from the Universal Characteristic, the physics of motion, solving the problem of longitude, the construction of a submarine, his calculating machine, and the defense of religious doctrines such as transubstantiation against the "insults of un-believers and atheists." On all these matters it is plainly stated or hinted that he has accomplished far more than anyone else.[9] Since this letter was functioning as a resumé, this hyperbole may be excusable. But the same kind of self-praise may be multiplied endlessly throughout his life in letters or in his many articles, pamphlets and reviews. These characteristics are in strong contrast to the unpretentiousness of a Newton or Boyle and are particularly evident in his writings on mathematics or physics: almost any result he discusses in these areas has, Leibniz suggests, already been done independently by him and often in a better way.[10] Sometimes this is not true as is the case of results by Gregory and Newton on series or his 1689 article *Tentamen de Motuum caelestium Causis* which, as we have remarked in Chapter 6, was probably

---

[7] Quoted in Stewart (2006), p 110–113.

[8] *Ibid.*, p. 112.

[9] *Ibid.*, p. 89f.

[10] To support priority claims Leibniz was not above occasionally backdating documents. As remarked in Chapter 5, he altered the date on a November 1675 mathematical manuscript to 1673.

not as independent of the *Principia* as Leibniz claimed.[11] We have also seen that he frequently promises more than he has actually delivered. This was especially evident in his writings on the Universal Characteristic or *Lingua Philosophica*. Admittedly this project motivated his unique approach to calculus and logic, which in itself is a singular achievement. Yet as far as the general project is concerned little had been accomplished beyond the production of a vast amount of propaganda. Over and over again, he describes the wonderful things he expects from the Characteristic, while strongly hinting that it is already is in his possession.

Sometimes also Leibniz will use the work of others (such as Barrow's transmutation theorems) without acknowledgement or fail to give due credit to his predecessors. But if he does mention other mathematicians in his letters or articles there is seldom overt disparagement or belittling; he often calls competitors "illustrious" or "brilliant." They may have fallen short, but they did their best and have something positive to offer which Leibniz will both exhibit and improve. But when confronting a competitor he deems a threat or during a quarrel, however, Leibniz can be nasty and sometimes devious or dishonest. This behavior is usually shown in "anonymous" reviews and pamphlets more than in his correspondence or signed articles; we see examples of it in his treatment of Newton. One can argue that such features of Leibniz's character have to be understood in the light of the intense competition for recognition and reward from the aristocratic establishment among seventeenth century intellectuals. He was by no means the only offender against modern conceptions of scholarly rectitude—Robert Hooke was *much* worse. And certainly, excessive modesty little became his very real achievements. Leibniz seems to feel that he is merely stating the obvious facts which every right minded person ought to recognize.

Despite his obvious "extroversion," social abilities—especially in relation to powerful people, and self-promotion there are opposite aspects to Leibniz's personality. Even when he is successful, Leibniz can exhibit insecurity. It is psychologically interesting that the self-perceptions, found in letters to close personal friends, are often in strong contrast to the public display of boastful self-confidence he shows in writing to the political and intellectual elite. For example, in a letter of 1673 to his friend Lichtenstern Leibniz relates that:

[11]Newton judged that the contents of this paper "Errors and Trifles excepted are Mr. Newton's (or easy Corollaries from them." The quotation is from p. 209 of the *Account* in Hall (1980), p. 299.

> I feel myself burdened with a deficiency that counts for a great deal in this world, namely, that I lack polished manners and thereby often spoil the first impression of my person. Whenever people attach great weight to such things, which I do not, and whenever it is a question of "socialize if you want to succeed", there I am out of place.[12]

In another note he mentions that a Parisian book dealer mistook him for a country bumpkin and sarcastically asked him if he did not even know the difference between metaphysics and logic.[13]

For several decades Leibniz enjoyed resounding professional success in Europe. We recall that he was elected to the Royal Society at age 27 in 1673. He was made a foreign member of the French Royal Academy of Science in 1700 and shortly afterwards was appointed president for life of the new Berlin Academy of Sciences which he had persuaded the Elector of Brandenburg to found. By the 1690s he was probably the most celebrated living philosopher, mathematician, and scientist in Western Europe. Yet in Hanover he suffered a slow but steady decline in esteem, especially under George Ludwig. A major part of the reason for this, of course, must have been the irritation caused by his inability to finish the History of the Guelfs. But even under the earlier Dukes the atmosphere had been inferior to that of Paris—which probably explains his constant desire to travel. He had little in common with the philistine aristocrats of a minor German court. But as he grew older the situation became worse. He began to be scoffed at for being overly French in his manners. "Late in life he became an object of ridicule for his old fashioned overly ornate clothes and huge black wig." The younger brother of George Ludwig, also named Ernst Augustus, once called him an "archeological find." He joked that Peter the Great (whom Leibniz had met in October 1711) must have confused him with the Duke of Wolfenbüttel's clown.[14]

Given his reduced status in Hanover, it was fortunate that he did develop enduring friendships with three high ranking women. One was Sophia of the Palatinate (1630–1714). She was the granddaughter of James I, King of great Britain,[15] and she was married to Ernst Augustus, the successor of Johann Friedrich as Duke of Hanover by whom she had six sons and one

---

[12]Mates (1986), p. 32.
[13]*Ibid.*
[14]Coudert (1995), p. 12. Also see Antognazza (2009), p. 469.
[15]Her parents were Frederick V, Elector Palatine (1596–1632) whose brief rule as King of Bohemia touched off the Thirty Years War and Elizabeth of Bohemia (1596–1662), the daughter of James I.

daughter. She was quite unlike her husband who was primarily a politician and had no intellectual interests. Sophia was intelligent and well educated and had great respect for Leibniz. The two used to walk in her beloved formal gardens of the Ducal palace Herrenhausen discussing philosophy. Her daughter Sophia Charlotte (1668–1705), Queen of Prussia from 1701 to her premature death, shared her mother's intelligence and was likewise a good friend of Leibniz. The third female friend was Caroline of Brandenburg-Ansbach (1683–1737), the daughter of the Elector of Brandenburg. She married George Ludwig's son George August (1683–1760) who was to succeed to the British throne as George II in 1727 and reign for 33 years. There is an extensive correspondence between Leibniz and these three women. Leibniz feels free to discuss political or even philosophic issues with them. He was especially close to the older Sophia who was often upset at his many absences and in letters, asking that he come home, expressed her fondness for him. Leibniz was devastated when Sophia Charlotte died in 1705 and her mother in 1714. There exist letters expressing his sorrow. Unfortunately, the passing of the older Sophia meant that Leibniz was completely isolated in Hanover during the final two years of his life.

Leibniz's last two decades in Hanover were further darkened by the vicious dispute with Newton and Newton's followers over priority in the invention of calculus.[16] The quarrel developed rather slowly. Leibniz seems never to have met Newton during his two trips to London in 1673 and 1676. We recall that after the second visit he received two letters: the so-called *Epistola Prior* and *Epistola Posterior* from Newton sent through Oldenburg. Both letters concerned Newton's methods of handling series and their use in quadratures. Neither letter explicitly mentioned fluxions, although in the *Epistola Posterior* a sentence relating fluxions and fluents was concealed by an anagram. During the second trip Leibniz also had the opportunity to read some of Newton's manuscripts in Collins' possession, principally *De Analysi*. These papers also related mainly to series, a subject that had attracted Leibniz's interest since his earliest work with Huygens; but as we have pointed out in Chapter 5, *De analysi* also contained statements equivalent to the FTC giving the inverse relationship between area and ordinate. Although Leibniz had long since found his series for $\pi/4$ or "arithmetic quadrature" of the circle, it was clear as late as 1676 that he was still far behind Newton and James Gregory concerning series.[17]

[16]This subject over the centuries has been written into the ground; extended treatments may be found in Hall (1980), Westfall (1980), and Hofmann (1974a). There is an excellent short summary in a review by Blank (2009).

[17]By early 1676 Leibniz had twice received letter of Collins sent by Oldenburg giving

Historians are divided over whether the tone of Newton's two letters indicated any hostility to Leibniz on Newton's part. In the writer's opinion they are quite friendly. Although Newton probably regarded Leibniz as a beginner, he is clearly impressed by his talent.[18] Relations remained satisfactory for nearly the next 20 years. As late as 1693 there were occasional exchanges of polite letters. Frequently Leibniz praised Newton's mathematical ability to others, and in 1687 Newton in a Scholium just after Lemma 2 of Book II of the first edition of the *Principia*, where it is shown that the "moment" of x^r is proportional to rx^{r-1} (or in differential notation $dx^r = rx^{r-1}dx$), essentially gives Leibniz credit as co-inventor of his methods for tangents and quadratures.[19] Leibniz in turn was impressed with Newton's achievement in the *Principia*. However, like most scientists on the continent who were influenced by Descartes, he disapproved of the unexplained gravitational attraction across empty space which Newton had introduced. This he felt was a reintroduction of the "occult qualities" of the scholastics and incompatible with a true mechanical philosophy in which all motion depends on contact. We have already seen how this objection motivated his attempt to explain Kepler's laws via vortex arguments in the *Tentamen* article in 1689. An even more fundamental objection for Leibniz to Newton's theory was theological: if the Universe could possess such an "immaterial" property as gravitation, then it was dangerously close to the self-sufficient monism of Spinoza. If matter could move by itself, then perhaps it could acquire thought, and life itself might then be a natural aspect or mode of the Universe.[20] Nor did Leibniz approve of a solution to the problem of gravitation by one of Newton's followers Richard Bentley (1662–1742), a noted theologian and classical scholar who later (1700) became master of Trinity College. Bentley argued that since gravity could not be an inherent property of matter, it must be directly implanted by

---

Newton's series for $\sin x$ and $\arcsin x$ as well as Gregory's series for $\tan x$ and $\arctan x$. After the first letter Leibniz said he had "no time" to compare these results with what he had found several years previously. After the second letter he asked for the derivations offering to share the $\pi/4$ series in return. This occasioned the *Epistola Prior* which contained the derivations.

[18]We share Hall's (1980) opinion (p. 33f, pp. 65–68) that the letters are polite and friendly. On the other hand, Hofmann (1974a), Westfall (1980), and Blank (2009) detect suspicion, hostility, and even paranoia concerning Leibniz in Newton's two letters. We are unable to detect this. Newton's tone seems perfectly polite and even complimentary to Leibniz.

[19]In later editions this was removed. See Newton (1999), p. 649 and Hall (1980), p. 33.

[20]This point is clearly stated in a 1714 letter to a French ally quoted in Stewart (2006), p. 274.

God. To Leibniz this called into question God's workmanship in the same manner as Newton's idea that periodic divine intervention was required to correct instability in the solar system.

These objections, as we have seen, led to Leibniz's attempt to construct a workable vortex theory in his *Tentamen* of 1689. Although Newton was in later years contemptuous of Leibniz's mathematics in his dynamical papers of 1689, viewing their results (aside from the mistakes) as either the same as his own or easily deducible from them, Newton probably did not read Leibniz's cosmological arguments before 1710. Hence as far as Newton personally was concerned, they initially played no direct role in launching the quarrel. However, Leibniz's criticisms probably did contribute to a general climate of bad feeling on the part of Newton's followers, reinforcing motives on their part to attack him on the calculus issue; and eventually this philosophical quarrel did become fused with the priority dispute.[21] Also for twenty years or more Newton seems not have paid any attention to what was going on mathematically on the continent. He seems to have only become aware of the rapid European development of the calculus and the credit being given Leibniz for it in 1695 when John Wallis warns him in a letter that "your Notions (of Fluxions) pass there with great applause, by the name of *Leibniz's Calculus Differentialis*" and strongly advised Newton to publish "lest others carry away the Reputation that is due you."[22] Even Leibniz in 1694 and again in 1697 urged Newton get his fluxions into print.[23] Even so, Newton remained passive. He disliked controversy and wanted mostly to be left alone. However, Newton seemed already aware of Leibniz's rising reputation and as early as 1685 had complained to people he knew intimately about the credit being given to Leibniz.[24]

The first claim that Leibniz had plagiarized the calculus came from a young Swiss mathematician Fatio de Duillier (1664–1753) who visited Newton in 1693. In 1699 he published a treatise on the brachistochrone problem published with the imprimatur of the Royal Society.[25] He asserted that Leibniz was the second inventor of the calculus and might be a plagiarist. A second attack on Leibniz came from John Keill (1671–1721). Keill, a Scottish mathematician, was an ardent Newtonian and opponent

[21] Hall (1980), p. 164.
[22] Quoted in Hall (1980), p. 112.
[23] *Ibid.* p. 113.
[24] Westfall (1980), p. 718f.
[25] That is the problem of finding a curve between two points such that an object sliding on it from the higher to lower point without friction in a gravitational field has the quickest descent.

of continental Neo-Cartesian philosophy. In 1710 there appeared a paper *On the Laws of Centripetal Force* by Keill in the *Philosophical Transactions* of the Royal Society for the year 1708[26] In the last third of the paper in commenting on some of his results involving integration Keill writes:

> All these things follow from the nowadays highly celebrated arithmetic of fluxions, which Mr Newton beyond any shadow of a doubt first discovered, as anyone reading his letters published by Wallis will readily ascertain, and yet the same arithmetic was afterwards published by Mr Leibniz in the *Acta Eruditorum* having changed the name and the symbolism.[27]

Keill's motives in doing this are uncertain. Like John Wallis he may have felt that Newton was indeed the first discoverer and resented the success that the Leibnizian calculus was having in Europe without any recognition being given to Newton. Perhaps he could not accept that since Newton had never published his work, continental mathematicians might be quite unaware of it. Then too, he might have resented the carping criticism of Cartesian philosophers against the "occult" quality of gravitation (Keill himself had been the object of such criticism).[28]

After Keill's publication matters got much worse. Fatio de Duillier had been more or less ignored. Newton had dropped him and when Leibniz had complained about him, Wallis replied that de Duillier had gained the Royal Society's imprimatur only through trickery. Further Wallis said that neither he or the Royal Society approved of Fatio's attack, and that Newton himself was annoyed by it.[29] In fact, de Duillier seems to have been a bit mentally compromised. He later became a religious fanatic and was made to stand in the pillory for some offense in Britain. As Johann Bernoulli joked, he required medical treatment with Hellebore for his mental condition rather than criminal prosecution.[30] The Keill affair was much more serious for the article had passed muster in the Royal Society's own journal. When Leibniz heard of Keill's action, he formally protested in his capacity as a Fellow and demanded an apology. Had it not been for a previous action by Leibniz this protest might have succeeded. In 1704 Newton at the urging of his friends had finally published *De Quadratura de Curvarum*, and it was reviewed by Leibniz anonymously in the *Acta*. After several flattering references

---

[26]But published two years later.
[27]Quoted in Hall (1980), p. 145. Also see Antognazza (2009), p. 487.
[28]*Ibid.* p. 145.
[29]*Ibid.*, p. 121 and Antognazza (2009), p. 429.
[30]Hall (1980), p 141.

to Newton he hints that his work derives from Leibniz's differences and compares Newton to Honoré Fabri (1607–1688) who was a French Jesuit theologian and minor mathematician. Newton's fluxions Leibniz writes:

> *are almost the same as the increments of the fluents generated in the least equal portions of time* He has made elegant use of these both in his *Principia Mathematica* and other publications since, just as Honoré Fabri in his *Synopis Geometrica* substituted the advance of movement for the method of Cavalieri.[31] [Emphasis in original]

Leibniz then explains his method using his symbolism and says that Newton "has labored very usefully at these tasks,"[32] but for further details Leibniz advises consulting a recent textbooks published by two other Scots: George Cheyne (1671–1743), a well known physician who dabbled in mathematics and John Craige (d. 1731), probably a student of David Gregory. Whether or not an insult or put-down was intended, Newton became quite angry when shown the review by Keill and said that he had more cause to complain of the review than "Mr. Leibniz has to complain of Mr. Keill."[33] Keill defended himself in a letter read at a meeting in late May of 1711 of the Royal Society in which he stated that Newton was clearly the first inventor of calculus and while not directly accusing Leibniz of theft claimed that it was possible that a man of Leibniz's intelligence could have derived calculus from the many hints in Newton's two letters of 1676. He further protested that the *Acta Eruditorum* had attempted to rob Newton of credit for his discovery by hinting that he had appropriated the work of others. This letter was transmitted to Leibniz with a barely polite cover letter by Hans Sloane (1660–1753), the secretary of the Royal Society. Leibniz's dignified reply to Sloane several months later (December, 1711) defended the *Acta* saying that it had given proper credit to everyone and claimed his own right to an independent discovery, while asserting that he had always granted that Newton had arrived at "basic principles similar to our own." But Newton could not claim to have forestalled him as he had made his discovery as far back as 1675. Leibniz also called for Newton himself to correct Keill.[34]

Probably influenced by Keill, Newton gradually became convinced not only that he was prior to Leibniz, but that Leibniz had stolen his fundamen-

[31] *Ibid.*, p. 138.
[32] *Ibid.*, p. 139.
[33] *Ibid.*, p. 169.
[34] *Ibid.*, p. 176.

tal ideas and disguised the theft by inventing a new notation. As Newton was president of the Royal Society it was easy for him to convene a committee to look into the dispute. Its report (written by Newton) was completed in April, 1711. Its title was *Correspondence of John Collins and others about the development of Analysis*; abbreviated in Latin, it was known as the *Commercium Epistolicum.* It consisted of all relevant correspondence. The reports concluded by asserting that Leibniz did not have a calculus before June 1677, more than a year after the *Epistola Posterior* "in which Letter the Method of Fluxions was sufficiently described to any intelligent person. Leibniz was not directly accused of plagiarism, but of concealing his knowledge of Newton prior work in the 1660s. In support of this thesis it mentioned Leibniz's attempt to claim credit for his 1673 results on series which were the same, as John Pell had pointed out, as work done by Mouton.[35]

A few years after the publication of the *Commercium Epistolicum* Newton published anonymously a 52 page work with the title *Account of the Book entitled Commercium Epistolicum* which was issued also in French and Latin. Here he further blackened Leibniz's reputation. Not only was much of Leibniz's earliest work in series stolen from Mouton, but his own inventions such as the summing of the series with terms $1/n(n+1)$ were fairly trivial. "See the Mystery!" Newton contemptuously exclaims before explaining Leibniz's results. Also, while the proof of his series for the circle quadrature may have been in some sense his own, he hints that the series itself was taken from Gregory, who had discovered it in 1669 and sent it to Collins; he also notes that the "Theoreme for transmuting of Figures" upon which Leibniz's argument depends is "like those of Dr. Barrow and Mr. Gregory."[36] Likewise, Newton had sent to Collins his series for the sine and arcsine, and Collins was given permission to communicate these results freely, i.e., probably also to Leibniz. Leibniz then had the effrontery to ask Oldenburg to send him the proofs of these two sine series, apparently forgetting that he had earlier claimed them as his own! In the case of the calculus Newton claims that Leibniz certainly could have easily deduced his method from the information in a letter of 1672 to Collins explaining his method of tangents together with his first and second letters which were all sent to Leibniz by Oldenburg; for had he not said that he had a method to find:

[35] *Ibid.*, p. 178f.
[36] *Account*, pp. 183, 185 in Hall (1980), pp. 273, 275.

> Tangents, Areas, Lengths, Solid Contents, Centers of Gravity, and Curvities of Curves, and curvilinear Figures Geometrical and Mechanical, without sticking at Surds; and that the Method of Tangents of *Slusius* was But a Branch or Corollary of this other Method.[37]

There is moreover according to Newton no evidence that Leibniz had any idea of differential calculus before his receipt of Newton's second letter, since his first explanation of it was in his letter to Oldenburg in June of 1677. Also an earlier letter of Leibniz (November, 1676) to Oldenburg speaks only of a method generalizing Sluse's rule so that "The Improvement by the differential calculus was not yet in his Mind."[38] Anyway, Leibniz's method of differentials is simply Barrow's except that a and e are written as dx and dy[39]

The quarrel over priority became ever more intense and occupied Leibniz for the rest of his life. Friends of both Newton and Leibniz made the situation even worse by egging them on. In Leibniz's case the principal actor was Johann Bernoulli who had already criticized the *Principia* in the *Mémoires* of the Paris Academy of Science by pointing out what he considered to be various mathematical mistakes in it. He now convinced Leibniz in a letter written in June 1713 that although Newton may have known of fluxions he "did not so much as dream of" Leibniz's calculus which served "as an algorithm in the manner of arithmetical and algebraic rules" until he had appropriated the idea from Leibniz.[40] Keill served the same role for Newton. At the end of July in reply to the *Commercium* and infuriated by Bernoulli's letter, Leibniz broadcast Bernoulli's accusation of plagiarism in a leaflet labeled in Latin *Charta Volans.*[41] which was published in a French translation in the *Journal Literaire* for November/December 1713. Leibniz included Bernoulli's comments in his June letter attributed not to Bernoulli but to a "leading mathematician." The leaflet was also purportedly written by a third person, but it was evident that the true author was Leibniz himself; and it was also easy to guess that the "leading mathematician" was Bernoulli. The *Charta Volans* further enraged Newton and his followers who continued attacks on Leibniz even after his death in 1716.

---

[37] *Ibid*, p. 191 in Hall (1980), p. 281f.
[38] *Ibid.*, p. 283.
[39] *Ibid.*, p. 286.
[40] *Ibid.*, p. 199. Yet in spite of these damming comments about Newton, Bernoulli seems to have had a healthy fear of the Newtonians, for in the same letter he requests Leibniz not to mention his name in connection with the dispute.
[41] This means "Flying Paper" or a leaflet.

The quarrel also began to go beyond questions of priority and plagiarism to become a fight between the metaphysics of Leibniz and Newton. Leibniz kept harping on the "occult" nature of Newton's gravitational force and ridiculed Newton's views that God had occasionally to step in and adjust the solar system to keep it stable and that space, which Newton believed to be real and absolute, was somehow the "sensorium" or organ of God. These questions were debated in a hammer and tongs correspondence from November 1715 until a month before Leibniz's death (November, 1716) with the English divine Samuel Clark (1675–1729) who was a friend of Newton and rector of his parish.

What are we to make of this dispute? Although Newton by nature was inclined to neurosis and paranoia, he did have reason to be suspicious of Leibniz's claims and behavior. Aside from the insulting anonymous review of 1704 comparing him to Fabri, in the 1684 *Acta* article on differential calculus, Leibniz failed to acknowledge him except (possibly) in the most oblique terms. In this paper Leibniz said that his new method applied to the most difficult problems which could not be solved without it *or a method like it* ("*aut simili*" in Latin). Newton (probably correctly) thought that this backhanded remark applied to him, and he considered it a veiled insult. In the *Account* he writes:

> What he meant by the words *AUT SIMILI* was impossible for the *Germans* to understand without an Interpreter. He ought to have done Mr. Newton justice in plain intelligible Language, and told the *Germans* whose was the *Methodus SIMILIS*, and to what Extent and Antiquity it was, according to the Notices ne had received from *England*; and to have acknowledged that his own Method was not so ancient.[42] [Emphasis in original]

Newton felt that Leibniz had surely been aware of his progress in quadrature and tangent problems, either from the *Epistola Posterior* of 1676 or the *De Analysi* which Newton knew through Collins that Leibniz had read. Moreover, since Leibniz failed to respond to that letter until June 1677 and being unaware of its late delivery, Newton not unreasonably suspected that Leibniz had studied it for many months and that consequently it must have influenced his reply to Oldenburg where (see Chapter 6) he explained his system for the first time.[43] And like other British mathematicians, Newton was aware that Leibniz's work in series (even his arithmetical quadrature

[42]p. 220 in Hall (1980), p. 310. Also quoted in Westfall (1980), p. 719.

[43]This perception may have been the initial motivation behind Newton's complaints to his friends in 1685.

of the circle) and some of his transmutation theorems were not new (and possibly therefore not even original). Over the decades what he may have at first regarded as the slips of a talented amateur began to assume a more sinister aspect, given the repeated warnings of de Duillier, Wallis, and Keill and his own perceptions of Leibniz's behavior. Newton's psychological makeup may explain his almost insane rage at Leibniz once he was provoked, but in the intensely competitive seventeenth century atmosphere which bred priority disputes *ad infinitum* some resentment would have been natural in any mathematician finding himself in Newton's position.

From our analysis of Leibniz's work in Chapters 5 and 6 we have been forced to conclude that the modern consensus of historians that Leibniz had arrived at the key ideas of his calculus as early as the fall of 1675 before his second visit to London and before Leibniz's communications with Newton is not quite correct, thus confirming the claim of the *Commercium Epistolicum* and Newton's *Account* that Leibniz's calculus in the form with which we are familiar was only perfected *after* his second trip to London in 1676, not before. Prior to it, as we have tried to show, Leibniz had "calculus-like" results like various "transmutations," where one quadrature could be reduced to another, or the generalization of Pascal's moment theorem. He had introduced the notation "$\int$" and "d" and was familiar with the characteristic triangle, but the meaning of some of his notation was different than in his later work. For example, $\int l$ was usually a sum of lines in the sense of Cavalieri and raised the dimension.[44] He was in possession of such formulas as $\int x^r = x^{r+1}/r + 1$ for certain r (which for the most part were not new) and what we would call integration by parts. But his reasoning depended on ideas from Cavalieri, moment arguments, or transmutation. He could also solve the occasional inverse tangent problem, given that the subtangent satisfied some geometrical condition. His discoveries relating to series in part depended on the realization that the sum $\sum_{k=1}^{n} t_k$ where the term t_k is a difference of the of the form $a_{k+1} - a_k$ is $a_{n+1} - a_1$. Except for a remark in a manuscript of November 11, 1675 he does not yet exploit this insight systematically in the infinitesimal case. His pre-London discoveries were praiseworthy and original (the *Commercium* and Newton are quite unfair here to suggest otherwise), but they were on a different track from what followed. They seem "modern" only from hindsight and if we "translate" them using modern notation. Doing this, however, without paying attention to Leibniz's underlying patterns of thought is as

[44]It is true, however, that by the summer of 1676 that $\int$ signaled a sum of strips or differences and did not raise the dimension.

un-historical as doing the same with the *Lectiones Geometricae* of Barrow. To summarize the situation, one feels that Leibniz was maddeningly close to his subsequent algorithm, but had by the summer of 1676 not quite arrived at it. What may have happened then in London to accelerate the changes in Leibniz's infinitesimal analysis that occurred soon afterwards? Let us begin by excluding one possibility. Contrary to Newton's assertions the *Epistola Prior* and *Posterior* contained nothing explicit about fluxions; they were devoted to Newton's various manipulations of series. And the famous anagram in the *Epistola Posterior* could not have been a clue as it was indecipherable; furthermore, Leibniz received this letter only very shortly before his own reply to Oldenburg in June of 1677 where he announced his own results. However, this conclusion may not hold for *De Analysi* which Leibniz read and made notes upon during his visit to London, since that work contained a clear statement of Newton's version of the FTC and its geometric implications. Yet Leibniz's notes on this work deal only with its results on series, not on the FTC. A charitable hypothesis is either that Leibniz did not notice this part of *De Analysi* (which was near the end anyway of a fairly complex work[45]) or it was already familiar to him, so he felt no need to take notes on it. If we wish to be uncharitable, we might suppose that Leibniz at once realized the implications of Newton's result for his own work, but did not take notes because, given his new understanding, he did not need to; perhaps also he wished to leave no "smoking gun" in the form of written evidence. It is also certainly possible that whatever caused Leibniz to accelerate his work on calculus in 1676–77 *may* have been independent of his second visit to London; perhaps he was inspired by conversations with Hudde, a very talented mathematician, whom he visited in November 1676 enroute to Hanover. Yet we are inclined to doubt this. It is a very common situation for a research mathematician (at least in the personal experience of this writer) to hear at, say, a conference that someone else has a method or has proved something that he had been working on. In these circumstances a sense of pride and competition will likely make him redouble his efforts. (Also the slightest hint or something unconsciously absorbed can produce a sudden epiphany and make the difference.) One can imagine (and certainly can't rule out) Leibniz, after reading something or after a conversation with someone (perhaps Collins?) in London explaining some detail of Newton's work, or even after some random remark realizing

[45]It is true that *De Analysi* begins with the standard formulas for the quadrature of x^r, but these were already quite familiar to Leibniz. The *De Analysi* itself and a reproduction of Leibniz's notes on it may be found in Newton (1967), II, pp. 206–259.

in a flash: "Aha! So that's what Newton is doing. Well, I can do the same with my differences." Or Leibniz may have been merely informed again (as he had been in letters from Oldenburg in April 1673, and in December 1674[46] by Collins or some other a third party in London that Newton had a method for quadratures, tangents, etc. Given this news with no details of the method, he might have supposed that the same was also true of several other people including Fermat, Barrow, Roberval, Huygens, and Sluse. Perhaps Newton's tangent "method" was only an improvement of Sluse's rule and his quadratures amounted only to the repeated quadrature (integration) of the terms of a power series. But in any case, seeing the enthusiasm of the British for Newton, he might have felt that it was time to finish his own work from which he had been distracted the myriad of his interests and diversions (some of which we have mentioned in Chapter 5). He may even have had his algorithm in mind at the time of his visit, but given the incredible number of his interests had not written it down until hearing of Newton's work. However, we will probably never know exactly where the truth lies unless some new evidence is discovered; except that conscious plagiarism is unlikely although it cannot be absolutely ruled out.

Whatever information Leibniz received in London, his own calculus is conceptually quite different from Newtonian fluents and fluxions. As we have tried to show, it was motivated (and this took some time!) by his first work on series in 1673 and is founded on the ideas of differences and sums. dy is a difference between two nearly equal values of y and $\int y\,dx$ can be evaluated provided $y\,dx$ can be written as a difference, thus generalizing the trivial case that $\int dy = y$. As for tangents, although dy/dx on occasion makes its appearance, Leibniz has little interest in it as an independent object, function, or rate of change. For him it is simply a device to calculate the subtangent t from the equation $dy/dx = y/t$. This equation was familiar (in some form) to many contemporary mathematicians including Sluse. Leibniz's only aim which he shared with others was to construct the tangent from a knowledge of t.

Had Leibniz been deeply influenced by Newton one would expect to see a different treatment of dy/dx, anticipating Euler's functional point of view. For Newton's fluxions are rates of change. If y is a quantity changing in time $\dot{y}$ is the velocity with which it is changing; conversely, given what

[46]See Hofmann (1974a), p. 42f. The first letter contained a long draft by Collins on the achievements of British mathematics by 1673. The second letter is given in and Newton (1959), II, p. 330f.

amounts to the fluxion $\dot{y}$, we want to find the fluent y.[47] Also even if the content of Leibniz's work was influenced by Newton's specific results, Leibniz is somewhat justified in his later claims that Newton did not invent a *calculus*.[48] After all, fairly trivial results from Leibniz's point of view such as

$$dx/dy = \frac{1}{dy/dx} \quad \text{or} \quad dz/dx = (dz/dy)(dy/dx)$$

would be difficult to express in the language of fluxions. These differences between the systems of Newton and Leibniz are persuasive evidence for a fundamental independence between the two.

The dispute over the invention of calculus did not harm Leibniz's reputation in Europe. He received support from continental mathematicians such as Varignon (1654–1722), Christian Wolff (1679–1754),[49] the Marquis de L'Hospital (1661-1704), and Johann Bernoulli who had enthusiastically embraced Leibniz's calculus and continued to develop it. Nevertheless it was a depressing episode for Leibniz in the last few years of his life. It may also have further damaged his prospects with the new king George I. Leibniz very much wanted to join the King in London and to be appointed Royal Historiographer. He was summary rejected and told to stay in Hanover. The main reason no doubt was his irritating failure to complete the history of the Guelfs. But even had Leibniz finished this task, the extreme hostility towards him in England would have made any such appointment unwise, especially given the fact that George was unpopular to begin with. Other setbacks followed. In 1715 Leibniz's emolument of 600 talers per annum as president of the Society of Sciences in Berlin which he helped to found was

[47]Both these points of view are still embedded in a modern calculus course. Leibniz's approach gives the slope of a tangent line to a point on the graph of a curve when we calculate dy/dx. Newton's leads to the concept of instantaneous velocity, or if we take the "second fluxion" instantaneous acceleration.

[48]In a letter of June 28, 1713 to Johann Bernoulli Leibniz writes of Newton's supporters that:

> They would maintain Newton in possession of his own invented calculus and yet it appears that he no more knew our calculus than Apollonius knew the algebraic calculus of Viète and Descartes. He knew fluxions,but not the calculus of fluxions which (as you rightly judge) put together at a later stage after our own was already published.

Quoted in Antognazza (2008), p. 496.

[49]Wolff, a professor of mathematics and natural philosophy at Halle, is also considered the greatest German philosopher between Leibniz and Kant. His philosophy was a modification of Leibniz's.

first reduced by 50% and then eliminated on the grounds that he had done no work for the Society and had not even visited Berlin for several years. In addition to this financial loss, George Ludwig refused to pay for most of his expenses in Vienna. Because he had done some work for the Electorate in Vienna he was offered compensation for three months spent there, but not for the remaining eighteen months.[50]

During Leibniz's forty years in Hanover, he pursued a vast correspondence with practically every major (and minor) intellectual figure in Europe and countless projects. In 1895 when an attempt was made to catalogue his work 15,000 letters were found to over 1000 recipients. Leibniz's manuscripts and printed work which survive amount to some 200,000 pages in seven languages. Most of this material which amounts to more than 57,000 separate items continues to be preserved in the State Library of Lower Saxony in Hanover. The effort to catalogue and print it began in 1901; however, because of delays caused by two wars, the Nazi dictatorship, the Holocaust, and the division of Germany the project was still incomplete in the 1980s.[51] Even during the last years in Hanover when much of his energy was spent trying to finish the history of the Guelfs he found the time to develop the intricate philosophical system for which he is remembered today. In 1710 he published an extensive philosophical work, one of the few he ever finished, entitled the *Theodicy*, which attempted to reconcile God's goodness, human freedom, and the existence of evil and argued that this is necessarily the best of all possible worlds. This is one of the few books he ever published, others being the youthful *De Arte Combinatoria* of 1666 and the very short *Hypotheses physica nova/ Theoria motus abstracti* of 1670. Around 1714 he also finished (but did not publish) an untitled essay known later as the *Monadology* which asserted that the ultimate reality is a collection of non-interacting spiritual substances (monads) existing outside of space and time and each mirroring the universe.[52] Aside from these works on philosophy or science he published three volumes of medieval documents relating to the Guelfs in 1707, 1710, and 1711.[53] There are substantial but unpublished treatises such the *Discours de Metaphysique* (1686), and *Nouveaux Essais sur l'Entendement* (1707),

[50] Antognazza (2009), p. 523 and 526.

[51] Mates (1986), p. 34.

[52] Intermediate versions were only published after his death in German and Latin translations in 1720 and 1721. The final version was not published until 1840 when Leibniz's philosophical works were edited by J. E. Erdmann. See Antognazza (2009), p. 500f.

[53] But these were unappreciated by George Ludwig who was expecting a short well organized history of the family. See *Ibid.*, p. 464.

an attack on John Locke's (1632–1704) *Essays on Human Understanding*, and an outline and several essays attempting to reconcile Catholic and Protestant theology, laying the ground work for church reunification which he called the *Demonstrationes Catholicae* (1671–). Leibniz also published more than two dozen journal articles together with a multitude of pamphlets (many anonymous), reports, and various schemes for his employers, but no other major publications. Most of his achievements are known only from his correspondence or the immense collection of manuscripts he left. Almost all his major projects were incomplete at his death. His secretary Eckhart who was trying to gain control of the Guelf project wrote:

> ... he is far too much distracted; and since he wants to do everything and to be involved in everything, he can finish nothing, not even if he had angels as assistants.[54]

And Leibniz himself wrote in a letter:

> I cannot tell you how extraordinarily distracted and spread out I am. I am trying to find various things in the archives; I look at old papers and hunt up unpublished documents. From these I hope to shed some light on the history of the [House of] Brunswick. I receive and answer a huge number of letters. At the same time, I have so many mathematical results, philosophical thoughts, and other literary innovations ... that I often do not know where to begin.[55]

Such deficiencies were the inevitable product of his incredible creativity. In spite of them he is clearly one of the greatest and most varied thinkers in human history.

Leibniz died in Hanover November 14, 1716 at the age of 70. His funeral and burial was held on December 14 in the Neustädter Kirche. Although the testimony we have may be unreliable, no one from the court was present (even though George I was in the area) except his secretary Johann Georg Eckhart, who had frequently complained about Leibniz behind his back to the Duke and who was to be his first biographer. No official notice was take by Berlin Society of Sciences. More understandably, given the mutual enmity, the same was the case with the Royal Society in London. Only in 1717 was an eulogy of Leibniz read by Bernard de Fontenelle (1657–1757), an important French mathematician and scientist, at the Royal Academy

[54] *Ibid.*, p. 546.

[55] Letter to Vincent Placcius in September 1695. Quoted from the Wikipedia article "Leibniz." An equivalent but slightly different translation is given in Antognazza (2009), p. 321.

of Sciences in Paris. At Leibniz's death the History of the Guelfs was still unfinished. However, Leibniz had almost reached the year 1024, the year of the death of the German Emperor Henry II. This text was only published in the nineteenth century in three volumes and is a remarkably well written and learned history.

Chapter 11

Some Concluding Remarks on Mathematical Change

One of our major theses is that Leibniz's calculus combined with Descartes analytic geometry represented the beginnings of a mathematical analog of a Kuhnian "paradigm shift," a phrase which has become a cliché not only in the history of science but in journalism, corporate management theory, or automobile advertisements (one wonders what its inventor, a rigorously trained Harvard physicist, would have made of it all). But as we have argued in Chapters 1 and 2 the analogy with scientific change goes only so far. Rejected scientific theory is viewed as *false*. Its claims are no longer seen to correspond to the way the world is. Despite the assertions by sociological proponents of the "Strong Programme" or various "postmodern" theorists of science, this situation has been rarely the case in mathematics. The conclusions of Viète or Apollonius are as valid today as they were when they were made. On the other hand, mathematics can certainly abandon a previously dominant "thought style" or subject matter and go off in new directions in such a profound way that the previous work is essentially forgotten or even made inaccessible, and this was the fate of seventeenth century geometry.

There are also other suggestive similarities between the mathematics of the seventeenth century and revolutionary epoches in the history of other scientific disciplines as interpreted by Kuhn. Recall that Kuhn held that in the sciences there is little real communication between the supporters of old and new paradigms. The two parties may argue, but they talk past each other. They appeal to standards and values which are alien to or not even understood by their opponents. Moreover, adherence to the new paradigm resembles a "conversion" experience or Gestalt shift rather than a purely rationally grounded change of view. To borrow an analogy from a well known psychological test, what had looked like an array of rabbits now becomes a flock of ducks. This shift cannot be communicated to the opposing party; it has to be directly experienced. In these senses the two paradigms are "incommensurable." Just as in the case of the hypotenuse of an isosceles right triangle compared with a unit side, there is no "common measure" between the two. From this Kuhn drew the implication that paradigm change can be more a matter of a political or sociological than rational argument. The Old Believers are simply defeated and die out without being won over. Kuhn goes on to argue that this phenomenon also means that the victorious paradigm cannot be "truer" than the other in the sense of correspondence with a "reality" which has the fundamentally unknowable status of a Kantian noumenon. Each paradigm in a sense creates its own world for its believers. At the same time, however, we can regard the new paradigm as "superior" to the old in the sense that it can explain puzzles that were beyond the capacity of the other.

Passing aside from the issue of truth—which as we have repeatedly remarked is rarely appropriate in mathematical change—and whatever the applicability of Kuhn's analysis to science, it is is obviously a bit too strong in the case of mathematics since Leibniz, Descartes, the Bernoulli brothers, *et al.* could understand and use the seventeenth century geometric paradigm; after all, they were trained in it. Yet we can also see in the transition to the Leibnizian calculus some of the phenomena Kuhn noted. Leibniz's relationship, for instance, with his ex-tutor and mentor Christiaan Huygens is similar in certain respects to what is seen in a period of revolutionary scientific change. The two, although polite to each other, sometimes seem to be unable to communicate because they lack a common frame of reference. Not only does Huygens, one of the ablest mathematicians of his time, until nearly the end of his life reject most of Leibniz's ideas, but he does not even seem to understand them or see them as "mathematics"[1]

It is clear that Huygens has a high regard for Leibniz's mathematical precocity. He had been impressed by Leibniz's series summations, the generalization of Pascal's computation of the surface area of a sphere, and the circle quadrature via series. Nevertheless for a long time he could not bring himself to quite accept the calculus, much less Leibniz's other "characteristics" such as the *Analysis situs.* Their letters are conventionally polite. In some cases Huygens is quite complimentary, in others (particularly to third parties) somewhat cool and distant regarding Leibniz. In a letter of 1690 Leibniz had sent a letter to Huygens announcing his new methods:

[1]This is amazing phenomenon from our point of view since today Huygens geometrical methods are nearly impenetrable, while calculus is a subject studied by college freshman.

> I do not know, Sir, whether you have seen in the Acts of Leipzig a method of calculating ... which will bring into the realm of analysis what even Mr. Descartes had excepted. Whereas the indices of quantities, used up to the present in calculating, were only roots or powers, I now use the sums and differences, such as $\overline{dy}$, $\overline{ddy}$, $\overline{dddy}$, that is to say the differences or increments of the quantity y, or the differences of the differences, or the differences of the differences of the differences, etc. ... By means of this calculus it occurred to me to give the tangents and to solve the problems *de maximis et minimis* when the equations are highly encumbered with roots and fractions, and it often saves me immense calculations.[2]

Leibniz goes on to say that his calculus applies the mechanical curves such as the cycloid which Descartes had barred from geometry. In reply Huygens replies that he had "seen something" of the new calculus, "but finding it obscure, I have not studied it sufficiently to understand it."[3] In a subsequent letter (August 1690) Huygens finds the basis of Leibniz's method "very good" but thinks that his own geometric methods will give equivalent results and challenges Leibniz to solve the catenary problem submitted by James Bernoulli which Huygens had already solved. Later, after examining Leibniz's solution, Huygens rejects the sort of "super transcendant lines" which Leibniz had introduced "in which unknown quantities enter into the Exponents." They appear "so obscure to me that I am not of the opinion that they should be introduced into geometry."[4] Huygens continues to have trouble understanding Leibniz's calculus and begs Leibniz not to "take it for granted" that he and Fatio de Duillier understand it.[5] In correspondence between Huygens and Fatio de Duillier both writers show a certain skepticism about Leibniz. Huygens again complains about Leibniz's obscurity and wonders if some of his quadratures are not impossible. In return Fatio doubts Leibniz's sincerity and originality, sentiments Huygens also shares in correspondence with the Marquis de l'Hospital.[6] When Huygens makes marginal notes on the pages of Leibniz's articles in *Acta Eruditorum* he can be much blunter: there are comments like: "I don't know whether he will solve the problem, although he wishes to appear to have solved it."

[2] Huygens (1899), IX, pp.450–52. Quoted in Dugas (1958), p. 496f.
[3] *Ibid.*, p. 471f and Dugas *op. cit.*, p. 497.
[4] *Ibid.*, p. 532f and Dugas *op. cit.*, p. 499.
[5] *Ibid.* p. 93 and Dugas *op. cit.* p. 500.
[6] Dugas *op. cit.*, pp. 502–504.

"This has long been easy to do, and we have done it." "A grandiose title, but in substance nothing," or just "Nonsense!"[7]

Huygens general reaction towards the calculus has been summarized by Hofmann:

> Whoever is able to acquire this technique will be superior to a degree undreamt of by the initiated, even when he has no particulary deep insight into the structure itself: the formalism will think for him It was this very possibility that Huygens deemed undesirable. He was ... the last of the old school of mathematicians who came to their results through truly ingenious but simple isolated conclusions: his bent was... to the pure geometrical chain of argument... Leibniz' procedure appeared to him as a more or less capricious game which threatened to restrict his own freedom of thinking, and so he rejected it.[8]

But this interpretation may be too simple. Part of the problem shown by Huygens comments lies in the differing character of the two men. Huygens was cautious, businesslike, and valued clarity. Like others who dealt with Leibniz, Huygens was alternately impressed by his talent and actual accomplishments and irritated (as others also were) by his tendency towards bombast, samples of which we have already seen particularly relating to the Universal Characteristic: there are frequent suggestions and hints of astounding results and powerful methods, but their exact statement and derivation are on some excuse postponed and not given. As others did Huygens sometimes felt that Leibniz promised more than he actually delivered.[9] For these reasons and his own allegiance to the classical tradition, Huygens also may not, at least at the beginning, realized the power of Leibniz's calculus. Like Newton, Huygens thought of mathematics aesthetically in terms of the elegant arguments of geometry. What Leibniz was doing appeared to him to be mere symbol mongering. Leibniz's employment of geometry moreover rarely went beyond elementary similarity arguments involving the characteristic triangle. Except for the notation there was nothing very new, for instance, in the observation

$$\frac{y}{t} = \frac{dy}{dx}$$

where t is the subtangent. Also from the tone of his letters and comments, Huygens had concluded—especially in the case of the *Analysis Situs* that

[7] "Nescio an solverit problema, etsi vult videri solvisse", "Hoc longe facilius est et factum habemus", "Speciosus titulus in re nihili," "O NUGAS!" [Huygens (1899), p. 809].

[8] 1974a, p. 299.

[9] Bos (1978), pp. 59–68.

Leibniz's efforts to develop "calculi" were often pretentious demonstrations of the obvious. Only after 1693 near the end of his life, when it became evident that Leibniz and others had solved many difficult problems using the new methods, did Huygens relent and begin to acknowledge the power of calculus and to learn it himself through explanations provided by the Marquis de l'Hôpital (1661–1704) and Johann Bernoulli. But he still remained uncomfortable with it.[10] The fact that Leibniz's fundamental paper in the *Acta Eruditorum* on the differential calculus in 1684 was initially more or less ignored shows a similar uncomprehending reaction on the part of the general mathematical community. We suspect too that Newton's grounding in and sympathy with the classical geometric tradition may have ill-disposed him to appreciate the importance of Leibniz's ultra-symbolic methods and so became one more psychological factor in his quarrel with Leibniz over priority.

We now see Leibniz as the bringer of progress. But in successful revolutions there is loss as well as gain. Few other cultural artifacts are as quite as dead as pre-Leibnizian seventeenth-century mathematics. Who can really appreciate Barrow or Huygens today? Barrow's construction of a tangent to a curve at a point by first fitting a conic section "tangent" to the curve at the point and then constructing a tangent to the conic section using theorems of Apollonius or Huygens' analysis of the cycloid are miracles of geometric ingenuity, but they can only be understood by historically trained specialists, not by modern mathematicians; and even when they are understood these techniques can no longer be used creatively to solve new problems or even be fully appreciated.[11] It is quite true, of course, that there has been a real gain in power in *some* aspects of mathematics such as the ability to churn out formulas like (1.7) in Chapter 1. As we have pointed out, such things are now quite within the reach of (some) college freshmen, and it is certainly true that the equivalent quadrature problem would have challenged Barrow or Huygens. Also the meaning of a "primitive" or indefinite integral or its derivative was quite outside the range of ideas they wished to consider. On the other hand, countless elegant arguments, a whole geometrical *weltanschauung*, has been effectively lost except for a dusty existence in rare book collections.

---

[10] Bos (1980), p. 143.

[11] The National Science Foundation has a special category of "arcana" to designate grant proposals dealing with obsolete techniques of problems no longer of interest. Anyone exhibiting a new and non-obvious ratio or locus property of some "mechanical curve," elegantly proved by geometrical techniques, and proposing to find new ones would surely fall into this category.

The crux of the matter is that these early mathematicians seem retrospectively to be only slightly more accessible to us than the mixture of hermeticsm, Llullism, and magic, characterizing so much of Renaissance philosophy. In a formal way we can "understand" the mathematics and bend it by hindsight into something whose "valuable" part anticipates or is incorporated into the modern while the remaining part is consigned to oblivion or as Philip J. Davis and Reuben Hersh have remarked:

> The doors of the mathematical past are often rusted. If an inner chamber is difficult of access, it does not mean that treasure is to be found therein.[12]

But they fail to add that what is and what is not a "treasure" depends on the frame of reference of those making the judgments and that this is as historically conditioned as the judgments in any other cultural field.

The Leibniz episode suggests that the history of mathematics is more complicated than the story of linear progress that is often narrated. One cannot doubt that the *quantity* of mathematics increases although at a varying rate. This (in a sense) is certainly evidence of cumulative progress, which we don't wish to deny. But much of it—even the dominating results of a period—can be forgotten nearly as fast. And we would suggest that the course of its development has been extremely contingent. A myriad of cultural, political, personal, sociological, or what might be called "irrational" factors have influenced mathematics over time besides purely "mathematical" considerations. Often "rational" causation like "internal logic", "unsolved problems", or anything resembling a Lakatos research program are prominent only in what we might call (in analogy to science) "normal mathematics" which can be thought of as "micro-development" of a mature mathematical paradigm, and even here less frequently than one would expect. At least as often the underlying mathematical paradigm may come into being and pass away for essentially non-mathematical reasons—the focus of practicing mathematicians on the current state of their discipline guaranteeing the evaporation of both these reasons and the previous paradigm from consciousness. To give two historical examples of this phenomenon: C. S Fisher (1966) has argued that the decline of the once vigorous algorithmic invariant theory whose prime representative was Paul Gordan (1837–1912) and its replacement by the existential nonconstructive methods of Hilbert (1862–1943) reflects an essentially ideological shift within the community of algebraists, involving sociological factors

[12] Davis and Hersh (1981), p. 59.

which are independent of the actual content or "merit" of the rival approaches. Secondly, H. A. Edwards (1999) has diagnosed similar factors in the rivalry between Kronecker's (1823–1891) constructive and Richard Dedekind's (1831–1916) nonconstructive treatment of ideal theory:

> Dedekind's approach carried the day not because of its intrinsic merit, but because his ideology meshed with that of Hilbert, the most influential mathematician of the next generation, and in no small degree, because Dedekind was such a good writer.

The disagreement was a significant one, for Dedekind was a harbinger of the twentieth-century triumph of nonconstructive set theory with its many foundational problems. Today Kronecker's entire view of mathematics except among a few intuitionists (mainly in philosophy rather than mathematics departments) is almost as dead as the mathematics of Huygens or Barrow. Any working research mathematician can think of similar recent cases where an area of research or method flourishes or dies for reasons that are sometimes random and are arguably independent of purely mathematical considerations. In the modern era such factors can include the presence (or lack of) personal charisma, marketing and networking skills and political ability of individual mathematicians, NSF funding policy,[13] the failure to produce (for many reasons) students capable of carrying on good research, death or retirement of key figures in the field[14], the demands of government for new areas of research, or just plain strife. We suspect (simply because it is a human enterprize) that factors of this general kind also influenced the development of seventeenth century mathematics including the calculus.

To give one fairly recent (global) example of the kind of phenomena we are discussing, one needs only go back to the post-Sputnik era of the 1960s and compare its mathematics with that of the present. Then the government, reacting to an exaggerated and irrational fear of Russian dominance in the sciences, gave enormous support to mathematics. Rather paradoxically, however, the highest prestige was enjoyed by the ultra-pure areas. This period saw the flourishing of point-set and algebraic topology, logic and model theory, category theory, algebra, homology theory, arrow chasing, etc. Very little of this kind of mathematics had any practical application

[13]Such as the refusal to fund an entire area of mathematics based on the dominance of competing fields in the relevant NSF policy committees.

[14]For instance, the passing of Abraham Robinson and its effect on non-standard analysis.

(certainly not to defense), and one can easily recognize the excessively abstract textbooks and papers just as one can the finned cars of the period.[15] Faculty in such fields were kings on campus, enjoying enormous NSF and DOD grants.[16] By contrast, the applied fields were considered as distinctly second-rate suitable only for those who lacked the ability to do "real mathematics." For various historical reasons (including the present lack of funding for higher education) the situation is almost reversed today. Mathematicians in the fashionable areas of the 1960s are now almost unemployable. In hiring, at least at the great middle range of institutions, more and more emphasis is being put on super applied forms of mathematics in the hope of interacting with engineers or obtaining financial support by consulting with industry. Computer models of the internal flow of phlegm or the mathematics of credit default swaps are increasingly preferred to Grothendieck sheaves or fiber bundles. The result is that many once trendy areas—particularly point-set topology—are now nearly dead. Many profound and elegant results (as judged from "internal" mathematical factors) were produced in the earlier period, and this represents genuine progress. Some may still be taught in courses, but the whole focus of present research of an ever larger segment of the mathematical community has largely changed, and the earlier work, embalmed in unread journals, is gradually being forgotten. It is not quite yet in the position of the seventeenth century geometry of Barrow or Huygens, but in the relatively near future (say, twenty years), given present trends, it almost certainly will be.

We can also see ingredients of a sociological kind at work in the destiny of Leibniz's mathematics, where they produced radically different results in England and on the Continent. In England the unique status of Newton, his admiration of the geometric classicism of the Greeks, the powerful propaganda of the *Commercium Epistolicum*, the loyalty and xenophobia of his disciples, and persuasive textbooks like as Colin MacLaurin's (1698–1746) 1742 *Treatise on Fluxions* ensured the triumph of Newton, with Leibniz being cast into the outer darkness. By contrast, Johann Bernoulli, Nicolas Malebranche (1638–1715), Pierre Varignon (1654–1722), Leibniz's own disingenuous polemic skill, and especially the Marquis de l'Hôpital's *Anal-*

[15]One amusing satire of the mathematical fashions of the time was Carl E. Linderholm's 1972 book *Mathematics Made Difficult* (World Publishers).

[16]As an impecunious graduate student in History at Berkeley around 1963, this writer can recall an acquaintance who was a graduate student in pure mathematics. He was so wealthy from consulting with local defense contractors that he had his own private plane!

yse des infiniments petits first published in 1696, ultimately ensured a European triumph for Leibniz. The complete story of this process has not yet been written, but as Hall (1980) has shown it is a tale involving political intrigue as well as purely mathematical factors. We don't wish to suggest, however, that there was anything "unfair" about the eventual victory of the Leibnizian version of calculus. As developed by his immediate followers and later by Euler, it became an extraordinarily powerful mathematical tool. Nevertheless a close look at how this was achieved can be disturbing to mathematical Platonists.

Turning from sociological to cultural factors, Leibniz's invention of calculus was probably unique: we can think of no other mathematical era where cultural forces actually helped determine the actual character of mathematics in the way that the Renaissance mind-set in which Leibniz was immersed did. Its preoccupation, which historically owed much to Llull, with emblems, "Real Characters," and philosophical languages, became in his hands both a precondition and a formative element in his calculus. Of course, there have been other examples where cultural or technical needs influenced the creation or development of certain mathematical areas—for example, one can think of high speed computers and numerical analysis, military needs and operations research, epidemiology and statistics. But the influence of such factors was more peripheral and had less impact on the actual *form* of the resulting mathematics than the cultural aspects we have described in Chapter 9.

To digress a bit from Leibniz and expand on some of our previous observations, we suggest again that the same kind of patterns permeating other branches of culture, have also been present in mathematics, and while they have played little role in the actual discovery of theorems, have helped to shape its development right up until the present. These factors have been largely invisible in part because—given the sheer difficulty of the subject and its remoteness from the humanities—interest or competence in the sociology of mathematics is in its infancy; for obvious reasons there is unlikely to be a widely accepted "Frankfurt School" interpretation of mathematical research anytime soon! Moreover, the history of mathematics although it *seems* to have produced an accurate chronicle of discrete and precise events is still far from the Rankean ideal of discovering the past "*wie es eigentlich gewesen.*" We suggest that in many cases accepted accounts of mathematical change are—if not gross simplifications of the chaos of events—are basically Whig interpretations, i.e., the propaganda of the beneficiaries of such change after the fact. The only unambiguous features in the practice of mathematics—as in all human things—are perpetual Heraclitian change, much of the time accompanied by loss of memory, together with a fractal-like complexity: the more one "magnifies" past mathematical events the more bewildering structure is revealed.

But even at the risk of excessive repetition, we cannot emphasize too strongly that it is not a question of extra-mathematical sociological forces determining the truth of mathematical results. The history of mathematics (aside possibly from the Chinese Cultural Revolution) does not contain the methods of mathematical persuasion used by O'Brien in his final interrogation of Winston in *1984*. The bulk of extant mathematics may become ignored or forgotten as a result of cultural or political change, but—*pace* occasional mistakes (sometimes spectacular) in the literature or the changing standards of rigor—it is unlikely to be falsified. However, the fact that a nontrivial mathematical proposition is true, although certainly pleasing to its discoverer, in the long term can be one of its least socially significant properties. Of greater interest and importance is the question of why the proposition does or does not have "influence", of why at one time it is embedded in and at another time cast out of a received canon; more generally, why does one intricate and difficult research program replace another? It is in the consideration of these questions that we argue that purely "internal" *mathematical* considerations may not provide a sufficient explanation.

Relativistic sociological interpretations of change have become commonplace in the analysis of parts of science other than mathematics. The fact that competing scientific theories have different truth claims, helps to highlight their mutability, and makes the kind of analysis that began with the Luther-like rebelion of Thomas Kuhn and which was soon carried to crazed Münzer-like extremes by Paul Feyerabend (1924–1994), British sociologists, and Bruno Latour seem natural. However, as we have repeatedly pointed out even dead or supplanted mathematics—for example, such as theorems on prosthaphaeresic multiplication,[17] remains "true"—in contrast to, say, the *Physics* of Aristotle; further some of this mathematics, although it may be almost forgotten and hardly ever studied even in universities, remains "useful", e.g., spherical trigonometry. These differences have made it historically more difficult to come to grips with the idea that conflicting

[17]That is, the use of trigonometric formulas as an aid to multiplication. For example, $\cos A \cos B = (\cos(A+B) + \cos(A-B))/2$. Hence after making adjustments in the decimal place, if $N = \cos A$ and $M = \cos B$, we can calculate NM by using a table of cosines and replacing multiplication by addition and subtraction in a manner similar to the use of logarithms.

paradigms may exist in mathematics as they do in other scientific fields; the inaccessibility of the discipline also makes it harder to see them.[18]

In the historical development of sciences like astronomy, physics, or chemistry we can often identify "anomalies" or occasions when a mature paradigm is failing to solve its "puzzles" or is producing contradictions, and although contemporary practitioners are aware of the problems, their response—often stubborn—in the short term is to push the accepted paradigm harder. Have anomalies existed in mathematics? As we have pointed out in Chapter 1 historians of mathematics have out suggested several candidates including the post-Cantor foundational crisis, the Pythagorean discovery of incommensurability, the dissatisfaction among mathematicians concerning the status of the Euclidian Parallel Postulate, etc. But we should beware of pushing Kuhn's conceptions too far in mathematics. In so far as the actual practice of mathematicians, as distinguished from that of philosophers or sociologists, is concerned anything resembling anomalies have been few and far between; nor are anomalies engines of change. At least in the modern university almost all mathematical research is "normal." Each research group is satisfied with its "paradigm"; unsolved problems are solved, or if too hard bypassed. Sometimes there is a genuine sense of fatigue among the practitioners, a sense that the area is "mined out" and no longer has decent problems, but this does not resemble the consciousness of a failure of an existing paradigm to solve "puzzles."[19] While one can speak of "revolutions" in mathematics (Leibniz's calculus being one) in the sense of radical departures from existing practice, they do not resemble in any simple way a response to a "crisis" in the accepted paradigm as Kuhn believed to be the case in science. Their development has no common structural cause; each is a special case. They are more analogous to a change in what Ludwik Fleck calls a "thought style," often reflecting the complicated and highly contingent causal factors described above in addition to purely "internal" aspects of mathematics.

Paradigm change in Mathematics also differs from science in one other respect. Supplanted scientific paradigms die (although this may take some time). No one practices Ptolemaic astronomy today. The fate of supplanted research areas or paradigms in mathematics is more complex. They tend to survive but become "non-mainstream" and suffers downward mobility

[18]This is also true of other sufficiently arcane areas of twentieth century science aside from standard examples such as quantum mechanics.

[19]But a certain skepticism is merited here: such a discourse is often fabricated after the fact—when one research area is replaced by another.

in the hierarchy of institutions, surviving for a time in minor "teaching mills", or Third World universities, or becoming embalmed in textbooks as material to be learned in order to be forgotten by beginning students. The theory of conic sections, for instance, customarily covered in a few pages in a beginning calculus course now has this status.[20] Plane geometry survives in an attenuated and debased form in high schools, but has almost vanished from universities. On a higher level, as we have already indicated, classical point set topology while still embedded in the curriculum is now nearly dead as a research area, while in the 1960s it was one of the most trendy areas of mathematics.

We can conclude that the process of mathematical change may be far more complicated at the macro-level than that described in *The Structure of Scientific Revolutions.* But if this is so, it is an ironic but inescapable fact about a discipline that in its micro-structure is the *ne plus ultra* of human rationality. Upon reflection we should not find this surprising. Since most mathematics has *nothing* to do with the world (at least initially), nothing is at stake except academic status. Therefore it is more open to change for all sorts of internal or external reasons than a field that is more "constrained" by physical reality; it follows that we can simultaneously admit that mathematics is one of the most intellectually demanding of subjects—and may indeed be an unfolding revelation of eternal truth as mathematical Platonists maintain, but find that its actual history can resemble those of literary criticism, art, cookery, philosophy, or clothing fashion more than that of physics or chemistry.

Perhaps a more usable set of metaphors in the history of mathematics than those furnished by Kuhn may be found in current interpretations[21] of evolution. Complex mathematical structures arise in random non-purposeful ways as a result of many unpredictable external and internal influences,[22] sometimes very rapidly—one can think of these periods

---

[20]Anyone doubting the reality of noncumulative aspects of mathematical change should consult George Salmon's (1819–1904) *A Treatise on Conic Sections* (1854). Salmon was an Anglican divine and Regius Professor of Divinity in the University of Dublin who also wrote many theological texts. The fourth edition of his textbook on conic sections published in 1863 was once a mainstay of the University curriculum and the source of terrifying problems for British Civil Service Examinations. To get a flavor of the work the reader may wish to tackle Exercise 4 of Article 227 (Sixth Edition) by whatever modern methods he chooses: "the area of the triangle formed by three tangents (to a general conic) is half that of the triangle formed by joining their points of contact".

[21]e.g., Gould (1977).

[22]As an illustration of the nonmathematical influences that may affect research areas of mathematics we may consider number theory. In a 1915 speech before British secondary

such as Leibniz's era as cases of "punctuated equilibria"; in other times there is the stately and logical unfolding of theorems implicit in a subject just as a fossil pattern for a given species remains stable. And since the results of Archimedes, Barrow, Huygens, Leibniz, Euler, Cauchy, or Weierstrass (with a few exceptions) remain "true", obsolete mathematics unlike obsolete science in a sense "survives". But this survival is only in a "fossilized" form. It is true that there is cumulative progress simply because more mathematics has been created just has the number of species that have once flourished increases. But as in the case of zoological species old theories become "extinct" as living intellectual forces. The destiny of most mathematics is to be forgotten, no longer studied, effectively lost to the current mathematical culture. Non-mathematicians (even perhaps if they are historians or philosophers of mathematicians) do not realize the fact that 95% or more of the mathematics that has been created is *much* more inaccessible than the products of pre-Galilean science or philosophy. One still finds an occasional specialist in Nifo (1473–1538 or 1545) or Zabarella (1533–1589). But most of the thousands of volumes of mathematical journals in a university library simply cannot as a practical matter be read. Even articles written as late as the 1970s may have forgotten vocabulary, symbolism, concepts, or be in a radically changed field. In principle, given infinite time, they may be understood. But almost no one has the interest or the leisure to do this. Even if the authors are still celebrated in textbooks, their papers are unread. Many are of interest only to antiquarians just as biological fossils are to paleontologists. In both cases (Jurassic Park excepted) the vital characteristics are difficult or impossible to reconstruct although practical uses may sometimes be found for the fossils. Prior concepts that are of later interest are simply reinterpreted and pulled from their environment just as dinosaurs "survive" by becoming birds. Moreover, and here the situation is similar to Whiggish accounts in the history

school teachers G. H. Hardy said:

> A science is said to be useful if its development tends to accentuate the existing inequalities in the distribution of wealth or more directly promote the destruction of human life. The theory of prime numbers satisfies no such criterion.

Further it "has always been regarded as one of the most obviously useless branches of Pure Mathematics" [Hardy (1915)]. Without the political developments of the twentieth century this judgement of Hardy would have remained true, and the area would have remained useless, perhaps comparable to algebraic geometry. Now, however, the theory lies at the heart of military codes controlling ICBM forces. For this and other reasons it has long been an area cultivated by the CIA and NSA.

of science, the history of the subject is constantly being rewritten so that it points to the current *Zeitgeist*; the result is tautologically "progress". In both evolutionary theory and the history of mathematics consequently there is the temptation of teleological illusions.

We are perhaps stating a pessimistic relativism in too extreme a way and have come full circle around to a position not differing in certain respects from that of Richard Rorty (1931–2007) and other postmodernists. But we think such ideas need to be explored, especially if one wishes to refute them. In particular, if the claims put forth above have any validity, they imply that the judgment of G. H. Hardy also quoted at the beginning of this essay about the eternity of mathematics and its superiority in this respect to other forms of cultural production is subject to qualification.

Appendix A

A Transmutation Theorem of Leibniz

At the end of a manuscript dated October 29, 1675 entitled *Analyseos Tetragonisticae pars seconda*, some of whose other results we have described in Chapter 5, Leibniz gives a transmutation result whose derivation by Hofmann has been criticized in the Preface. In Figure A1 $T(L)$ is tangent to the curve $\mathcal{C}$ at the point L. I is the intersection of the tangent with the ordinate AQ and N is the midpoint of AI. Leibniz claims that omn.(QN) is equal to the area of the triangle ABL. Whether or not this result is entirely Leibniz's own is unknown to me. It is possible that it is a modification of something in Barrow since it is similar in spirit to several of the theorems in Lecture XI. James Gregory's (1638–1678) *Geometriae Pars Universalis* is another possibility. In any event, transmutation theorems of various sorts where the area of one region is shown to be equal to another, perhaps simpler, region were very much in vogue in the 1660s and 1670s.

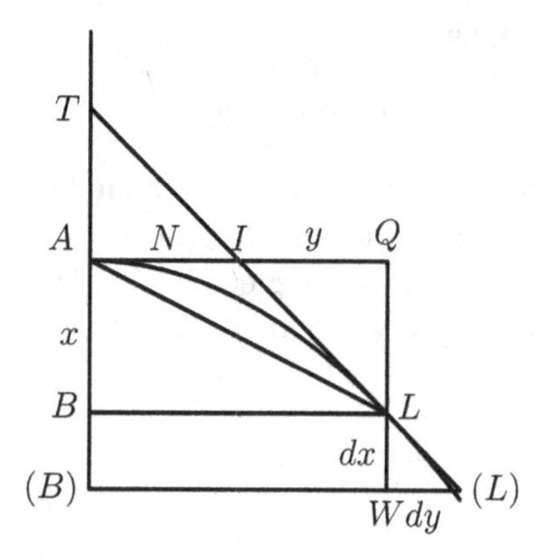

Figure-A1

Earlier in the manuscript Leibniz had introduced the "integral sign" $\int$. But this was not yet an integral in our sense, but rather a symbol signifying the aggregate of lines l in the sense of Bonaventura Cavalieri defining an area. Previously he has denoted this by omn.l. Now he writes $\int l$. Therefore the proof given by Hofmann which we have reproduced in the Preface using modern technology cannot have been true to Leibniz's way of thinking in 1675, although one can conceive of him giving a similar argument in the period 1684–1716 (but without using definite integrals). How did Leibniz prove his result? We don't know for no proof is given. Instead he says that it "can easily be shown by what I have written in another place."[23] The following is an reconstruction of an argument Leibniz might have used which is consistent with his mathematical development in 1675. Denote AI by c. Although Leibniz has not yet introduced this notation, for convenience we call $W(L)$ dy and $LW \equiv B(B)$ dx. Then since the triangle IQL is similar to the characteristic triangle

$$QI : dy = x : dx.$$

Hence $QI = x(dy/dx)$, and

$$c = y - QI = y - x(dy/dx),$$

so $2y - c = y + x(dy/dx)$. Multiplying through by dx gives

$$(2y - c)dx = ydx + xdy.$$

For infinitesimal dx and dy, the rectangular strips with infinitesimal base can be considered lines, or like Roberval and others Leibniz may have already thought of areas being made up of the sum of infinitesimally thin rectangular strips. Therefore

$$\int 2y - c = \int y + \int x.$$

The first sum of lines (or strips) is the area bounded by AB, BL and the curve C. The second sum is the complement of this area in the rectangle $ABLQ$. Therefore $\int 2y - c =$ rectangle($ABLQ$). Dividing by 2 gives that the sum of the segments QN is the area of the triangle ABL.

[23] Leibniz (1920), p. 83.

Appendix B

Leibniz's Series Quadrature of a Conic

We consider a conic with equation

$$\lambda\frac{x^2}{a^2}+\frac{(y-b)^2}{b^2}=1 \qquad (B.1)$$

where $\lambda = \pm 1$. Figure B1 is an illustration of the arc RFA of the conic when it is an ellipse or circle. The diagram for the hyperbola will be omitted, but our reasoning will also cover that case. Let A be the vertex and Q the center. We take the positive y axis to be AQ and the positive x axis to be the continuation QR to the left. Under these conventions $A = (0, 0)$, $Q = (0, b)$, and $R = (a, b)$.

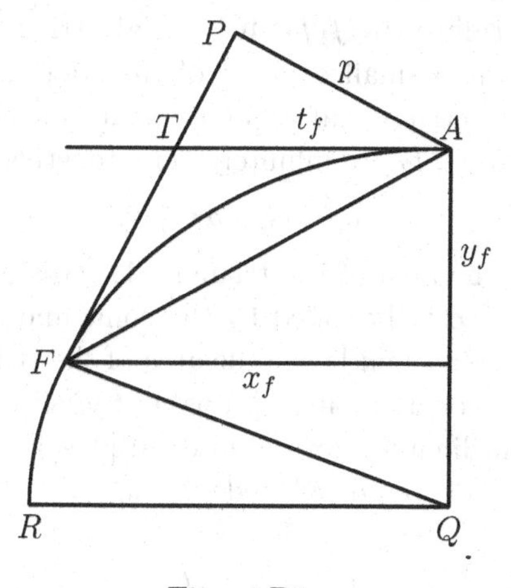

Figure-B1

The semi-latus rectum of the conic is b^2/a and the semi-latus transversum is a. Since Leibniz is assuming that the rectangle having these sides is of unit area, $b^2 = b = 1$. So $P = (a, -1)$, $Q = (0, 1)$, and $A = (0, 0)$. Let F have coordinates (x_f, y_f). We want to represent the area of the sector QFA by a series in $TA = t$ where TA is a tangent to the ellipse (hence parallel to the x axis at A). In Figure B2 let Af_1f_2 be a sector of QFA where f_1 and f_2 are very close together and between F and A.

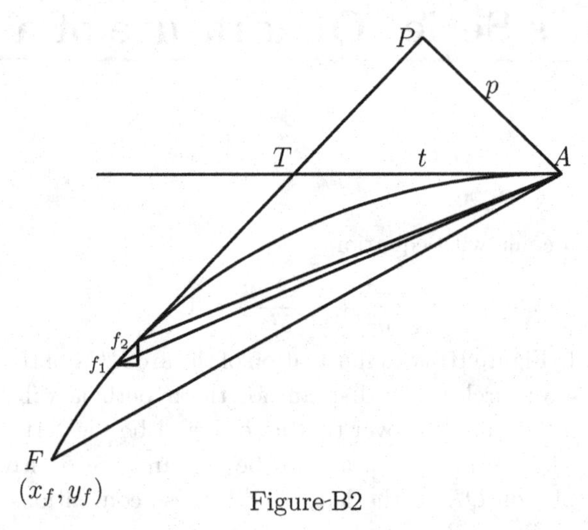

Figure-B2

This is nearly a triangle with third side $f_1f_2 \equiv \delta s$. Consider the characteristic triangle determined by f_1f_2. It is a right triangle with hypotenuse $f_1f_2 \approx \delta s$ where δs is a small element of the conic, and horizontal and vertical sides we respectively call δx and δy. It is similar to the triangle ATP. Assuming that f_1, f_2 are infinitely close together, $f_1f_2 = \delta s$. Hence

$$\delta y : \delta s = p : t,$$

so $t\,\delta y = p\,\delta s$. Now the area of the triangle Af_1f_2 is $(1/2)p\,\delta s$. It follows that the area of the sector bounded by the conic and chord AF ought to be the sum of these areas, and hence the area of the sum of the strips with height t and base δy, or as Leibniz probably thought of it the aggregate of "indivisibles" or ordinates t on the interval $[0, y_f]$. In modern notation (which Leibniz had not yet developed—he would have used the "omn" notation)

$$\frac{1}{2}\int p\,ds = \frac{1}{2}\int t\,dy.$$

Since t is the x-intercept of the secant line f_1f_2TP, it is given by

$$t = x - y(\delta x/\delta y) \qquad (B.1)$$

where (x, y) are the coordinates of f_1. If we expand the equation for the conic we get

$$\lambda\frac{x^2}{a^2} = 2y - y^2. \tag{$B.2$}$$

Hence (probably using Sluse's rule),

$$\frac{\delta x}{\delta y} = a^2\left(\frac{1-y}{\lambda x}\right). \tag{$B.3$}$$

Therefore from $(B.2)$ and $(B.3)$, we get that

$$t = a^2\frac{y}{\lambda x}, \tag{B.1}$$

or

$$\lambda x = a^2\frac{y}{t}.$$

Substituting this into $(B.1)$ and simplifying gives that

$$y = \frac{2\lambda t^2}{a^2 + \lambda t^2}. \tag{$B.4$}$$

Hence also,

$$x = \frac{2a^2 t}{a^2 + \lambda t^2}. \tag{$B.5$}$$

We have seen that the area of the sector with chord AF is $(1/2)\int_0^{y_f} t\,dy$. Using Leibniz's moment arguments (Chapter 5) to get what we now think of as integration by parts gives that

$$(1/2)\int_0^{y_f} t\,dy = (1/2)\left(y_f t_f - \int_0^{t_f} y\,dt\right).$$

After expressing y in terms of t, we get that the area of the sector AFA is given by

$$(1/2)\left(y_f t_f - \int_0^{t_f} \frac{2t^2\lambda\,dt}{a^2 + \lambda t^2}\right).$$

Now the area of the total sector QFA in Figure B1 is the sum of the above area and the triangle QFA which is $x_f/2$. However, from $(B.4)$ and $(B.5)$

$$(1/2)(x_f + y_f t_f) = t_f.$$

The proof is completed by expressing the integrand as a series using Mercator's technique of long division and "integrating" term by term. Specifically, if $\lambda = 1$, we must integrate the series on $[0, t_f]$

$$-\left(\frac{t}{a}\right)^2 + \left(\frac{t}{a}\right)^4 - \left(\frac{t}{a}\right)^6 + \cdots + (-1)^n\left(\frac{t}{a}\right)^{2n} + \ldots,$$

obtaining

$$a\left(-\frac{z^3}{3}+\frac{z^5}{5}-\frac{z^7}{7}+\cdots+(-1)^n\frac{z^{2n+1}}{2n+1}+\ldots\right)$$

where $z = t_f/a$. So the area of the sector is

$$a\left(z-\frac{z^3}{3}+\frac{z^5}{5}-\frac{z^7}{7}+\cdots+(-1)^n\frac{z^{2n+1}}{2n+1}+\ldots\right).$$

If $\lambda = -1$ we get

$$a\left(z+\frac{z^3}{3}+\frac{z^5}{5}+\frac{z^7}{7}+\cdots+\frac{z^{2n+1}}{2n+1}+\ldots\right).$$

If $a = 1$ we get Leibniz's result in his August 17, 1676 letter to Oldenburg as stated for the circle and the hyperbola. In the case of a circle or a hyperbola if $F = (1, 0)$, $t_f = 1$, and the series is the same as given in his arithmetical quadrature for the quarter circle.[24]

[24]The above has been my summary of Leibniz's quadrature arguments. The original versions which are quite lengthly may be found in his manuscript *De quadratura arithmetica circuli ellipseos et hyperbolae.* See Leibniz (1993).

Appendix C

Syllogistic Logic

The following material is a sketch of the classical logic Leibniz and his fellow students would have been trained in from secondary school onward. It was a core subject and absolutely familiar to educated people.

Types of categorical propositions and their relations[25] Recall that categorical propositions can be of four kinds: (1) universal affirmative, (2) universal negative, (3) particular affirmative, and (4) particular negative which have the respective forms (1) All S is P, (2) No S is P, (3) Some S is P, and (4) Some S is not P. To denote these types of propositions in a convenient manner, we use the letters A, E, I, and O. Thus A denotes a universal affirmative, E a universal negative, I a particular affirmative, and O a particular negative proposition. According to traditional logic certain logical relations between these types are indicated by an ancient diagram (Figure C1) dating from the Middle Ages called the Square of Opposition.

In this diagram propositions of forms A and O or E and I negate or contradict each other. So if one is true, the other is false and *vice-versa*. For example, the negation of "All men are wise" (A) is "Some men are not wise" (O) and the negation of "No men are wise" (E) is "Some men are wise" (I). A proposition of form A or E implies its subaltern I or O provided we assume that the propositions refer to a class of existing objects. Thus "No men are wise" implies that "Some men are not wise," and "All men are wise" implies that "Some men are wise."

[25] We are indebted to Eaton (1931) for the material presented here.

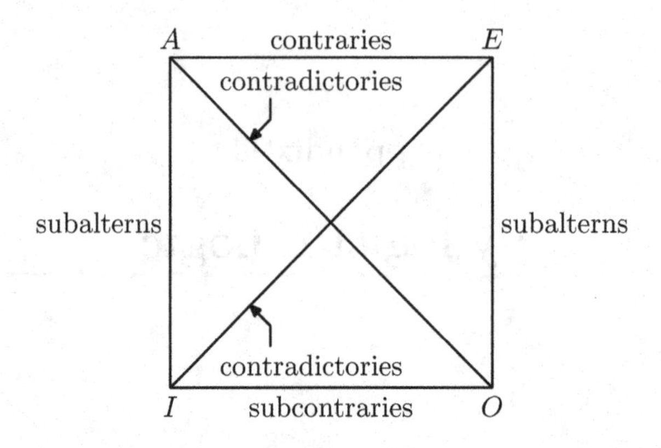

Figure-C1

Finally, pairs of propositions of the form A and E are called contraries. These cannot be true together although both may be false. To see this, if examples of both A and E are true, the subalterns I of A is true, but this contradicts E, so E is false, contrary to assumption. However it is possible that "All men are wise" and "No man is wise" can be both false: for "All men are wise" is false, if, and only if, "Some man is not wise" is true, and similarly the falsehood of "No man is wise" is equivalent to the truth of "Some man is wise." Both are possible states of affairs. This rule is reversed for subcontrary pairs of the form I and O. "Some man is wise" (I) and "Some man is not wise" (O) cannot be simultaneously false. For if the first is false its contradiction "No man is wise" is true, meaning the same is the case for its subaltern "Some man is not wise," contradicting our assumptions.

Syllogisms A syllogism consists of three categorical propositions, such that the third or "conclusion" is a consequence of the first two or "premises." Let S and P stand for the subject and predicate of the conclusion where P is called the *major term* and S the *minor term* We suppose there is a term M which occurs once in each of the two premises, the remaining term in each premise being either S or P, and call the premises where P and S occur respectively the *major* and *minor* premises, and we suppose that each of S, P and M can be either subject or predicate of the premises. It is customary to divide the possibilities of this arrangement into four classes called "figures." In each figure the first two lines are the premises and the third is the conclusion. In all cases "quantity" (the modifiers "all" or "some") and "quality" (affirmation or negation) are ignored. Under these conventions we have:

I	II	III	IV
M is P	P is M	M is P	P is M
S is M	S is M	M is S	M is S
S is P	S is P	S is P	S is P

Now each of the of the premises and the conclusion can have two qualities and two quantities, i.e., the forms A, E, I and O. Hence each figure contains 64 possible syllogisms for a total of 256. Most of these are invalid since the conclusion will not follow from the premises. For example,

Some men are saintly

Some men are murderers

Therefore some murderers are saintly

is in the third figure, being of the form

M is P

M is S

S is P,

but is plainly invalid. It turns out that there are exactly 24 valid syllogisms among the 256.[26] How can they be determined? Let begin by calling a subject term S "distributed" if it is signifies the entire class that the term names; equivalently this means that S is not restricted by the modifier "some" or equivalently, its quantity is signaled by "all." Otherwise S is said to be "undistributed." For a predicate term P the question whether or not it is distributed is a little more complicated. Consider, for example, "Some S is P." P is not explicitly restricted, but we take it to be undistributed since we cannot conclude "All P is S," although this might be the case. However, in the proposition "No S is P" both S and P are distributed because we are saying something about the entire class S and "No S is P" is equivalent to "No P is S, so that P signifies the entire class. In the case of a particular negative proposition "Some S is not P" we have to consider the entire class P to determine any S which does not belong to it. We can summarize the possibilities for propositions of the form A, E, I, O as follows:

[26]Note this is exactly half the number (512) that Leibniz determined in his *De Arte Combinatoria*. This is because he also took into account the order in which the premises are stated.

- **A**: All S is P. S is distributed; P is undistributed.
- **E**: No S is P. S is distributed; P is distributed.
- **I**: Some S is P. Neither S or P is distributed.
- **O**: Some S is not P; S is undistributed. P is distributed.

Having made these definitions, the following rules apply to determine valid syllogisms:

RULE I. The middle term M must be distributed in at least one premise. Note that in the example given above, this rule is violated. The middle term "men" is undistributed in each premise and hence the groups referred to may be disjoint.

RULE II. No term in the conclusion can be distributed if it not distributed in one of the premises. For example:

all brave warriors control their fear

all brave warriors do battle

therefore all who do battle conrol their fear.

Neither the class which does battle or the class that controls its fear are distributed in the premises. The two classes have a common intersection, but the doers of battle may not be a subclass of those who control their fear. The syllogism would be valid, however, if "some" were substituted for "all" in the conclusion.

RULE III. A syllogism with two negative premises has no valid conclusion. To see this consider

Italians are not Turks,

Turks are not Buddists,

therefore Italians are not Buddists.

This is of the first figure, but is plainly invalid. (Nor does it follow that Italians are Buddhists.)

RULE IV. Two affirmative premises cannot have a negative conclusion.

RULE V. If one premise is negative, so is the conclusion. If one premise is particular so is the conclusion.

RULE VI. Nothing follows from two particular premises.

Using these rules we can list the 24 valid syllogisms or "moods" in the four figures. In the Middle Ages schoolmen devised artificial names as an aid to memory, the vowels signifying propositions of the form A, E, I, O. Using these names we have:

FIRST FIGURE: *Barbara, Celarent, Darii, Ferio, Barbari, Celaront.*

SECOND FIGURE: *Cesare, Camestres, Festino, Baroco, Cesaro, Camestros.*

THIRD FIGURE: *Datisi, Disamis, Ferison, Bocardo, Felapton, Darapit.*

FOURTH FIGURE: *Calemes, Dimatis, Fresion, Calemos, Fesapo, Bamalip.*

For example, in Figure 1 *Barbara* has the form

All S is M

All M is P

Therefore all S is P

while in the third *Bocardo* would have the structure

Some M is not P

All M is S

Therefore some S is not P

and so forth.

Not all of these syllogisms are independent of each other. Subaltern syllogisms are those whose conclusions follow by particularizing the conclusions of other syllogisms in the same figure with the same premises. Thus in the first figure *Barbari* and *Celaront* follow in this way from *Barbara* and *Celarent.* By the same token, we can dispense with *Cesaro*, *Camestros* in the second figure and *Calemos* in the fourth figure. We are left with nineteen valid moods (out of 256 possibilities). But we can go even further since many of these moods are equivalent to each other. Consider all the syllogisms of the form $(EI)O$ where the parentheses indicate the premises. There are four of these *Ferio*, *Festino*, *Ferrison*, and *Fresison* in Figures 1 to 4. All are equivalent to each other. To see this *Ferio* which is of the form

No M is P

Some S are M

Therefore some S are not P

has the same premises as *Festino* in the Figure 2 or

No P is M
Some S are M
Therefore some S are not P.

This in turn has premises equivalent to *Ferison* in Figure 3 or

No M is P
Some M are S
Therefore some S are not P

and to *Fresison* (Figure 4)

No P is M
Some M are S
Therefore some S is not P.

In the same way we can show that

Celarent (1) = *Cesare (2)* = *Camestres (2)* = *Camenes (4)*

and that

Darii (1) = *Datisti (3)* = *Disamis (3)* = *Dimaris (4)*

where the numerals indicate the Figure. Finally, *Felapton (3)* or

No M is P
All M are S
Therefore some S is not P

is equivalent to *Fesapo (4)* or

No P is M
All M are S
Therefore some S is not P.

The remaining five moods *Barbara (1)*, *Bramantip (4)*, *Darapti (3)*, *Bocardo (3)*, and *Baroco (2)* are not equivalent to any other mood. In this way the nineteen moods can be reduced to nine.

Appendix D

The *Vis Viva* Dispute

In 1686 in an article published in the *Acta Eruditorum* whose title we abbreviate as "Brief Demonstration of a Notable Error of Descartes and Others Concerning a Natural Law."[27] Leibniz launched an attack on Descartes' dynamical ideas that initiated a quarrel concerning conservation laws and the nature of force that lasted more than fifty years and indeed may not have been quite over by the beginning of the nineteenth century. In his *Principles of Philosophy* Descartes had argued that extension was the sole defining property of matter: "For in fact the extension in length, breadth, and depth which constitute the space occupied by a body, is exactly the same as that which constitutes the body."[28] From this view certain consequences followed: Since extension characterizes matter there is only one kind of matter. Furthermore, because what ever is extended is divisible *ad infinitum* matter is continuous and there cannot be indivisible atoms of matter. Also there can be no void.[29] Since"space does not in fact essentially differ from material substance", it must be a plenum completely full of matter.[30]. Indeed, a completely empty space or vacuum consisting only of immaterial extension, Descartes felt would be a logical contradiction. Turning to motion, Descartes derived a conservation law from his idea of the perfection of God. God is the source of all motion. Since His actions are perfect, He need not intervene in the Universe after His creation of it. Therefore the amount of motion initially created cannot diminish, for in

[27]*Acta Eruditorum*, 1686, pp. 161–163. The full Latin title of the article was *Brevis demonstratio erroris memorabilis Cartesii et aliorum cira naturalem, secundum quam volunt a Deo eandum sember auantitatem motus conservari; qua et in re mechnanica abutuntur.*

[28]Descartes (1983), II.10. For the *Principles* we refer to part and section.

[29]*Ibid.*, II.16.

[30]*Ibid.*, p. 44.

this case the Universe might "run down" and require Divine repair. To Descartes and his followers this fact entails a conservation law: If we calculate the quantity of motion of a body by $m|v|$ where m is the "size" or volume of the body and $|v|$ is its speed (the direction is immaterial) the sum of this quantity over all bodies in the Universe must be constant over time. Hence if $m|v|$ of a particular body increases or decreases, the total quantity of motion of the remaining part of the Universe must decrease of increase by the same amount. This conservation law may look like a version of the conservation of momentum principle, but it is different as Descartes did not consider motion to have vector properties; all that mattered was the "speed" or magnitude of velocity. Also Descartes had no real concept of mass—the quantity of a matter in a body is measured by its volume. He explained the apparently different kinds of matter having differing of density, hardness, color, etc. in bodies in terms of the sizes and motion of the particles composing them. As for the "force" associated with a body in motion, Descartes identified it with $m|v|$.

Leibniz was sympathetic with very little of this. As early as 1680, probably expressing thoughts that he had by the 1670s he wrote to a certain Filippi that:

> M. Descartes physics has a great defect: it is that his rules of motion or laws of nature, which are to serve as the basis, are for the most part false. And his great principle, that the same quantity of motion is conserved in the world is an error.[31]

As we shall see Leibniz had a much more complicated view of matter and its defining properties than Descartes. But in the "Brief Demonstration" he concentrates his attack on Descartes' view that $m|v|$ is the correct measure of force. He agrees that what he calls "motive force" is conserved "since we never see force lost by one body without being transferred to another—or augmentes," and this conservation implies the impossibility of perpetual motion machines.[32] But Leibniz argues that this motive force cannot be the same as Descartes quantity of motion. The two concepts are very different. It follows that whatever the general truth of a conservation law of force, in the form stated by Descartes it is false. To show that Descartes concept of force is in error he proposes a simple thought experiment. He begins with two assumptions. The first is:

> ... a body falling from a certain altitude will have exactly the

[31] Dugas (1958), p. 467.

[32] Leibniz (1969), p 296.

> same force which is necessary to lift it back to its original altitude if its direction were to carry it back and if nothing external interfered with it. For example a pendulum would return to exactly the height from which it falls except for air resistance and other similar obstacles which absorb something of its force ...[33]

Secondly, the force needed to lift a body weighing one pound four yards is exactly the same as that required to lift a body of four pounds one yard. Leibniz observes that both assumptions are accepted by "Cartesians as well as other philosophers of our times"[34] A consequence of both assumptions that the force acquires by the four pound body falling one foot is the same as that acquired by the one pound body falling four feet. This follows because by the first assumption each of the two bodies, having fallen, will acquire the force needed to raise themselves back to their original height, and by the second assumption the forces needed to raise the objects are the same. If this is true, Leibniz, by appealing to Galileo's laws concerning falling bodies, shows that Descartes' identification of force with quantity of motion cannot be true. Galileo had demonstrated that the velocity of the body falling four feet is twice the velocity v of the body falling one foot.[35] This would imply according to Descartes that for the one pound body the quantity of motion is $2v$ while for the four pound body it is $4v$. On the other hand, as has just been seen, the forces of the two bodies should be equal; "thus there is a big difference between motive force and quantity of motion, and the one cannot be calculated by the other, as we undertook to show."[36]

Nine years later Leibniz continued the attack in a more detailed form in the *Specimen Dynamicum*, Part I of which was published in April, 1695 in the *Acta* while Part II was discovered by Gerhardt in Leibniz's unpublished manuscripts preserved in Hanover. Leibniz begins Part I with an elaborate classification of types of forces. There are two fundamental kinds: active and passive, and each can be either primitive or derivative. His fundamental thesis is that the behavior of bodies must be explained by something more than their extension. Bodies are active and contain a "force" or principle of activity implanted by God. Leibniz calls this force active and primitive and identifies it with the substantial form of the Scholastic-Aristotelian

[33] *Ibid.*

[34] *Ibid.* The two assumptions are discussed in additional detail and proofs given in a "Supplement" included at the end of the article. See *Ibid.*, pp. 298–301.

[35] That is, if h is the height fallen v is proportional to $\sqrt{h}$ and does not depend on the mass or weight of the body.

[36] *Ibid.*, p. 297.

tradition. It rather than Cartesian extension is fundamental to a material object. It:

> ... must constitute the inmost nature of the body, since it is the character of substance to act, and extension means only the continuation or the diffusion of a striving or counter-striving already presupposed by it, i.e., the diffusion of a resisting substance. So far is extension itself from comprising substance.[37]

The second kind of active force or derivative force Leibniz vaguely defines as "exercised in various ways through a limitation of primitive force resulting from the conflict of bodies with each other."[38] Turning to passive forces, passive primitive force explains inertia, resistance to motion and qualities such as impermeability or resistance to penetration by another body. This to Leibniz corresponds to Aristotelian *materia prima* or prime matter, i.e., the underlying substrate on which the properties of a substance are imposed. Although these properties may change, the prime matter lacking all positive qualities remains unchanged. The second kind of passive force or derivative force "or the force by which bodies actually act and are acted upon by each other" is associated with "local motion" which explains "all other material phenomena." Motion, in turn, "is the continuous change of place and thus requires time." What we would call the instantaneous velocity v considered as a vector quantity, Leibniz calls "*conatus*" while "*impetus*" is mv or our momentum.[39] This, of course, is close to the Cartesian notion, but Leibniz thinks of the "total quantity of motion" as a time integral of instantaneous impetuses or

$$\int_0^t m\frac{dv}{dt}\,dt.$$

Derivative passive force is also of two kinds: "dead" and "living" force or *vis viva.*[40] The first does not involve actual motion; it is only a "solicitation to motion" where motion does not yet exist as, for instance, in the outward striving felt by someone whirling a sling before releasing the projectile. *Vis viva*, on the other hand, is the force actually associated with motion. Again, Leibniz divides this kind of force into a further three subtypes: (i) total

[37] *Ibid.*, p. 435.

[38] *Ibid.*, p. 436.

[39] Leibniz's concept of mass seems different from Newton's. It is not the quantity of matter, but rather a measure of the passive force of inertia found in the "secondary matter" of a body or that aspect of a body that participates in physical interactions.

[40] This may be the first published instance of this term. It is not present in the *Brevis Demonstratio.*

force, (ii) relative partial force, and (iii) directive partial force. Briefly, (ii) and (iii) refer respectively to the force by which members of an aggregate of bodies act on each other and to the *vis viva* of the aggregate directed externally, while (i) is the force that is the union of (ii) and (iii). Dead force he claims was known to the ancients as it is associated with statics and simple machines such as the lever, pulley, and inclined plane. These forces collectively are what is "really real" in a body. Extension itself is sort of a phenomenal result of them, and motion being relative is not fully real.[41]

Beyond these rather arid definitions there is very little mathematics in the *Specimen Dynamicum* and what there is (as in the earlier *Brevis Demonstratio*) is an application of Galileo's law of free fall. For example, Leibniz argues, let us consider two pendulums with the bobs of both having equal weights. Suppose the first pendulum is given a unit velocity and ascends to a height of 1 foot above the rest position. Suppose the second has twice the velocity. Then from Galileo's laws it will ascend 4 feet. Both pendulums must expend all their "force" in achieving their maximum amplitude. But the activity of the second pendulum is equivalent to lifting the 1 pound bob 1 foot four times, so its force is 4 times the other and is proportional to the square of the initial velocity.

> In the same way we can conclude in general that the forces of equal bodies are proportional to the squares of their velocities and that the forces of bodies in general are proportional, compositely, to their simple masses and the squares of their velocities.[42]

In a second argument Leibniz points out that Descartes identification of force with $m|v|$ leads to the possibility of a perpetual motion machine. For suppose a body A of magnitude 2 and speed 1 is dynamically equivalent to a body C of magnitude 1 and speed 2, so that one can be substituted for the other. By Galileo's analysis of falling bodies C thrown upward will ascend a height of 4 feet. Now if the two bodies in motion are equivalent, then the 2-pound body A ought also to ascend 4 feet, then being dropped it will also have speed 2 and quantity of motion 4. In this way, we constantly gain quantity of motion or Cartesian force and therefore could build a perpetual motion machine to take advantage of this.[43]

---

[41]That is, in the absence of forces it is impossible to tell whether one is in motion and the environment at rest or the reverse. See a quotation in Garber (2008), p. 299: "... il n'est pas possible de determine par la seule consideration de ces changements à qui entre eux le mouvement ou le repos doit estre attribué."

[42]*Ibid.*, p. 443.

[43]*Ibid.*, p. 444.

Neither the *Brevis Demonstratio* or the *Specimen Dynamicum* directly attack Descartes' conservation principle, although its falsehood is a reasonable inference, especially from the second work. However, in an unpublished manuscript of 1691 *Essay de Dynamique sur les Loix du Mouvement, où il est monstré qu'il ne se conserve pas la même quantité de mouvement mais la même force absolue ou bien la même quantitë de l'action motrice* Leibniz is more direct. He remarks that Descartes' law has been accepted as an "incontestable Axiom" of dynamics for a long time. "But now we are beginning to lose our illusions" about it, especially since it is being abandoned by some of its older defenders including Malebranche. Leibniz, however, feels that natural philosophers have gone to the other extreme and do not recognize the conservation of anything that might take the place of $m|v|$. But a better conservation principle involving motion is what Leibniz calls the "Quantity of progress" where the direction of motion is taken into account. Here Leibniz is speaking of what we would call the conservation of momentum or mv where v is a vector quantity. But this is not "absolute" since as we have seen motion to Leibniz is plain velocity is relative and thus not fully real: "*Motion in so far it is phenomenal consists in a mere relationship.*"[44] Instead Leibniz propounds as an absolute conservation principle the conservation of *vis viva* which is proportional to the squares of the velocities.[45]

Very early on, just after the *Brevis Demonstratio*, Huygens remarked that Leibniz was somewhat unfair to Descartes. He agreed that Descartes laws of collision were wrong as they contradicted his, Wren's and Wallis' work. But he denied (correctly) that Descartes ever identified his quantity of motion with motive force, and even had he done so Leibniz's arguments do not prove that the quantity of motion is not conserved. To prove this, additional arguments would be needed.[46] As usual Huygens was being judicious and fair, but very quickly Leibniz's criticisms of Descartes provoked a firestorm. Soon after it was published the *Brevis Demonstratio* was answered by the Abbé Catelan who was a Cartesian. Catelan argued that Leibniz did not take time into consideration. In Leibniz's example the 1 pound weight ascends to height 4 in time 2, but the 4 pound weight reaches a unit height in time 1. Since the times differ it is natural that the Carte-

[44]Leibniz (1969), p. 445. Emphasis in original.
[45]*Ibid.*, pp. 473–477. Leibniz seems to be thinking here that the square of the (magnitude) of the velocity does not depend on the reference system. It is the same whether or not the observer or his environment is moving at a constant velocity.
[46]Dugas (1958), p. 468.

sian quantities of motion differ. If the two weights are suspended from the fulcrum of a lever at distances reciprocal to their and we imagine the lever to move, then the quantity of motion of the two weights would be equal! Leibniz replied that time had nothing to do with the issue. To say that one force is greater than another because it took longer to acquire the force is like thinking that a man who took longer to earn his money is wealthier than if he had taken less time.[47] A similar objection to Catelan's was given by Denis Papin (1647–1712) in 1689 article in the *Acta*. For a falling body the force $mv \propto t$. Then:

> If the times are equal no more or less force can be added by making the space traversed linger or shorter. Thus a measure of force estimated by the spaces traversed cannot be correct[48]

Leibniz answers Papin by agreeing that one can define "force" as one wishes, but the issue is whether the Cartesian view of conservation of quantity of motion leads to perpetual motion. Suppose a body of weight 4 descends from a height 1 on an inclined plane. After it reaches the horizontal let it interact with a body of weight 1. If all the force of the larger body is transferred to the smaller one, the latter body would have to be given a speed of 4 in order for the Cartesian conservation law to hold. Then by Galileo's law this body could ascend to a height of 16 on another inclined plane. If we now think of the body at the right end of a lever slightly more than a distance 4/5 of the length of the lever from the fulcrum at a height of slightly less 16/5 above the horizontal with the 4 pound weight on the left end of the lever and release the lever, then 1 pound weight will fall to the horizontal and elevate the 4 pound weight to a height of 4 instead of its initial state of 1 which is absurd. Papin's answers this argument by insisting that no way is known to transfer all the force of larger body to the smaller one; and if Leibniz can find a way to do this short of a miracle he will concede defeat.[49]

The polemics Leibniz's criticisms of Cartesian dynamics touched off were not limited to Catelan and Papin. They lasted for most of the eighteenth century, and perhaps even longer. To give its history in reasonable detail is beyond the scope of this book which centers on Leibniz's mathematics and logic. An entirely new book, in fact, could easily be written on this topic.

---

[47] Iltis (1971), p. 28f.

[48] Quoted in *Ibid.*, p. 30.

[49] *Ibid.*, p. 30f. I have added the implied details of Leibniz's argument and supplied the location of the fulcrum for it to be correct; the number 4/5 is consistent with the diagram in the *Acta* reproduced in *Ibid.*, p. 31.

But good short accounts have been given in the papers we have cited by Iltis, Hankins, and Laudan.[50]

In Britain Leibniz's doctrine was psychologically conflated with his criticism of gravitation, his attempt at a vortex theory, and the controversy over priority in the calculus. Hence like all his work it should be consigned to the flames. Newton himself thought that Leibniz's concept of *vis viva* applied to accelerated motion led to absurdity. For if a body is accelerated uniformly equal increases in momentum are produced in equal intervals of time. However, the same is not true for *vis viva*, for if a body is accelerated from 10 feet/second to 15 feet/second the gain in visa viva $\propto$ 125 feet^2/second^2. But if it accelerated from 5 feet/second to 10 feet/second the gain $\propto$ 75/feet^2/second^2 even though the "force" as measured by the changes in velocity is the same. The only way force could increase in this manner would be if the weight of the falling body increased as it fell which is absurd. Moreover, *vis viva* is not conserved since in a totally inelastic collision of two bodies M_1 and M_2 of equal mass M with velocities $v_1 = -v_2$, motion of both M_1 and M_2 ceases, but the total momentum $Mv_1 + Mv_2 = 0$ is conserved, while the total *vis viva* of the universe diminishes since after collision $v_1^2 = v_2^2 = 0$. Aside from these observations Newton did not participate actively in the dispute;[51] but, as we shall see, this did not prevent his followers from doing so. A typical British example was Colin Maclaurin's *Account of Sir Isaac Newton's Philosophical Discoveries* (1748). Maclaurin argued that mv is the only correct measure of motive force and mv^2 was neither conserved or a measure of force.[52] On the continent Jean d'Alembert (1717–1783) was willing to accept *vis viva* in the study of elastic collisions and in certain other circumstances as a convenient mathematical aid, but believing in the existence of perfectly hard and inelastic particles, he accepted Newton's objection to its conservation while accepting the conservation of momentum and the instantaneous vanishing of motion after collision.

On his side as a reply to objections based on inelastic collisions, Leibniz had denied that perfectly inelastic bodies exist. If they did the motion of two bodies of equal mass colliding with velocity of equal magnitude and opposite direction would vanish *instantaneously* violating the principle of continuity. A corollary is that there can be no perfectly hard fundamental

[50] Also see the excellent survey of Terrall (2004).
[51] Hankins (1965), p. 289 and n. 20.
[52] Laudan (1968), p. 133.

units of matter or atoms. All matter is to some extent elastic.[53] Leibniz explained the apparent loss of *vis viva* in some partially inelastic collisions by arguing that it was transferred to the colliding bodies by exciting motion of the particles composing them.

On the continent Leibniz fared better than in Britain since at least two prominent European physicists sided with with him. In 1724 the Paris Academy of Sciences posed an essay competition on the nature of the laws of collision between perfectly hard bodies in a vacuum or a plenum. Colin Maclaurin won the prize with an essay backing the Cartesian/Newtonian position. Although praised by the Academy, Johann Bernoulli's submission did not win. But when he published independently, it became central to the debate for many years afterwards. He defended Leibniz's denial of perfect inelasticity and in considering springs proved that the *vis viva* created or diminished by a body stopped by compressing a spring or set in motion by the release of a compressed spring is proportional to the distance the spring is compressed or extends.[54] Willem 's Gravesand (1688–1742), a Dutch member of the Royal Society, and professor in Leiden initially opposed Leibniz. However, he became convinced by the doctrine of *vis viva* when he designed an experiment where balls of different mass fell on soft clay. Much to his surprise, he found that if the heights from which they fell are inversely proportional to their masses, the indentations made by the balls were the same. But since the height was proportional to the square of velocity v of impact, it followed that the depth of impact $\propto mv^2$, and hence to the "force."[55] Yet despite Johann Bernoulli and 'sGravesand, a significant fraction (possibly a majority) of European natural philosophers still did not accept *vis viva* as a measure of motive force. To give just two examples (out of many) in 1747 Kant argued that momentum or mv was the only correct measure of force. And in 1758 the Croatian Jesuit astronomer and physicist Roger Boscovich (1711–1787), agreeing with Kant's position, further argued that all moving forces are dead forces and living forces or *vires vivae* do not exist; "there are no living forces in nature. The only forces associated with motion are dead forces of one body acting on another when they collide.[56]

What accounts for the length of time (nearly a century) taken by this

---

[53]Leibniz (1969), p. 446f.
[54]Terrall (2004), p. 193.
[55]Hankins (1965), p. 287.
[56]Laudan (1968),p. 132 and Hankins (1965), p. 292; the quotation of Boscovitch may be found in Hankins, p. 291.

dispute. A central problem seems to have been that the notions of force, work and energy were not given clear mathematical definitions, not even by Newton in the *Principia.* For instance, at the very beginning of the work he speaks of the "inherent force of matter" as "the power of resisting by which every body 'so far as it is able perseveres in its state either of resting or of moving uniformly straight forward."[57]. In our terminology therefore Newton identifies one kind of force with inertia, A second kind is "impressed force" which is the action exerted on a body to change its state either of resting or moving straight forward."[58] In particular, the law (or definition) that $F = ma$[59] where a is the acceleration is not found in the *Principia.* Newton's version of the second law is "a change in motion is proportional to the motive force and taken along the straight line in which that force is impressed," which suggests that "force" varies as the change in momentum. So if we view mv as a change from zero momentum the force *ought* to be measured by mv.[60] Leibniz, on the other hand, although he has an intuitive feeling for the relation between what we would call kinetic and potential energy, identifies mv^2 with "force" which to him is the ability to do work.

The peculiar intensity and duration of the quarrel, however, resulted from more than disagreement over technical matters. It involved fundamental metaphysical and theological disagreements over the laws of nature. For instance, do perfectly hard inelastic bodies exist. Are there atoms or is matter infinitely divisible as Leibniz maintained? What conservation principle did God introduce to prevent the world running down? Emotion, personal rancor, and politics were also factors. As we have already pointed out bad feelings between continental and British mathematicians and physicists dating from the priority dispute over calculus were involved. These continued long after the death of both Newton and Leibniz. For instance, in 1728 Samuel Clark argued in a bitter polemic that those adhering to *vis viva* theory were attempting to tarnish the memory of the great New-

[57] Newton (1999), p. 404

[58] *Ibid.*, p. 405. Still a third kind of force seemingly independent of the other two is "centripetal" which he defines just after the impressed variety.

[59] "Force" is still a slippery concept. The writer can recall being puzzled by $F = ma$ in an elementary physics class. Was this a *definition* of a convenient mathematical concept called "force" or something which *measured* an otherwise familiar physical property? But if the latter, what *was* force? After wasting time thinking about this issue (and also why the angular velocity ω should be a vector quantity) until the disastrous first midterm examination, he concluded that physics was a far too difficult a subject for him.

[60] *Ibid.*, p. 416.

ton.[61] Then too the issue was mixed up with contemporary Enlightenment political feelings. Voltaire, for example, felt that Newton and the honor accorded to him in Britain represented all that was the best in that political system, as contrasted with the corruption and authoritarianism of the Old Regime. Hence he viewed himself as Newton's representative in France and he argued that *vis viva* represented a reactionary assault on the reputation of Newton. Hence without knowing any physics, he attacked Leibniz (incidently opposing the pro-Leibnizian position of his mistress, the talented physicist and mathematician, Émilie du Châtelet (1706–1749).

It has been conventionary maintained that the *vis viva* dispute came to an end soon after 1743 when Jean d'Alembert (1717–1783) in the preface of his *Traité de dynamique* that this was a dispute about definitions and empty of real physical content. "Force" could be described in two ways, either by mv and mv^2, and both quantities could be conserved. However, as Laudan has shown that it went on for some time after this date, D'Alembert's opinion being discounted or ignored. For example, besides Kant (writing in 1747) and Boscovich (writing in 1758) who we have already mentioned. Euler in 1745 recognized that the quarrel over *vis viva* was still a heated one. A physics textbook published in 1744 by Desaguliers (1683–1744), a French natural philosopher and member of the Royal Society, devoted almost fifty pages to it.[62] The dispute seems to have gradually ceased not because one of the parties won but mainly through exhaustion. One can still find traces of it in the late eighteenth century. George Atwood (1745–1807), for example, a physicist writing in 1784 still felt it necessary to defend Newton against Leibniz and Charles Hutton (1737–1823) in his *Mathematical and Philosophical Dictionary* of 1795 devoted several pages to the dispute and did not regard it as settled;[63] but most natural philosophers by this time had come around to d'Alembert's position.

[61] *Ibid.*, p. 282.
[62] Laudan (1968), p. 132.
[63] Laudan (1968), p. 134.

Appendix E

Some Applications of Conics, Mechanical Curves and Neusis in Greek Geometry

Greek geometry was one of the finest creations of the human mind, and in its emphasis on proof and rigorous argument it (although it perhaps is no longer politically correct to say this) stands alone among the mathematics created by the civilizations of the ancient world, and was the main inspiration for almost all European higher mathematics.

The following material is not intended to be a survey of the subject which would require an entire book and has been done by others.[64] Instead it is intended to illustrate a small part of the subject matter with which a sixteenth or early seventeenth century mathematician in a university would have been completely familiar. By the middle of the seventeenth century it was probably no longer inspiring work at the research frontier, but it would have been imprinted in his mental DNA as firmly as calculus or linear algebra would be among his descendants today. Added to this would have been a mastery of all the books of Euclid, as well as the surviving works of Apollonius, Archimedes, and Pappus. The problems and solutions we sketch below should not be thought of a primitive version of our own mathematics. Instead they are part of a mathematical tradition which for complex historical reasons we have more or less abandoned, but which in depth and insight is equal to our own.

Conic Sections Menaechmus (fl. 350 B.C.) was the brother of Dinostratus, another mathematician whose use of the quadratrix to square the circle will be described below.[65] Menaechmus studied under Eudoxus of

[64]The best such survey is in two volumes by Sir Thomas Heath (1981). Also see Allman (1889) and Gow (1968).

[65]There is an anecdote concerning an exchange between Menaechmus and Alexander the Great. "O king," Menaechmus is supposed to have told Alexander, "for traveling

Cnidus (c.410–355 B.C.) who himself was a student of Plato, the creator of a theory of proportion that applied to incommensurable magnitudes, and the inventor of a system of 27 celestial spheres to explain the motion of the stars, sun, moon, and planets. Menaechmus is said to have discovered the conic sections and showed that they suffice to solve two of the classical problems of antiquity: duplication of the cube and angle trisection. Here is Menaechmus' solution of the problem of finding two mean proportionals which can be easily shown to be equivalent to cube duplication. Suppose y and x are mean proportionals between line segments a and b, i.e.,

$$a : y = y : x = x : b.$$

In Figure E1 where $AO = a$, $BO = b$, $x = OX$, $y = OY$ complete the rectangle $OXPY$

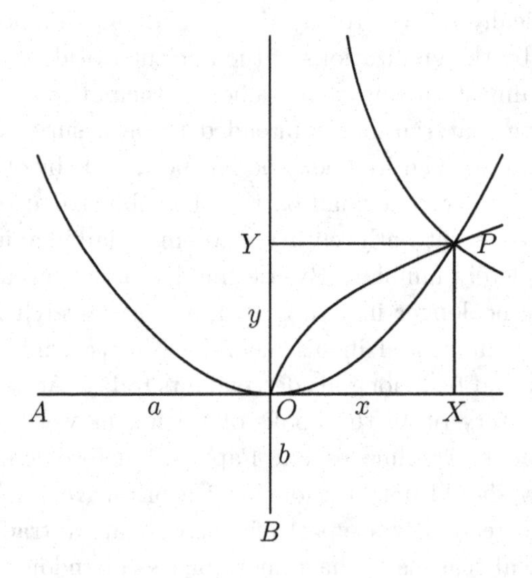

Figure E1

Since $a : y = a : PX = y : x = PX : x$, we have that $y^2 = PX^2 = ax = aOX$. Secondly, $OY \cdot OX = yx = ab$. Menaechmus knew that the first equation implies that P lies on the parabola with axis the prolongation of AX and latus rectum a and that the curve corresponding to the last equation is a hyperbola with asymptotes the prolongations of OX and OY.

over the country there are royal roads and roads for common citizens, but in geometry there is one road for all." [Heath (1981), I, p. 252.]

He also gave a second solution by observing that $x^2 = yb$ so that P also lies on a parabola with axis the prolongation of OY.

Here is one of two solutions given by Pappus to trisect an angle using a hyperbola. It is a good example of the analysis procedure recorded by Pappus.

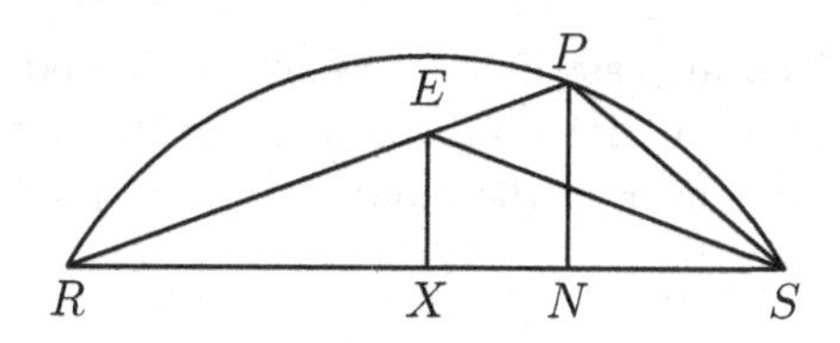

Figure-E2

In Figure E2 $\widehat{RPS}$ is the arc of a circle we want to trisect. Suppose $\widehat{PS}$ is one-third of this arc. Draw RP and PS. Let SE bisect $\angle RSP$ and draw XE and PN perpendicular to RS. Now

$$\angle PRS = (1/2)\widehat{PS} = (1/6)\widehat{RPS}.$$

Since triangles REX and RPN are similar, $RE : RP = RX : RN$. But $RP = RE + EP$ and $RN = RX + XN$. Hence

$$(RE + EP) : RE = (RX + XN) : RX \implies RE : EP = RX : XN.$$

Since $\angle ESX$ bisects $\angle RSP$ and $\angle RSP = (1/2)(2/3)\widehat{RPS}$,

$$\angle ESX = (1/2)\angle RSP = (1/6)\widehat{RPS} = \angle PRS.$$

Hence triangles REX and SEX are congruent, $RE = ES$, and $\angle PRS = \angle ESR \equiv \angle PSE$. We conclude that triangles PRS and PSE are similar. Therefore

$$\begin{aligned} EP : PS = ES : RS \equiv RE : RS &\implies RS : SP \equiv ES : EP \\ = RE : EP = RX : XN. \end{aligned}$$

and so

$$RS : RX = SP : XN.$$

But $RS = 2RX$ so that $SP = 2XN$. We know that a hyperbola is the locus of points P such that the ratio of the distance of P to a fixed point to the distance of P to a fixed line is a number $e > 1$. Hence P lies on a hyperbola with EX being the line or directrix and S the fixed point or focus. If we reverse the reasoning by constructing the hyperbola with these characteristics its intersection with the arc $\widehat{RPS}$ at P will trisect $\widehat{RPS}$.

Mechanical Curves The earliest invented curve used to solve one of the classic problems was the quadratrix. This curve was constructed in the late fifth century B.C., possibly by the Sophist Hippias of Ellis (born c. 460 B.C.) who lived at least as long as his contemporary Socrates and who is the subject of two unflattering dialogues of Plato, the *Hippias Major* and *Minor*.[66] The definition of the quadratrix is as follows. In the square $ABCD$ (Figure E3)

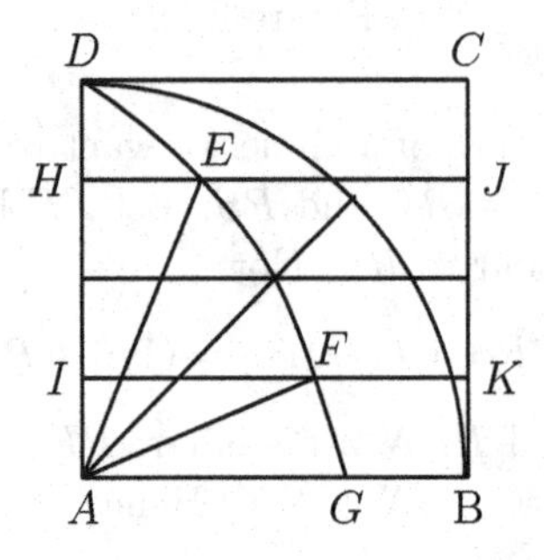

Figure-E3

let DC descend with constant velocity parallel to AB. At the same time let the radius AD of the quarter circle rotate clockwise about A so that AD falls on AB at the same time DC does. The intersection of the descending segment DC and the rotating radius AD defines the quadratrix $DEFG$.

Let us see how we can use the quadratrix to trisect an angle. Let $\angle EAG$ be the given angle where E lies on the quadratrix. Draw HJ through E parallel to AB. Let IA be one third of HA. Draw IK parallel to AB intersecting the quadratrix at F. Then $\angle FAG = \angle EAB/3$. The Greeks would have used a formal ratio argument, but an algebraic one is easier. If y is the height above AB of a point E on the quadratrix at time t and ϕ is the angle swept out by the radius at time t and ω is the angular velocity of the radius we must have $y = s(1 - 2\omega t/\pi)$ where s is the length of AB.

[66]The attribution of the quadratrix to Hippias is, however, disputed. See Allman (1889), pp. 92–94 or Heath (1981), I, pp. 225–230.

Then if θ denotes the given $\angle EAG$, $\omega t = \pi/2 - \theta$, we get that

$$\begin{aligned} y &= s(1 - 2(\pi/2 - \theta)) \\ &= \left(\frac{2s}{\pi}\right)\theta. \end{aligned}$$

Now if $y' = IA = y/3$ and $\theta' = \angle FAG$, by the previous reasoning

$$\begin{aligned} y' &= \left(\frac{2s}{\pi}\right)\theta' \\ &= \left(\frac{1}{3}\right)\left(\frac{2s}{\pi}\right)\theta \\ &= \left(\frac{2s}{\pi}\right)\left(\frac{\theta}{3}\right), \end{aligned}$$

so that

$$\theta' \equiv \angle FAG = \theta/3 \equiv \angle EAG/3.$$

Obviously the quadratrix will also serve to divide an angle into n equal parts.

Another Greek geometer Dinostratus (c. 390 B.C.– c.320 B.C.) used the quadratrix to square the circle. Here it is first shown that

$$\frac{\widehat{DB}}{AB} = \frac{AB}{AG}.$$

We pass over the argument as it is a bit intricate. The idea is to show, using the definition of the quadratrix, that the ratio on the left cannot be greater or less than the ratio on the right. The details are given in Proposition 26 of Book IV of Pappus' *Collection*.[67] However, once this Lemma is proven, the quadrature of a circle is nearly trivial. If L is a line segment such that

$$\frac{L}{AB} = \frac{AB}{AG},$$

then $L = \widehat{DB}$, and the circumference C of the circle is $4L$. But a result of Archimedes shows that the rectangle with sides C and AB is twice the area of the circle.[68] This rectangle then can be transformed into a square.[69]

Other curves that can be used to solve all or some of trisect angles, duplicate the cube, or square the circle include the spiral of Archimedes or

[67] Pappus (2008), p. 134ff.
[68] *Measurement of a circle*, Archimedes (2010).
[69] *Elements*, Bk II, Proposition 14.

the cissoid of Diocles (fl. 100 B.C.).[70] The use of such curves to solve the classic problems is to give what Pappus calls a "linear" solution. In the case of angle trisection or cube duplication, he feels that this method represents bad taste in geometry, since these are"solid" problems; conic sections will suffice to solve them.

Neusis Constructions Recall that a neusis construction amounted to placing a given line segment between two given lines such that the line determined by the line segment passed through a given point. (See Figure 2.1.) This is impossible using a ruler and compass. But in the third century B.C. a mathematician named Nicomedes (c. 280 –c. 210 B.C.) of whom nothing is known devised a special curve called the conchoid to do this.[71] In Figure E4

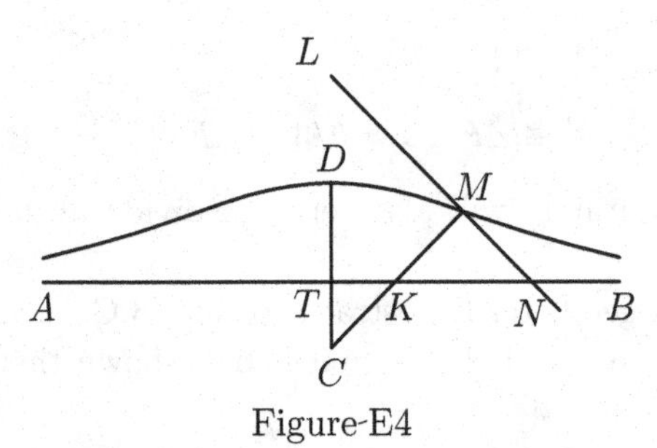

Figure-E4

let AB be a given line, C a fixed point, and DC perpendicular to AB in such a way that DT is equal to a given line segment. Then the line CM rotates clockwise about C such that KM is always equal to DT. Then if LN is the second given line, KM accomplishes the required neusis by the definition of the conchoid.

Neusis could be used to solve two of the classical problems of antiquity: angle trisection and the duplication of the cube. Here are two neusis constructions to trisect an angle. The first is recorded by Pappus[72] and the second is due to Archimedes and is known from an Arabic version of his book *Lemmata*. In Figure E5 let DAC be a given angle. Complete the rectangle $DABC$ and extend BC sufficiently far to H. Next draw GE intersecting the line BH at E such that GE is twice AC. We claim that the angle $DAG = \beta$ is $\angle DAC/3$.

[70] Details may be found in Heath (1981) or in Klein (1956).
[71] The conchoid is recorded in Pappus (2008), Proposition 26, p. 126.
[72] *Ibid.*, Proposition 32, p. 148.

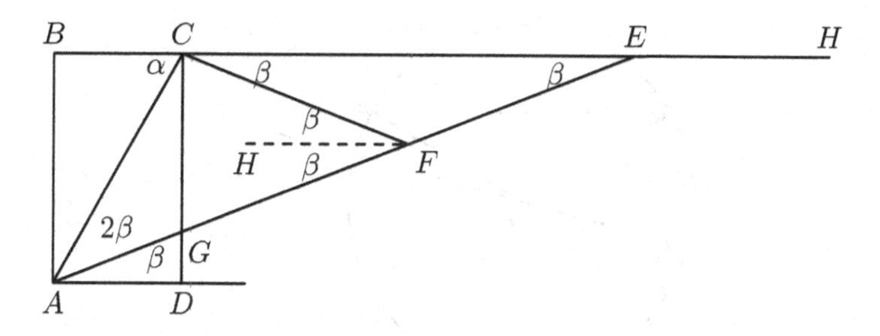

Figure-E5

To see this, let F be the midpoint of GE. Draw CF and JF perpendicular to BH and HF parallel to BH. Then $GF = FE = AC$. Now triangles CGE and JFE are similar, so that

$$CG : JF = CE : JE = GE : FE = 2.$$

Hence $CJ = JE$ and triangles CJF and EJF are congruent, and CFE is isosceles. So also triangle CAF is isosceles since $CF = FE = AC$. It follows that $\angle CAG = \angle CFG = 2\beta$. Since $\angle GAD = \beta$, we conclude that $\angle CAD = 3\beta$.

Our second example is Proposition 8 of the *Lemmata* and involves neusis between a circle and a line instead of two lines. It states (see Figure E6) that:

> If a chord AB of a circle be produced until the part produced, BC is equal to the radius; if then the point C is joined to the center of the circle, which is the point D, and if CD which cuts the circle in F be produced until it cut [sic] it again in E, the arc AE will be three times the arc BF.[73]

The application of this proposition to angle trisection is fairly obvious: We start with the angle ADE and extend EF indefinitely. Then the chord AB is chosen such that when its extension cuts the extension of EF at C $BC = BD$. Then it will follow that $\angle BDF = \angle ADE/3$. To prove the proposition, first note that $\angle BDF = \angle BCF$ since by construction triangle DBC is isosceles. Call one of these angles β. Then $\angle ABD = 2\beta = \angle DAB$. Hence $\angle ADB = 180° - 4\beta$. Then $\angle ADE = 180 - (\angle ADB + \angle BDC) = 180 - ((180 - 4\beta) + \beta)$ which is 3β.

[73]Quoted in Allman (1889), p. 91.

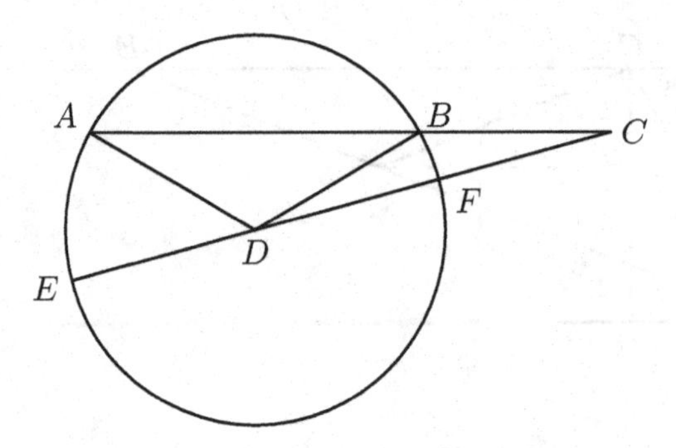

Figure-E6

To duplicate the cube by neusis we solve the equivalent problem of finding two mean proportionals between two given line segments. Here is a remarkably ingenious construction recorded by Pappus.[74] We construct Figure E7 as follows:

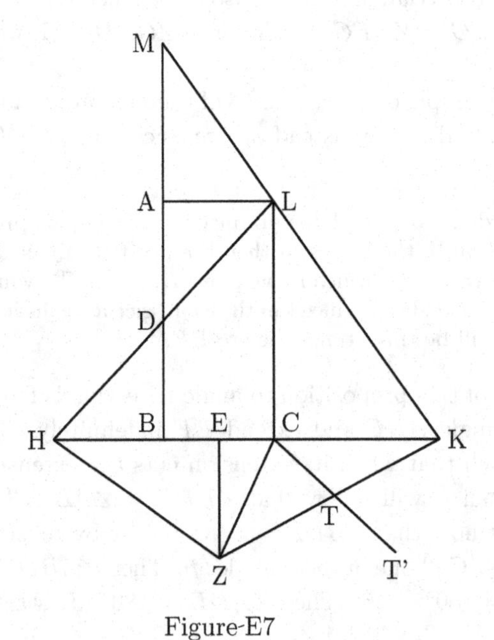

Figure-E7

[74]Pappus (2010), Book IV, Proposition 24, p. 129f. Also see Heath (1981), I, p. 260f.

Let the given lines be AB and BC. Make AB perpendicular to BC and complete the rectangle $ABCL$. Next let D bisect AB and draw LH through D so that it intersects the prolongation of BC at H. Also let E bisect BC and construct the right triangle EZC so that the hypotenuse $ZC = AD$. Next draw CT' parallel to HZ and by neusis let K be the point on the prolongation of BC so that $TK = AD = DB$. Also draw KL and let it intersect with the prolongation of AB at M. Then we claim that

$$AB : KC = KC : MA = MA : BC.$$

To prove this, consider the rectangle with sides BK and CK. If we add the square with side CE to it, the result is equal to the square with side EK,[75] that is

$$BK \cdot KC + (CE)^2 = (EK)^2. \qquad (E.1)$$

Next add the square with side EZ to Both sides. By the Pythagorean theorem we get that

$$BK \cdot KC + (CE)^2 + (CF)^2 = (KF)^2. \qquad (E.2)$$

Since triangles AML and CLK are similar

$$MA : AB = ML : LK = BC : KC.$$

Now also $AB = 2AD$ and $BC = (1/2)HC$ since triangles DAL and DBH are congruent. Because triangles HKZ and CKT are similar

$$HC : CK = ZT : TK.$$

But also $MA : 2AD = HC : 2KC$, so that

$$HC : CK = MA : AD = ZT : TK.$$

But by the neusis construction $AD = TK$. Hence $MA = ZT$ and $(MD)^2 = (ZK)^2$. Now

$$(MD)^2 = BM \cdot MA + (DA)^2$$

by the same type of reasoning that gives the identity (E.1). Since $(MD)^2 = (FK)^2$ and using (E.2) we get that

$$BM \cdot MA + (DA)^2 = BK \cdot KC + CZ^2.$$

But $(DA)^2 = (CZ)^2$. Hence

$$BM \cdot MA = BC \cdot KC,$$

[75]This follows from the *Elements*, Book II, Proposition 6. It also can easily be proved algebraically. Let $x = BC$ and $y = CK$. Then $(x + y)y + (x/2)^2 = ((x/2) + y)^2$.

i.e.,

$$KC : MA = BM : BK = LC : KC.$$

Since also $BM : BK = MA : AL$ we have that

$$LC : KC = KC : MA = MA : AL.$$

Because $AB = LC$ and $AL = BC$ we conclude that KC and MA are the two mean proportionals between AB and BC.

Although the above arguments are extremely ingenious, if the conchoid is used to accomplish the neusis, they give "linear" solutions to the angle trisection and duplication of the curve problem. But in reality both problems are "solid." Angle trisection is equivalent to the solution of a cubic equation and hence can be solved using conic sections, in particular as shown above using a hyperbola, while Menaechmus' solution to the duplication of the cube problem involves the intersection of two parabolas or a parabola with a hyperbola. But in spite of the apparent need to use the conchoid of Nicomedes which is a mechanical curve, it can be shown that neusis constructions are also "solid" provided only lines are involved. One way to see this using modern analytic geometry is to set up a coordinate system with x axis parallel to the one of the two lines and y axis perpendicular to it. Then we want to find a line with equation $y = mx$ such that the points of its intersection with the two lines are a given distance d apart. This leads to a (complicated!) quartic equation in m but such equations can be solved geometrically, as we have seen, using conic sections by the methods of Descartes (or even those of Viète).

It is interesting to contrast the point of view of the ancient geometers as revealed in some of these constructions with that of modern mathematics. There is an element of incommensurability between these two "thought styles." Take for instance the duplication of the cube problem. If s is the side of the original cube the doubled cube has side $\sqrt[3]{2}s$, and to a contemporary mathematician that would be the end of the matter. If questioned by his mathematical ancestor about "$\sqrt[3]{2}s$" he would say that this is usually an irrational real number; and if questioned further, there would be some hand waving about equivalence classes of Cauchy sequences or Dedekind cuts. Still the geometer might observe that all of this simply approximates $\sqrt[3]{2}s$ as closely as we please where the construction locates it *exactly*. Therefore we have not really solved the problem. The only possible reply to such criticism is that we are no longer interested in this kind of construction. Our view of what constitutes a worthwhile mathematical problem has simply changed.

Appendix F

Infinitesimals

The reader will have noticed that we have given almost no analysis concerning Leibniz's views on infinitesimals and the foundational disputes arising from them. The reason for this omission is that we feel that this subject both has been adequately dealt with elsewhere,[76] and also that it is a diversion from our main themes. To conflate calculus with the use and misuse of infinitesimals we feel gives a distorted and limited vision of the complexity of seventeenth century mathematical development. Nevertheless it seems worthwhile to give a brief sketch here of some of the issues raised by infinitesimal concepts in the late seventeenth and early eighteenth centuries, especially after the achievements of Newton and Leibniz.

It should be obvious from our account that infinitesimal ideas, at least on an intuitive level, had been used by many precursors of the calculus almost certainly going back to Archimedes and Apollonius to solve difficult problems. For Fermat, Pascal, Roberval, Huygens, Kepler, Cavalieri, Barrow and many others they were an aid to discovery. To find tangents, quadratures, or surface area it was tempting to regard the geometric object one was considering as composed of infinitely many infinitely small elements. For example, to construct tangents a curve could be thought of as having a polygonal structure made up of infinitesimal line segments. Similar considerations were employed for areas and surfaces. Such techniques were intuitively persuasive and gave answers, but they were not rigorous. It was a commonplace in philosophy since Aristotle that the "actual infinite" did not exist. Infinity was only "potential." Thus there is no largest or smallest positive number. All we can say is that for any number

[76]There has been an enormous literature on this subject and also on the same question concerning Newton. In volume it may equal or exceed all other work concerning the historical development of calculus. See, for example, the book *Infinite Differences* (2008) edited by Goldenbaum and Jesseph and the numerous references contained therein.

$n > 0$ there is a greater one, say, $n + 1$, and for any fraction $1/n$, the fraction $1/(n+1)$ is smaller yet positive. To make infinitesimal arguments logically sound, two strategies had been evolved. The first went back to Archimedes: After the desired conclusion was obtained using infinitesimals one then considered them to be finite but very small. Then a *reductio ad absurdum* argument (discussed in Chapter 2) could be given, by trapping the "true" answer between approximations using collections of these finite elements which respectively under and overestimated it and showing that the difference between these approximations could be made as near to zero as one pleased by making the elements small and numerous enough. The second and much more recent method consisted of the elaborate theory of "indivisibles" introduced by Cavalieri. As we have seen, Cavalieri thought of plane figures as made up of line segments. But this collection of segments was not the area of the figure. Instead the ratio of "all the lines" of a figure F to all the lines of another convenient figure, say a rectangle R, was equal to the ratio of the areas of the two figures. If this ratio could be obtained the area of F could be determined.[77]

By the last third of the seventeenth century infinitesimals began to assume a more undisguised and unapologetic role especially in the works of John Wallis, Leibniz, and Newton. To try to overcome traditional objections to them several points of view (some rather confused) emerged concerning their status. Some mathematicians like Leibniz's partisans the Marquis de la L'Hôpital, Nicolas Malebranche, the Bernoulli brothers, and Pierre Varignon strongly defended the thesis that infinitesimals actually existed and while smaller than any assigned number they were positive and not zero. Others, for example John Wallis, considered that an infinitesimal was the limit of a decreasing ultimately vanishing quantity so that it was actually "nothing" or equivalent to zero.[78] Newton in his early work around 1666 thought of the "moment" o (equivalent to Leibniz's dx or dy) as actually infinitely small. But in his mature formulation of the fluxional calculus he developed the doctrine of "prime and ultimate ratios" which did not explicitly involve infinitesimals and amounted to a rough and intuitive precursor of the modern limit concept. For example, given a particle with coordinates $(x(t), y(t))$ where t is time, Newton defines the instantaneous

[77] Cavalieri intended that his theory of indivisibles should satisfy Euclidean standards of rigor. However, his efforts to do this met considerable contemporary criticism. Several mathematicians including Galileo felt that he did not escape the paradoxes of the infinite and the continuum. See Giusti (1980) and de Gandt (1992).

[78] Wallis felt quite comfortable in using $1/\infty$ to denote the base of an infinitesimal rectangle in a quadrature argument.

velocity or "fluxion" p/q (recall that in Leibniz's notation $p = dy/dt$ and $q = dx/dt$) at t by first considering the ratio

$$\frac{y(t+o) - y(t)}{x(t+o) - x(t)}$$

and then thinking of p/q as the "ultimate" ratio *just* as o vanishes. This approach has the advantage that the moment o is always finite, but the rigorous meaning of "just as o vanishes" is unclear.[79]

All these positions were subject to criticisms, particularly since mathematicians sometimes seemed to adopt more than one of them at the same time. The idea, for instance, that quantities existed that were smaller than any number yet positive seemed to violate the basic laws of the number system, particularly the Archimedean principle that given two positive numbers a and b with $a < b$, there exists an integer n such that $na > b$. On the other hand, if infinitesimals were actually equivalent to zero as Wallis believed, it was difficult to see how anything useful could be obtained by multiplying, dividing by, or adding zero. Worse yet, in many arguments infinitesimals were first considered to exist and then when convenient discarded as equivalent to zero. As the popularity of Leibniz's calculus increased on the continent during the 1690s, especially after l'Hôpital's publication of what can be considered the first calculus textbook *Analysis des infiniments petits pour l'intelligence des lignes courbes*, its foundations began to attract critics raising these kinds of objections. The most prominent were the Dutch physician and amateur mathematician Bernard Nieuwentijt (1654–1718) and a group in the Parisian Academy of Sciences centering on Michel Rolle (1652–1718). These people were believers in traditional geometric rigor. To them many aspects of Leibniz's calculus were at least non-rigorous, and very possibly confused and incorrect.

The least offensive critic was Nieuwentijt. He praised the accomplishments of the calculus and the genius of Leibniz, but in a paper published in 1694 argued that its foundations had not been properly demonstrated. Sometimes according to Nieuwentijt Leibniz omits differentials in his arguments as infinitely small but at other times as zero without distinguishing between the two cases. If infinitesimals are not nothing, they or their higher powers cannot be ignored in calculations as Leibniz did. Doing so treats the differential as zero or nothing which is a contradiction. In particular, that $dx = 0$ is implied by the identity $x + dx = x$ used by both Leibniz and l'Hôpital since we can legitimately subtract x from both sides. "I declare this proposition is indubitable and carries with it most evidently the

[79]For an analysis of Newton's strategy for justifying fluxions, see Kitcher (1973).

certain signs of truth: Only those quantities are equal whose difference is zero or equal to nothing."[80] Ultimately, however, Nieuwentijt was prepared to retain first order differentials which he felt existed as a consequence of the infinite divisibility of line segments, but only if they were not discarded in calculus arguments. But even if first order infinitesimals existed, Nieuwentijt gave a rather murky proof that second or higher order differential infinitely smaller than dx or dy could not. Nieuwentijt also claimed that the differential dz of $z = y^x$ cannot be computed by Leibniz's methods. Attempting to do so leads only to the trivial identity $y^x - y^x = 0$.[81]

Rolle was much more vehement, publishing several articles denouncing Leibniz's calculus in the proceedings of the Academy of Sciences. Broadly speaking his objections were similar to Nieuwentijt's, but more detailed. He argued that Leibniz assumed the existence both of arbitrarily great and small orders of infinity which made no sense, that $x \pm dx = x$ implied that a part was equal to a whole, violating "common notion 5" of the *Elements*; and that in the same argument differentials were both zero and nonzero quantities. One proof of Rolle, which was rather more complicated than Nieuwentijt's, demonstrated that they must be absolute zeros. He reasoned as follows: given the parabola with equation $y^2 = ax$, according to the calculus $adx = 2ydy$. But since the point with coordinates $(x + dx, y + dy)$ is also on the parabola

$$ax + adx = y^2 + 2ydy + dy^2,$$

subtracting ax and y^2 from the left and right sides yields (since an equality is not changed if equals are subtracted from equals) that $adx = 2ydy + dy^2$. Hence given that $adx = 2ydy$, we conclude that $dy^2 = 0$ which implies that both dy and dx are also zero. Furthermore, Rolle maintained that Leibniz's methods gave no new results and sometimes even produced mistaken or incomplete conclusions. As an example he presented a curve for which he claimed Leibniz's procedure to find maxima and minima did not give results that could be found using the methods of Hudde. Suppose in Figure F1 we consider the curve having the equation

$$y - b = \frac{(xx - 2ax + aa - bb)^{2/3}}{a^{1/3}}.$$

[80] Quoted from Nieuwentijt's paper *Considerationes circs analyseos ad quantities infinite parvas applicatae principia & calculi differentialis usum in resolvendis problematibus Geometricis* in Jesseph (1998), p. 9.

[81] Nagel (2008), p. 201ff. Beginning calculus students sometimes make Nieuwentijt's mistake as well.

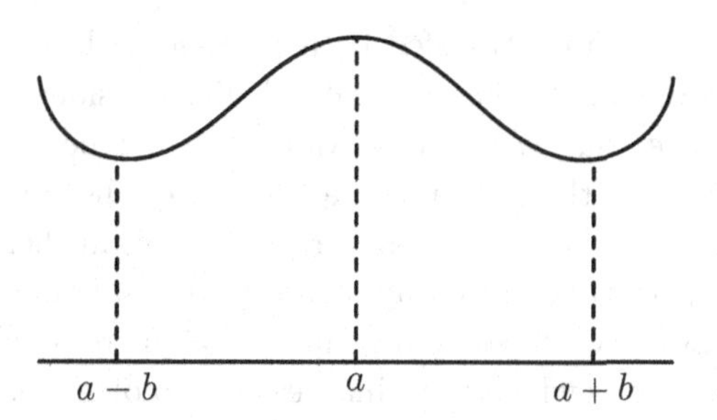

Figure-F1

Then according to Leibniz

$$dy = \frac{4xdx - 4adx}{3\sqrt{axx - 2aax + a^3 - abb}}$$

which yields that $dy = 0$ when $x = a$. This is a local maximum. However, using Hudde's method we also find besides the maximum at $x = a$ two minima at $x = a - b$ and $x = a + b$, "one can therefore be assured that this geometry leads to error."[82] Rolle's objections to Leibniz rapidly evolved into a nasty quarrel mainly within the Academy of Sciences. Supporting Rolle was a group principally composed of the Abbé Gallois (1632–1707), the Abbé Gouye and the astronomer and mathematician Phillipe de la Hire (1640–1718); while Leibniz's defenders included Joseph Saurin (1659–1737), Johann Bernoulli and Bernard Fontenelle (who had become secretary of the academy in 1697) besides Varignon and l'Hôpital. Soon the polemics on each side degenerated to invective and name calling. On the one hand, it was asserted that Rolle was motivated by his anger in not being able to understand the calculus, while for his part Rolle spoke of one supporter of Leibniz as "un pitoyable géomètre."[83] Ultimately in 1706 a committee of the Academy of Sciences which (although it was weighted in favor of Rolle) brought what Fontenelle later called the "peace of the infinitely small" by refusing to take sides and only requesting that everyone observe standards of academic politeness.

On the grounds of pure logic (excluding his false counterexamples) Rolle may have had the better of the argument. Postulates by l'Hôpital asserting the existence of infinitesimal quantities, Varignon's identification of a

[82]This counterexample of Rolle was refuted by Varignon who pointed out that at $x = a-b$ and $x = a + b$) there are indeed two minima. But here $dy = \infty$ so that the curve has cusps with *vertical* tangents which Rolle failed to perceive at these two values of x. See the discussion in Blay (1998), pp. 113–119 or Mancosu (1989) p. 233.

[83]Mancosu (1989), pp. 239, 242.

differential as the distance traveled by a particle with a given velocity in an "instant" of time, or statements that calculus considers differentials not before and not after but *just* as they vanish cannot be said to have been satisfactory. But after the action of the committee the supporters of Leibniz rapidly became victorious in spite of weaknesses in their arguments, in part because their opponents simply died out and in part because calculus "worked"—regardless of its soundness it became increasingly obvious that it could solve problems which were inaccessible to older techniques.

As is well known, criticisms similar to those of Nieuwentijt and Rolle were later (1734) echoed in England by Bishop Berkeley (1685–1753) in his book *The Analyst* and applied both to Leibniz's calculus and Newton's fluxions. Berkeley did not question the fact that calculus gave impressive result, but he thought differentials and Newtonian moments made no sense, being "ghosts of departed quantities." Berkeley was motivated, however, by more than purely mathematical concerns. His real aim was to defend religion against what he thought was the tendency to unbelief by mathematically oriented scientists, by arguing that their mathematical theories rested on weaker foundations than anything in Christian theology.

Where did Leibniz stand in relation to infinitesimals? His writings on the subject show several changes in points of view. In his early work on quadratures when he was using Cavalieri's indivisibles he seems to have accepted the reality of infinitely small quantities, especially in relation to the characteristic triangle which is said to have infinitely small sides.[84] But in the same period (and as late as the 1690s) possibly under the influence of Hobbes notion of *conatus* and geometric points[85] Leibniz also sometimes spoke of infinitesimals as "incomparably small" but finite objects whose magnitude could be ignored. Their size in relation to an ordinary line segment was similar, for example, to the size of the earth in relation to the heavens.[86] Either position was vulnerable, the first to the objections of Nieuwentijt and Rolle, and the second to the observation that if infinitesimals were in some sense finite then calculus offered only approximations (though very good ones). As early as 1676 Leibniz seems to have rejected the actual existence of infinitesimals, giving in that year a proof (which he still accepted in correspondence with Varignon in 1702) of the impossi-

---

[84]For a discussion of Leibniz's early views on indivisibles and infinitesimals see Probst (2008).

[85]See especially Jesseph (2008), p. 216ff. Recall from Chapter 4 that Hobbes thought of points as extended, but of a magnitude too small to be considered in demonstrations.

[86]Jesseph (1998) and (2008), p. 227.

bility of infinitely small line segments.[87] Leibniz's mature position which he had reached by 1700 was to regard infinitesimals merely as useful "fictions," much like imaginary numbers in algebra. They are convenient to use and properly employed lead to correct results. Moreover, whatever can be shown using them could (in theory) be proven rigorously using the Archimedean method of exhaustion. It is interesting and a bit amusing that this reasonable and diplomatic position dismayed supporters like l'Hôpital and Varignon, who in their enthusiasm to defend Leibniz against his Academy of Sciences critics argued for the absolute reality of infinitesimals. In a letter written in 1716 Leibniz recalled that l'Hôpital in particular accused him of betraying the cause of real infinitesimals and asked that he please shut up.[88]

While quarrels over infinitesimals are part of the mathematical history of the period and deserve mention, it is easy to make too much of them. Identifying infinitesimals with calculus itself (as some of the literature seems to do) diverts attention both from the complicated structure of seventeenth century mathematics and the techniques which were actually evolved. It should also be noted that Leibniz, Newton, (and many other mathematicians of the period) had a sound intuitive sense of limits. Their notions of infinitesimals were often confused and their rigor may not have satisfied either Weierstrass or modern calculus teachers who insist on the pedantic infliction of δ, ϵ arguments on hapless audiences of engineering students; but they were seldom misled, and they realized that Archimedean methods could be used to give rigorous proofs of their results. (If they often neglected to do so, it was because such proofs would have been boring and nearly unreadable.[89]) Moreover, they were comfortable using infinitesimals

---

[87] Basically Leibniz argued as follows: Suppose AB is an infinitely short segment and CD is a finite segment. Consider the mean proportional EF between AB and CD (which he assumes to exist) and let GH and IK be line segments such that $AB : EF = EF : CD = CD : GH = GH : IK$. Next extend GH to a line GL having the same length as IK. Now for these ratios to hold since AB is infinitely short GH, GL, and IK must be infinitely long. But $GH \subset GL$; hence GH is an infinite line with a right endpoint H which is impossible. Leibniz was fond of mean proportion arguments; for instance, he defines the second order differential ddx by considering dx to be the mean proportional between ddx and x. For further details see Bassler (2008), pp. 136–139 and Levey (2008), pp. 112–115.

[88] Mancosu (1989), p. 237.

[89] However in the *Principia* Newton gives a perfectly rigorous argument using what would later be called Riemann sums, that an area under a curve can be approximated to any degree of accuracy between sums of rectangles that respectively over and underestimate it. See Lemma 3, Corollary 4 in Newton (1999), p. 433 and the discussion in Arthur (2008), p. 17f. A similar argument was given by Leibniz in his *De quadratura arithmetica* in 1675–766. (*Ibid.*, pp. 20–24.)

because they thought that such techniques amounted to a rediscovery of the method by which the ancients actually found their results and then carefully concealed behind their elegant but unmotivated proofs .

While we we agree with Michel Blay (1998) that the seventeenth and early eighteenth century saw a turning away from a traditional geometric paradigm to a more sophisticated "mathematization" of natural phenomena, culminating in the work of Euler and Lagrange (1736–1813) we do not feel that the transition has much to do with any new admission of the infinite into mathematics. Whatever the status infinitesimals were believed to have in the seventeenth century, they were not the most critical novelty of the mathematics of the period—after all, as we have noted, the idea of the infinitely small had been used intuitively by many geometrically oriented mathematicians as a means of discovery. And in the case of Leibniz we have tried to show that the originality of his calculus does not rest on a novel treatment of the infinite. Instead its emergence represented a complicated interaction between Leibniz's exposure to contemporary mathematics and a much more general set of nonmathematical cultural beliefs (ultimately traceable to Lulle) concerning the possibility of constructing a genuine *Ars Inveniendi* which inspired him.

A Note on the Author

Born in 1939, Richard Brown is Professor Emeritus of Mathematics at the University of Alabama. He initially studied European history and history of science at the Berkeley campus of the University of California, receiving the BA in 1960 and MA in 1962. After teaching college level history for a few years, he earned a PhD in mathematics in 1972 from the Pennsylvania State University. Following a post-doctoral appointment at the Army Mathematics Research Center in Madison Wisconsin (1972–74), he taught at the University of Alabama in Tuscaloosa, retiring in 2002. He is the author of more than sixty research papers in mathematics and one previous book related to the history of science: *Are Science and Mathematics Socially Constructed? A Mathematician Encounters Postmodern Interpretations of Science* (World Scientific, 2009).

Bibliography

Aiton, E. J. (1972). *The Vortex Theory of Planetary Motion* (Macdonald, London and American Elsevier Inc., New York).

Aiton, E. J. (1985). *Leibniz: A Biography* (Adam Hilger Ltd, Bristol and Boston).

Aiton, E. J. (1986). The application of the infinitesimal calculus to some physical problems by Leibniz and his friends, *Studia Leibnitiana* **14**, pp. 133–143.

Allman, G. (1889). *Greek Geometry from Thales to Euclid* (Dublin University Press, Dublin).

Alvarez, C. (2008). François Viète et la mise en équation des problèmes solides, in P. Radelat-de Grave and C. Brichard (eds.), *Liber amicorum Jean Dhombres* (Brepols, Turnhout, Belgium), pp. 16–61.

Andersen, K. (1985). Cavalieri's method of invisibles, *Archive for History of Exact Science* **31**, pp. 291–367.

Anonymous (1910). *Barrow*, in *The Encylopaedia Britannica*, 11th edn. (The Encyclopaedia Britannica Company, New York), Vol. 3, p. 440.

Antognazza, M. R. (2009). *Leibniz: An Intellectual Biography* (Cambridge University Press, Cambridge, New York).

Apollonius (1961). *Apollonius of Perga, Treatise on Conic Sections*, editied in modern notation including an essay on the earlier history of the subject by T. L. Heath (W. Heffer & Sons, Cambridge).

Archimedes (2010). *The Works of Archimedes*, edited in modern notation with introductory chapters by Thomas L. Heath (Cambridge University Press, Cambridge, New York).

Arthur, R. T. W. (2008). Leery bedfellows: Newton and Leibniz on the status of infinitesimals, in U. Goldenbaum and D. Jesseph (eds.), *Infinitesimal Differences: Controversies between Leibniz and His Contemporaries* (Walter de Gruyter, Berlin, New York), pp. 7–30.

Baron, M. E. (2003). *The Origins of the Infinitesimal Calculus* (Dover Publications, Inc., Mineola, New York).

Barrow, I. (1916). *The Geometrical Lectures of Isaac Barrow*, edited and translated with a commentary by J. M. Child (Open Court, Chicago and London–Original work published 1670).

Bassler, O. B. (1999). Towards Paris: The growth of Leibniz's Paris mathematics and pre-Paris metaphysics, *Studia Leibnitiana* **31**(2), pp. 160–180.

Bassler, O. B. (2008). An enticing(im)possibility: infinitesimals, differentials, and the Leibnizian calculus, in U. Goldenbaum and D. Jesseph (eds.), *Infinitesimal Differences: Controversies between Leibniz and His Contemporaries* (Walter de Gruyter, Berlin, New York), pp. 135–151.

Beck, L. W. (1969). *Early German Philosophy: Kant and His Predecessors* (The Belknap Press of Harvard University Press, Cambridge, Mass.).

Beeley, P. (2004). A philosophical apprenticeship: Leibniz's correspondence with the secretary of the Royal Society, Henry Oldenburg, in P. Lodge (ed.) *Leibniz and his Correspondents* (Cambridge University Press, Cambridge), Chap. 3, pp. 47–73.

Beeley, P. (2009). *De abstracto et concreto*: Rationalism and empirical science in Leibniz, in M. Dascal (ed.), *Leibniz: What Kind of Rationalist?* (Springer, Berlin 2009).

Bell, E. T. (1965). *Men of Mathematics* (Simon & Shuster, New York).

Bernstein, H. R. (1980). *Conatus*, Hobbes, and the young Leibniz, *Studies in the History and Philosophy of Science* **11**, pp. 25–37.

Beyerchen, A. D. (1978). *Scientists under Hitler: Politics and the Physics Community in the Third Reich* (Yale University Press, New Haven).

Bird, A. (2000). *Thomas Kuhn* (Princeton University Press, Princeton).

Blank, B. E. (2009). Review of *The Calculus Wars: Newton, Leibniz, and the Greatest Mathematical Clash of All Time* by J. S. Bardi, *Notices of the AMS* **56**(5), pp. 602–610.

Blay, M. (1998). *Reasoning with the Infinite: From the Closed World to the Mathematical Universe* (The University of Chicago Press, Chicago and London).

Bloor, D. (1976). *Knowledge and Social Imagery* (Routledge & Keegan Paul, Boston, London and Henley).

Boehner, P. (1952). *Medieval Logic* (University of Manchester Press, Manchester).

Boole, G. (1951). *An Investigation of the Laws of Thought: On Which are Founded the Mathematical Theories of Logic and Probabilities* (Dover Publications, New York).

Bos, H. J. M. (1974). Differentials, higher order differentials and the derivative in the Leibnizian calculus, *Archive for History of Exact Sciences* **14**, pp. 1–90.

Bos, H. J. M. (1978). The influence of Huygens on the formation of Leibniz's ideas, *Studia Leibnitiana* **17**, pp. 59–68.

Bos, H. J. M. (1980). Huygens and mathematics, in H. J. M. Bos, M. J. S. Rudwick, H. A. M. Snelders, and R. P. W. Visser (eds.), *Studies on Christiaan Huygens* (Swets & Zeitlinger B. V., Lisse, Holland), pp. 126–146.

Bos, H. J. M. (1981). On the representation of curves in Descartes' Géométry, *Archive for History of Exact Sciences* **24**, pp. 295–388.

Bos, H. J. M. (1984). Arguments on motivatiòn in the rise and decline of a mathematical theory; the "construction of equations", 1637–*ca.* 1750, *Archive for History of Exact Sciences* **30**, pp. 331–380.

Bos, H. J. M. (1990). The structure of Descartes' Géométrie, in *il Metodo e i Saggi; Atti del Convegno per il 350° Anniversario della Publicazione del*

Discours de la Methode e degli Essais (Instituto della Enciclopedia Itali, Rome), pp. 349–369.

Bos, H. J. M. (2000). *Redefining Geometrical Exactness: Descartes Transformation of the Early Modern Concept of Construction* (Springer, New York).

Bos, H. J. M., and Mehrtens, H. (1977). The interactions of mathematics and society, in history: some exploratory remarks, *Historia Mathematica* **4**, pp. 7–30.

Bouwsma, W. J. (1957). *Concordia Mundi: The Career and Thought of Guillaume Postel, 1510–1561* (Harvard University Press, Cambridge, Mass.).

Boyer, C. (1959). *The History of the Calculus and Its Conceptual Development* (Dover Publications, Inc., New York).

Brandt, F. (1927). *Thomas Hobbes' Mechanical Conception of Nature* (Copenhagen, 1927).

Broad, C. (1949). Leibniz's predicate-in-notion principle and some of its alleged consequences, *Theoria* **15**, pp. 54–70.

Broad, C. (1975). *Leibniz: an Introduction*, C. Lewy (ed.) (Cambridge University Press, London).

Broughton, J. and Carriero, J. (eds.) (2011). *A Companion to Descartes*, Blackwell Companions to Philosophy (John Wiley and Sons, Inc., Chichester, West Sussex, UK).

Brown, R. (2009). *Are Science and Mathematics Socially Constructed? A Mathematician Encounters Postmodern Interpretations of Science* (World Scientific, London, Singapore, Beijing).

Burton, D. M. (1985). *The History of Mathematics, an Introduction* (Allyn and Bacon, Inc., Boston, Sydney, and Toronto).

Burton, D. M. (1998). *Elementary Number Theory*, 4th edn. (The McGraw-Hill companies, inc., New York).

Burtt, E. (1932). *The Metaphysical Foundations of Modern Sciences* (Doubleday and Co., Garden City, N.J.).

Cohen, J. (1954). On the project of a universal characteristic, *Mind*, New Ser. **63**, pp. 49–63.

Copenhaver, B. and Schmitt, C. (1992). *Renaissance Philosophy* (Oxford University Press, Oxford and New York).

Corry, L. (1993). Kuhnian issues, scientific revolutions and the history of mathematics, *Stud. Hist. Phil. Sci.* **24**(1), pp. 95–117.

Coudert, A. (1995). Leibniz and the Kabbalah, *International Archives of the History of Ideas* **142**) (Kluwer Academic, Boston).

Coudert, A. (1999). *The Impact of the Kabbalah in the Seventeenth Century: the Life and Thought of Francis Mercury van Helmont (1614–1698)*, Brill's Series in Jewish Studies **9** (Brill, Leiden, Boston).

Couturat, L. (1961). *La Logique de Leibniz d'après des Documents Inédits* (Georg Olms Verlagsbuchhandlung, Hildesheim—Original work published 1901).

Crowe, M. J. (1975). Ten "laws" concerning patterns of change in the history of Mathematics, *Historia Mathematica* **2**, pp. 161–166.

Crowe, M. (1992). A revolution in the historiography of mathematics? in D.

Gillies (ed.) *Revolutions in mathematics* Clarendon Press, Oxford), pp. 306–316.

Crowe, M. (1994). *A History of Vector Analysis* (Dover, Mineola, N.Y).

Dauben, J. (1984). Conceptual revolutions and the history of mathematics: two studies in the growth of knowledge, in E. Mendelsohn (ed.), *Transformation and Tradition in the Sciences, Essays in Honor of I. Bernard Cohen* (Cambridge University Press, Cambridge), pp. 81–109. Reprinted in Gillies (1992), pp. 49–71.

Davis, P. J. and Hersh, R. (1981). *The Mathematical Experience* (Houghton Mifflin, Boston).

De Gandt, F. (1992). Cavalieri's indivisibles and Euclid's canon, in *Revolution and Continuity; Essays in the History and Philosophy of Modern Science* (Catholic University Press, Washington D.C.) pp. 157–182.

De Gandt, F. (1995). *Force and Geometry in Newton's* Principia (Princeton University Press, Princeton, N.J.).

De Risi, V. (2007). *Geometry and Mondology: Leibniz's Analysis Situs and Philosophy of Space* (Springer, Berlin).

Descartes, R. (1954). *The Geometry of René Descartes*, translated from the French and Latin; by D. E. Smith with a facsimile of the first edition, 1637 (Dover Publications, New York).

Descartes, R. (1965). *Discourse on Method, Optics, Geometry, and Meteorology*, translated by P. J. Olscamp (Bobbs-Merrill Co., Inc., Indianpolis—Original work published 1637).

Descartes, R. (1983). *Principles of Philosophy*, translated with explanatory notes by V. R. and R. P. Miller (Dordrecht, Holland and Boston, Mass.; Reidel, Hingham, Mass.).

Dhombres, J. (1983). Is one proof enough? Travels with a mathematician of the baroque period, *Educational Studies in Mathematics* **24**, pp. 401–411.

Dugas, R. (1958). *Mechanics in the Seventeenth Century (From the Scholastic Antecedents to Classical Thought* (Central Book Company, New York).

Dugas, R. (1988). *A History of Mechanics*, translated by J. R. Maddox (Dover Publications, Inc., New York).

Dunham, W (1990). *Journey Through Genius: The Great Theorems of Mathematics* (Wiley Science Editions, John Wiley & Sons, Inc., New York).

Eaton, R. M. (1931). *General Logic: An Introductory Survey* (Charles Scribner's Sons, New York).

Echeverria, J. (1979). L'Analyse géométrique de Grassmann et ses rapports avec la caractéristique geometrique de Leibniz, *Studia Leibnitiana*, **11**, pp. 223–76.

Eco, U. (1995). *The Search for the Perfect Language*, translated by J. Fentrss (Blackwell, Oxford, UK and Cambridge, USA).

Edwards, C. H. (1979). *The Historical Development of the Calculus* (Springer-Verlag, New York).

Edwards, H. A. (1989). Kronecker's views on the foundations of mathematics, in *The History of Modern Mathematics* (Academic Press, Inc., New York), pp. 67–77.

Ernest, P. (1998). *Social Constructivism as a Philosophy of Mathematics* (State University of New York Press, New York).

Euclid (1956). *The Thirteen Books of the Elements*, Vols. I–III, 2nd edn., translated and edited by Sir Thomas Heath (Dover Publications, Inc., New York).

Fauvel, J. and Wilson, R. J. (1994). The Llull before the storm. *Bulletin of the ICA*, **11**, pp. 49–58.

Feingold, M. (1990). Isaac Barrow: divine, scholar, mathematician, in M. Feingold (ed.) *Before Newton: the Life and Times of Isaac Barrow* (Cambridge University Press, Cambridge), Chap. 1, pp. 1–104.

Feingold, M. (1993). Newton, Leibniz, and Barrow too, *Isis*, **84**, pp. 310–338.

Fisher, C. S. (1966). The death of a mathematical theory: A study in the sociology of knowledge, *Archive for History of Exact Sciences* **iii**, pp. 137–139.

Fleck, L. (1979). T. J. Trenn and R. K. Merton (eds.), *The Genesis and Development of a Scientific Fact* (The University of Chicago Press, Chicago).

Fowler, D. (1987). *The Mathematics of Plato's Academy: A New Reconstruction* (Clarendon Press, Oxford).

Freguglia, P. (2008). Les équations algébriques et la geométrie chez les algébrists du XVIe siècle et chez Vi'ete, in P. Radelat-de Grave and C. Brichard (eds.), *Liber amicorum Jean Dhombres* (Brepols, Turnhout, Belgium), pp. 148–161.

Gale, G. (1973). Leibniz's dynamical metaphysics and the origins of the vis viva controversy, *Systematics* **11**, pp. 184–207.

Garber, D. (2008). Dead force, infinitesimals, and the mathematization of nature, in in U. Goldenbaum and D. Jesseph (eds.), *Infinitesimal Differences: Controversies between Leibniz and His Contemporaries* (Walter de Gruyter, Berlin, New York), pp. 281–306.

Garber, D. (2009). *Leibniz: Body, Substance, Monad* (Oxford University Press, Oxford).

Gardner, M. (1982). *Logic Machines and Diagrams*, 2nd edn. (The University of Chicago Press, Chicago).

Gaukroger, S. (1995). *Descartes: An Intellectual Biography* (Oxford University Press, Oxford).

Gilles, D. (ed.) (1992). *Revolutions in Mathematics* (Clarendon Press, Oxford).

Giusti, E. (1980). *Bonaventura Cavalieri and the Theory of Indivisibles* (Edizioni Cremonese, Bologna).

Giusti, E. (1992). Algebra and geometry in Bombelli and Viète, *Boll. Storia. Sci. Mat.* **12**, pp. 303–328.

Goldenbaum U. (2008). Indivisibilia vera—how Leibniz came to love mathematics, in U. Goldenbaum and D. Jesseph (eds.), *Infinitesimal Differences: Controversies between Leibniz and His Contemporaries* (Walter de Gruyter, Berlin, New York), pp. 53–94.

Gould, S. (1977). *Ever Since Darwin: Reflections on Natural History* (Norton, New York).

Gow, J. *A Short History of Greek Mathematics* (Chelsea, New York.)

Gracia, J. J. E. (ed.) (1994). *Individuation in Scholasticism: The Later Middle*

Ages and the Counter-Reformation (State University of New York Press, Albany).

Grinell, G. (1973). Newton's *Principia* as Whig propaganda, in P. Fritz and D. Williams (eds.) *City and Society in the 18th Century* (Hakkert, Toronto), pp. 181–192.

Grosholz, E. (1991). Descartes geometry and the classical tradition, in Peter Barker and Roger Ariew (eds.) *Revolution and Continuity: Essays in the History and Philosophy of Early Modern Science* (Catholic University of America Press, Washington, D. C.), pp. 183–216.

Grosholz, E. (1992). Was Leibniz a mathematical revolutionary? in D. Gillies (ed.) *Revolutions in Mathematics* (Clarendon Press, Oxford), Chap. 7, pp. 117–138.

Grosholz, E. and Yakira, E. (1998). *Leibniz's Science of the Rational* Studia Leibnitiana Sonderheft 26 (Franz Steiner Verlag).

Gross, P. R. and Levitt N. (1994) *Higher Superstition: the Academic Left and Its Quarrel with Science* (John Hopkins University Press, Baltimore and London).

Guicciardini, N. (1999). *Reading the Principia: The Debate on Newton's Mathematical Methods for Natural Philosophy from 1687* (Cambridge University Press, Cambridge, New York).

Guicciardini, N. (2009). *Isaac Newton on Mathematical Certainty and Method* (MIT Press, Cambridge, Mass.).

Guhrauer, G. E. (1966). *Gottfried Wilhelm Freherr von Leibnitz. Eine Biographie* (Reprint of original 1842 edition, Georg Olms, Hildesheim).

Giusti, E. (1980). *Bonaventura Cavalieri and the Theory of Indivisibles*(Ediziones Cremonese, Bologna).

Giusti, E. (1992). *Algebra and geometry in Bombelli and Viète Boll. Storia Sci. Mat.* **12**(2), pp. 303–328.

Hall, A. R. (1976). Leibniz and the British mathematicians, in *Leibniz a' Paris 1672–1676: Symposium de la G. W. Leibniz-Gesellschaft (Hanover) et du Centre National de la Recherche Scientifique (Paris a' Chantilly (France) de 14 au 18 November (sic) 1976* (Steiner, Wiesbaden), pp. 131–152.

Hall, A. R. (1980). *Philosophers at War* (Cambridge University Press, Cambridge).

Hankins, T. L. (1965). Eighteenth-century attempts to resolve the vis viva controversy, *Isis* **56**(3), pp. 281–297.

Hardy G. H. (1915). Prime numbers, *British Association Report*, **10**, pp. 350–354.

Hardy, G. H. (1992). *A Mathematician's Apology* (Cambridge University Press, Cambridge).

Heath, T. (1981). *A History of Greek Mathematics* (2 Vols.) (Dover Publications, Inc., New York).

Heilbron, J. L. (2010). *Galileo* (Oxford University Press, Oxford, New York).

Henrici, O. (1910). "Calculating machines," in *The Encylopaedia Britannica*, 11th edn. (The Encyclopaedia Britannica Company, New York), vol. 4, p. 973.

Hersh, R. (1997). *What Is Mathematics Really?* (Oxford University Press, New York).

Hessen B. (1931). The social and economic roots of Newton's Principia, in *Science at the Crossroads* (Kniga, London).

Hillgarth, J. N. (1971). *Ramon Lull and Lullism in fourteenth-century France* (Clarendon Press, Oxford).

Himmelfarb, G. (1967). *Darwin and the Darwinian Revolution* (P. Smith, Gloucester, Mass.).

Hofmann, J. E. (1959). *Classical Mathematics, a Concise History of the Classical Era in Mathematics* (Philosophical Library, New York).

Hofmann, J. E. (1974a). *Leibniz in Paris 1672–1676* (Cambridge University Press, London).

Hofmann, J. E. (1974b). The mathematical studies of G. W. Leibniz on Combinatorics, *Historia Mathematica*, **1**, pp. 409–430.

Hogben, L. (1940). *Mathematics for the Million*, revised and enlarged edn. (W. W. Norton & Company, Inc., New York).

Huygens C. (1899). *Oeuvres Complètes de Christiaan Huygens, tome huitième: correspondance 1676–1684 publiées par la Societe Hollandise des Sciences* [Complete works of Christiaan Huygens, volume 8: correspondence 1676–1684 published by the Dutch Society of Sciences] (Martinus Nijhoff, The Hague).

Iltis, C. (1971). Leibniz and the vis viva controversy, *Isis* **62**(1), pp. 21–35.

Jesseph, D. M. (1981). Leibniz on the foundations of the calculus: The question of the reality of infinitesimal magnitudes, *Perspectives on Science* **6**, pp. 6–40.

Jesseph, D. M. (1999). *Squaring the Circle: The War between Hobbes and Wallace* (The University of Chicago Press, Chicago, London).

Jesseph, D. M. (2006). Hobbesian mechanics, in *Oxford Studies in Early Modern Philosophy* (Clarendon Press, Oxford) **3**, pp. 119–152.

Jesseph, D. M. (2008). Truth in fiction: origins and consequences of Leibniz's doctrine of infinitesimal magnitudes, in U. Goldenbaum and D. Jesseph (eds.), *Infinitesimal Differences: Controversies between Leibniz and His Contemporaries* (Walter de Gruyter, Berlin, New York), pp. 215–233.

Jungius, J. (1971). *Logicae Hamburgenis additamenta*, Wilhelm Risse (ed.) (Vandenhoeck & Ruprecht, Göttingen).

Keynes, J. M. (1947). Newton the man, in *The Royal Society Newton Tercentary Celebrations 15–19 July 1946* (Cambridge University Press, Cambridge).

Kirsanov, V. (2008). Leibniz in Paris, in Hartmut Hetch, et al. (eds.) *Kosmos und Zahl Beitrage zur Mathematik und Astronomie zu Alexander von Humbolt und Leibniz*; unter Mitarbeit von Katharina Zeitz (Steiner, Stuggart), pp. 137–151.

Kitcher, P. (1973). Fluxions, limits and infinite littleness: A study of Newton's presentation of the calculus, *Isis* **64**, pp. 33–49.

Klein F. (1956). *Famous Problems of Elementary Geometry: The Duplication of the Cube, the Trisection of an Angle, the quadrature of the Circle* (Dover Publications, Inc., New York).

Klein, J. (1968). *Greek Mathematical Thought and the Origin of Algebra*, Eva Brann (trans.) (M.I.T. Press, Cambridge, Mass.).

Kneale, W. and Kneale, M. (1962). *The Development of Logic* (Clarendon Press, Oxford).

Knecht, H. H. (1981). *La Logique chez Leibniz: Essai sur le Rationalism Baroque* (Editions L'Age d'Homme, Lausanne).

Kuhn, T. S. (1962). *The Structure of Scientific Revolutions* (The University of Chicago Press, Chicago).

Kuhn, T. S. Second thoughts on paradigms, in F. Suppe (ed.) *The Structure of Scientific Theories* (The University of Illinois Press, Urbana, Chicago, London, 1974), pp. 458–482; Discussion, pp. 500–517.

Latour B. (1985) *Science in Action: How to Follow Scientists and Engineers Through Society* (Cambridge University, Cambridge).

Laudan, L. L. (1968). The Vis Viva controversy, a post-mortem, *Isia* **59**(2), pp. 130–143.

Leibniz, G. N. (1671). *Hypothesis Physica Nova, Qua Phaenomenorum Naturae plerorumque causae ab unico quodam universali motu, in globo nostro supposito, neque Tychonicis, neque Copernicanis aspernando, repetuntur. Nec non Theoria Motus Abstracti.* (J. Martyn, Regiae Societatis Typographi, Londini).

Leibniz, G. N. (1920). *The Early Mathematical Manuscripts of Leibniz*, translated and edited by J. M. Child (Open Court, Chicago and London) (Original manuscripts and letters from 1673–1714).

Leibniz, G. N. (1961). *Opuscules et Fragments Inédits de Leibniz*, L. Couturat (ed.) (Georg Olms Verlagsbuchhandlung, Hildesheim—Original work published 1903).

Leibniz, G. W. (1966). *Leibniz: Logical Papers*; a selection translated and edited with an introduction by G. H. R. Parkinson (Clarendon Press, Oxford)

Leibniz, G. W. (1969). *Philosophical Papers and Letters*, translated and edited by L. E. Loemker (D. Reidel Publishing Company, Dordrecht—Holland).

Leibniz, G. N. (1971) *Leibnizens mathematische schiften, herausgegeben von C. Gerhardt* [Leibniz's mathematical writings collected by C. L. Gerhardt]. (Vols. 1–7) (G. Olms, Hildesheim and New York—paperback edition of original published in 6 volumes., 1849–1863).

Leibniz, G. W. (1989). *La naissance du cacul différentiel: 26 articles des* Acta Eruditoum, edited and translated by Marc Parmentier (Librarie Philosophique J. Vrin, Paris).

Leibniz, G. W. (1993). *De quadratura arithmetica ciculi ellipeos et hyperbolae cujus corollarium est trigonometria sine tabulis*, edited with a commentary by E. Knobloch (Vanden hoeck & Ruprecht, Göttingen).

Leibniz, G. W. (2001). *The Labyrinth of the Continuum: Writings on the Continuum Problem, 1672–1686*, translated, edited, and with an introduction by Richard T. W. Arthur (Yale University Press, New Haven and London).

Leibniz, G. W. (2008). *Sämtliche Schriften und Briefe*, Seventh Series, Mathematical Writings, Vol 5 (Berlin and Göttingen Academies of Science).

Levey, S. (2008). Arcimedes, infinitesimals and the law of continuity: on Leibniz's fictionalism, in U. Goldenbaum and D. Jesseph (eds.), *Infinitesimal Differ-*

ences: Controversies between Leibniz and His Contemporaries (Walter de Gruyter, Berlin, New York), pp. 107–133.

Llull, R. (1985). *Selected Works of Ramon Llull (1232–1316)*, Vols. I–II, edited and translated by A. Bonner (Princeton University Press, Princeton, New Jersey).

Maat, J. (2004). *Philosophical Languages in the Seventeenth Century: Dalgarno, Wilkins, Leibniz*, The New Synthese Historical Library, Vol. 54 (Kluwer Academic Publishers, Dordrecht/Boston/London).

Mackie, J. M. (1845). *Life of Godfrey William von Leibnitz on the Basis of the German Work of Dr. G. E. Guhrauer* (Gould, Kendell and Lincoln, Boston, Mass.).

Maclane, S. (1995). Mathematics at Göttingen under the Nazis. *Notices of the American Mathematical Society*, **42**(10), pp. 1134–1138.

Mahoney, M. (1968). Another look at Greek geometrical analysis, *Archive for History of Exact Sciences* **5**, pp. 318–348.

Mahoney, M. (1990a). Barrow's mathematics: between ancients and moderns, in M. Feingold (ed.), *Before Newton: The Life and Times of Isaac Barrow* (Cambridge University Press, Cambridge), Chap. 3, pp. 179–249.

Mahoney, M. (1990b). Infinitesimals and transcendent relations: The mathematics of motion in the late seventeenth century, in D. C. Lindberg and R. S. Westerman (eds.), *Reappraisals of The Scientific Revolution* (Cambridge University Press, Cambridge), Chap. 12, pp. 461–491.

Mahoney, M. (1994). *The Mathematical Career of Pierre de Fermat 1601–1665*, 2nd edn. (Princeton University Press, Princeton, New Jersey).

Mancosu P. (1989) The metaphysics of the calculus: a foundational debate in the Paris Academy of Sciences, 1700–1706, *Historia Mathematica* **16**, pp. 224–248.

Mancosu P. (1992). Descartes's *Géometrie* and revolutions in mathematics, in D. Gillies (ed.) *Revolutions in Mathematics* (Clarendon Press, Oxford), Chap. 6, pp. 83–116.

Mancosu P. (1996). *Philosophy of Mathematics and Mathematical Practice in the 17th Century* (Oxford University Press, New York and Oxford).

Mates, B. (1986). *The philosophy of Leibniz* (Oxford University Press, New York and Oxford).

McCullough, L. B. (1996). *Leibniz on Individuals and Individuation: The Persistence of Premodern Ideas in Modern Philosoiphy* (Kluwer Academic Publishers, Dordrecht/Boston/London).

Mehrtens H. (1976). T. S. Kuhn's theories and mathematics: a discussion paper on the "new historiography" of mathematics, *Historia Mathematica* **3**, pp. 297–320.

Mehrtens H. (1987). Ludwig Bieberbach and "Deutsche Mathematik," *Studies in Mathematics* **26**, pp. 195–241.

Meli, B. (1993). *Equivalence and Prioity: Newton versus Leibniz Including Leibniz's Unpublished Manuscripts on the Principia* (Oxford University Press, New York).

Mercer, C. (2001). *Leibniz's Metaphysics: Its Origins and development* (Cambridge University Press, Cambridge)

Molland, G. (1976). Shifting the foundations: Descartes transformation of ancient geometry, *Historia Mathematica* **3**, pp. 21–49.

Molland, G. (1993). Science and mathematics from the Renaissance to Descartes, in G. H. R. Parkinson (ed.), *The Renaissance and Seventeenth-century Rationalism: Vol. IV. Routledge History of Philosophy* (Routledge, London and New York), pp. 104–139.

Muir, T. (1910). "Circle," in *The Encylopaedia Britannica*, 11th ed. (The Encyclopaedia Britannica Company, New York) Vol. 6, p. 386.

Nagel, F. (2008). Nieuwentijt, Leibniz, and Jacob Hermann on infinitesimals, in U. Goldenbaum and D. Jesseph (eds.), *Infinitesimal Differences: Controversies between Leibniz and His Contemporaries* (Walter de Gruyter, Berlin, New York), pp. 199–214.

Newton, I. (1959). *The Correspondence of Isaac Newton* (Vol. I (1661–1675) & Vol. II (1676–1687)), H. W. Turnbull (ed.) (Cambridge University Press, London).

Newton, I. (1962). *Unpublished Scientific Papers of Isaac Newton: A Selection from the Portsmouth Collection in the University Library, Cambridge*, translated and edited by A. Rupert Hall and Marie Boas Hall (Cambridge University Press, London).

Newton, I. (1967). *The Mathematical Papers of Isaac Newton* (Vol. I (1664–1666) & Vol. II (1667–1669)), D. T. Whiteside (ed.) (Cambridge University Press, London).

Newton, I. (1999). *The Principia: Mathematical Principles of Natural Philosophy*, translated by I. B. Cohen and Ann Whitman (University of California Press, Berkeley, Los Angeles, London).

O'Briant, W. H. (1967). Leibniz's preference for an intensional logic (A reply to Mr. Parkinson), *Notre Dame Journal of Formal Logic* **VIII**(3), pp. 254–256.

Panza, M. (2008). Isaac Barrow and the bounds of geometry, in P. Radelat-de Grave and C. Brichard (eds.), *Liber amicorum Jean Dhombres* (Brepols, Turnhout, Belgium).

Pappus (1986).. *Pappus of Alexandria: Book 7 of the Collection*, (2 Vols.), Sources in the History of Mathematics and Physical Science **8**, edited and translated with a commentary by A. Jones (Springer Verlag, Berlin).

Pappus (2010). *Pappus of Alexandria: Book 4 of the Collection*, edited and translated with a commentary by Heike-Sefrin-Weiss (Springer, Berlin).

Parkinson, G. H. R. (1985). *Logic and Reality in Leibniz's Metaphysics* (Clarendon Press, Oxford).

Penrose, R. (1989). *The Emperor's New Mind: Concerning Computers, Minds, and the Laws of Physics* (Oxford University Press, New York).

Probst, S. (2008). Indivisibiles and infinitesimals in early mathematical texts of Leibniz, in U. Goldenbaum and D. Jesseph (eds.), *Infinitesimal Differences: Controversies between Leibniz and His Contemporaries* (Walter de Gruyter, Berlin, New York), pp. 95–106.

Rashed, R. (2008). Le concept de tangent dans les conoques d'Apollonius, in

Hartmut Hetch, *et al* (eds.) *Kosmos und Zahl Beitrage zur Mathematik und Astronomie zu Alexander von Humbolt und Leibniz*; unter Mitarbeit von Katharina Zeitz (Steiner, Stuggart), pp. 365–374.

Rivaud, A. (1914–1924). *Catalogue critique des manuscrits de Leibniz* (Poitiers).

Robinson, A. (1986). *Nonstandard Analysis*, Princeton Landmarks in Mathematics (2nd ed.) (Princeton University Press, Princeton).

Rorty, R. (1979) *Philosophy and the Mirror of Nature* (Princeton University Press, Princeton, New Jersey).

Rose, P. E. (1975). *The Italian Renaissance of Mathematics* (Librairie Droz, Geneva).

Rosenthal, E. (1986). *The Calculus of Murder* (St. Martin's Press, New York).

Rossi, P. (1961). The legacy of Ramon Lull in sixteenth-century thought, *Medieval and Renaissance Studies* **5**, pp. 182–213.

Rossi, P. (1979). Universal languages, classifications, and nomenclatures in the seventeenth century, *History and Philosophy of the Life Sciences* **6**(2), pp. 119–131.

Rossi, P. (1989). The twisted roots of Leibniz' characteristic, in P. Rossi & W. Bernardi (eds.), *The Leibniz Renaissance: Vol. 28. Biblioteca di storia della scienza* (Firenze: Leo S. Olschki Editore, Firenze), pp. 271–289.

Rossi, P. (2006) *Logic and the Art of Memory: The Quest for a Universal Language* (Continuum International Publishing Group, Limited, London).

Russell, B. (1964). *A Critical Exposition of the Philosophy of Leibniz* (George Allen & Unwin Ltd, London).

Salmon, G. (1954). *A Treatise on Conic Sections.* Sixth Edition (Chelsea Publishing, New York).

Scott, J. F. (1958). *A History of Mathematics From Antiquity to the Beginning of the Nineteenth Century* (Taylor & Francis Ltd, London).

Scriba, C. J. (1965). The inverse method of tangents a dialogue between Leibniz and Newton, *Archive for History of Exact Sciences* **2**, pp. 113–137.

Segal, S. (1986). Mathematics and German politics: the National Socialist experience, *Historia Mathematica* **13**, pp. 118–135.

Shapin, S. and Schaffer, S. (1985). *Leviathan and the Air Pump* (Princeton University Press, Princeton, New Jersey).

Shapiro, B. (1969). *John Wilkins 1614–1672, an Intellectual Biography* (University of California Press, Berkeley and Los Angeles).

Shimony, I. (2010). Leibniz and the *vis viva* controversy, in *The Practice of Reason; Leibniz and His Controversies*, Chap. 3, pp. 51–73.

Sluse, R. (1672). *Philosophical Transactions of the Royal Society* **7** (1665–1678), pp. 5143–5147.

Stewart, M. (2006). *The Courtier and the Heretic: Leibniz, Spinoza and the Fate of God in the Modern World* (Norton, NY).

Struik, D. J. (1969). *A Source Book in Mathematics, 1200–1800.* (Harvard University Press, Cambridge, Mass).

Terrall, M. (2004). *Vis viva* revisited, *History of Science* **xlii**, pp. 189–209.

Thorndike, L. (1923–1958). *A History of Magic and Experimental Science* (Macmillan, New York).

Wakefield, A. (2010). Leibniz and the wind machines, *Osiris*, second series **25**, pp. 171–188..

Walker, D. P. (1975). *Spiritual and Demonic Magic: from Fincino to Campanella* (University of Notre Dame Press, Notre Dame).

Wallis, J. (2004). *The Arithmetic of Infinitesimals* (Springer, New York).

Westfall, R. S. (1980). *Never at Rest: A Biography of Isaac Newton* (Cambridge University Press, Cambridge, New York).

Whiteside, D. T. (1961). Patterns of mathematical thought in the seventeenth century, *Archive for History of Exact Sciences* **1**, pp. 179–388.

Wilder, R. L. (1981).*Mathematics as a Cultural System* (Pergamon Press, Oxford, New York, Toronto, Sydney, Paris, Frankfurt).

Yates, F. A. (1954). The art of Ramon Lull: an approach to it through Lull's theory of the elements, *Journal of the Warburg and Courtauld Institutes* **17**, pp. 115-173.

Yates, F. A. (1964). *Giordano Bruno and the Hermetic Tradition* (Routledge and Paul, London).

Yates, F. A. (1966). *The Art of Memory* (University of Chicago Press, Chicago).

Yates, F. A. (1982). *Lull and Bruno* (Routledge and K. Paul, London and Boston).

Yoder, J. G. (1988). *Unrolling Time: Christiaan Huygens and the Mathematization of Nature* (Cambridge University Press, Cambridge).

Index